AAPG Treatise of Petroleum Geology
Reprint Series

The American Association of Petroleum Geologists
gratefully acknowledges and appreciates the leadership and support
of the AAPG Foundation in the development of the
Treatise of Petroleum Geology.

Geophysics IV

Gravity, Magnetic, and Magnetotelluric Methods

Compiled by
Edward A. Beaumont
And
Norman H. Foster

Treatise of Petroleum Geology
Reprint Series, No. 15

Published By
THE AMERICAN ASSOCIATION OF PETROLEUM GEOLOGISTS
TULSA, OKLAHOMA 74101-0979, U.S.A.

Library of Congress Cataloging-in-Publication Data

Geophysics IV.

(Treatise of petroleum geology reprint series; no. 15
Includes bibliographical references.
1. Petroleum—Prospecting.   2. Seismic interpretation.
I. Beaumont, E.A. (Edward A.)   II. Foster, Norman H.
III. Series.
TN271.P4G47   1989   622'.1828   89-18158

ISBN 0-89181-414-0
ISSN 1046-0144

AMERICAN ASSOCIATION OF PETROLEUM GEOLOGISTS FOUNDATION
TREATISE OF PETROLEUM GEOLOGY FUND*

Major Corporate Contributors
($25,000 or more)

Chevron Corporation
Mobil Oil Corporation
Oryx Energy Company
Pennzoil Exploration and Production Company
Shell Oil Company
Union Pacific Foundation

Other Corporate Contributors
($5,000 to $25,000)

Cabot Energy Corporation
Canadian Hunter Exploration Ltd.
Conoco Inc.
Marathon Oil Company
The McGee Foundation, Inc.
Phillips Petroleum Company
Texaco Philanthropic Foundation
Transco Energy Company

Major Individual Contributors
($1,000 or more)

C. Hayden Atchison
Richard R. Bloomer
A.S. Bonner, Jr.
David G. Campbell
Herbert G. Davis
Paul H. Dudley, Jr.
Lewis G. Fearing
James A. Gibbs
George R. Gibson
William E. Gipson
Robert D. Gunn
Cecil V. Hagen
Frank W. Harrison
William A. Heck
Roy M. Huffington
Harrison C. Jamison
Thomas N. Jordan, Jr.
Hugh M. Looney
John W. Mason
George B. McBride
Dean A. McGee
John R. McMillan
Rudolf B. Siegert
Robert M. Sneider
Jack C. Threet
Charles Weiner
Harry Westmoreland
James E. Wilson, Jr.

The Foundation also gratefully acknowledges the many who have supported this endeavor with additional contributions, which now total more than $12,000.

*Contributions received as of November 1, 1989.

Treatise of Petroleum Geology
Advisory Board

*Deceased

IN APPRECIATION...

The American Association of Petroleum Geologists and the AAPG Foundation gratefully acknowledge the contributions of the Society of Exploration Geophysicists to the Treatise of Petroleum Geology Reprint Series volumes on geophysics. The crucial role played by SEG in advancing exploration geophysics is universally recognized in the petroleum industry and is plainly documented by the many papers from *Geophysics* and *Geophysics: The Leading Edge of Exploration* reproduced in these volumes. The spirit of advancement and expansion in the science of geophysical exploration for hydrocarbons as well as the continuing synergism that melds the professions of geology and geophysics is exemplified by the permission granted by SEG to reproduce its papers in this series.

Although SEG and AAPG both have a long history of independent activities and autonomous operation, there have been, and there continue to be, cooperative efforts to improve the professionalism of both geologists and geophysicists. Previous activities, such as joint meetings, research conferences, and publications, exemplify the ongoing cooperative efforts which create results far more valuable than independent efforts by either group would have achieved. SEG's contributions to the earth sciences and to the Treatise of Petroleum Geology Reprint Series volumes on geophysics are gratefully acknowledged by AAPG, the AAPG Foundation, and the Advisory Board of the Treatise of Petroleum Geology.

INTRODUCTION

This reprint volume belongs to a series of that is part of the *Treatise of Petroleum Geology*. The *Treatise of Petroleum Geology* was conceived during a discussion we had at the 1984 AAPG Annual Meeting in San Antonio. When our discussion ended, we had decided to write a state-of-the-art textbook in petroleum geology, directed not at the student, but at the practicing petroleum geologist. The project to put together one textbook gradually evolved into a series of three different publications: the Reprint Series, the Atlas of Oil and Gas Fields, and the Handbook of Petroleum Geology; collectively these publications are known as the *Treatise of Petroleum Geology*. With the help of the Treatise of Petroleum Geology Advisory Board, we designed this set of publications to represent the cutting edge in petroleum exploration knowledge and application. The Reprint Series provides previously published landmark literature; the Atlas collects detailed field studies to illustrate the various ways oil and gas are trapped; and the Handbook is a professional explorationist's guide to the latest knowledge in the various areas of petroleum geology and related fields.

The papers in the various volumes of the Reprint Series complement the different chapters of the Handbook. Papers were selected on the basis of their usefulness today in petroleum exploration and development. Many "classic papers" that led to our present state of knowledge have not been included because of space limitations. In some cases, it was difficult to decide in which Reprint volume a particular paper should be published because that paper covers several topics. We suggest, therefore, that interested readers become familiar with all the Reprint volumes if they are looking for a particular paper.

Geophysics is an indispensable tool for geologists looking for and developing oil and gas fields. Because it lets us "see" into the subsurface, geophysics allows petroleum geologists to build better images of the subsurface than is possible using only surface geology and information from well bores. In the past, geophysics was the domain of the geophysicist, and the geophysicist alone acquired, processed, and interpreted geophysical data. During the past two decades, however, the technology of geophysics has exploded; at the same time, the petroleum industry has been forced to look for more and more subtle traps in more and more difficult terrain. This placed a tremendous burden on geophysicists, and they naturally looked to their colleagues, the geologists, for relief. At first, geologists only helped with interpretation. Today, however, geologists are also involved in helping geophysicists make decisions regarding acquisition and processing of data.

The choice of papers in these geophysics reprint volumes reflects this evolution. The papers were chosen to help geologists, not geophysicists, enhance their knowledge of geophysics. Math-intensive papers were excluded because those papers are relatively esoteric and have limited applicability for most geologists. Many of the papers included do contain mathematical equations, but they were selected because they are germane, and the math is presented at a level that, we trust, the majority of geologists are now comfortable with.

The number and distribution of the papers reprinted in these volumes reflect the current importance and uses of the different geophysical methods described in the papers. We have divided the topic of geophysics into four volumes. The first volume contains papers on Seismic Methods. Papers in this volume are concerned with seismic theory and are grouped into six sections: Seismic Methods, Seismic Rock Properties, Seismic Acquisition, Seismic Processing and Display, Seismic Velocities, and Migration. Volume II is subtitled Tools for Seismic Interpretation. Section titles in this volume are Synthetic Seismograms and Velocity Inversion; Seismic Modeling; Seismic Attributes: Amplitude, Frequency, Phase, Velocity; Shear Waves; Amplitude Variation with Offset; and Vertical Seismic Profiling. Volume III, Geologic Interpretation of Seismic Data, contains sections on Structural Interpretation and Stratigraphic Interpretation. The last volume is on Gravity, Magnetic, and Magnetotelluric Methods. It contains two sections: Gravity and Magnetic Methods, and Magnetotelluric Methods.

We would like to thank the various societies and publishers who gave us permission to reprint these papers, especially the Society of Exploration Geophysicists. We also wish to thank the members of Advisory Board of the Treatise of Petroleum Geology who suggested papers for these volumes, especially R. Randy Ray. Randy Ray is a geophysicist who was trained initially as a geologist. From a large list of proposed papers, he helped us select papers that would be both understandable and useful to a geologist exploring for and developing oil and gas fields.

Edward A. Beaumont
Tulsa, Oklahoma

Norman H. Foster
Denver, Colorado

TABLE OF CONTENTS

GEOPHYSICS IV
GRAVITY, MAGNETIC, AND MAGNETOTELLURIC METHODS

GRAVITY AND MAGNETIC METHODS

MAGNETOTELLURIC METHODS

TABLE OF CONTENTS

GEOPHYSICS I
SEISMIC METHODS

Predictive deconvolution: theory and practice.
K. L. Peacock and Sven Treitel.

Estimation and correction of near-surface time anomalies.
M. Turhan Taner, F. Koehler, and K. A. Alhilali.

Seismic signal processing.
Lawrence C. Wood and Sven Treitel.

SEISMIC VELOCITIES

Seismic velocities from surface measurements.
C. Hewitt Dix.

An analysis of stacking, rms, average, and interval velocities over a horizontally layered ground.
M. Al-Chalabi.

Time-depth and velocity-depth relations in western Canada.
C. H. Acheson.

Apparent velocity from dipping interface reflections.
F. K. Levin.

A velocity function including lithologic variation.
L. Y. Faust.

Seismic data indicate depth, magnitude of abnormal pressures.
E. S. Pennebaker, Jr.

Synthetic sonic logs—a process for stratigraphic interpretation.
R. O. Lindseth.

Velocity spectra—digital computer derivation and applications of velocity functions.
M. Turhan Taner and Fulton Koehler.

The effects of cracks on the compressibility of rock.
J. B. Walsh.

MIGRATION

Migration.
P. Hood.

Two-dimensional and three-dimensional migration of model-experiment reflection profiles.
William S. French.

A simple theory for seismic diffractions.
A. W. Trorey.

Migration of seismic data from inhomogeneous media.
Les Hatton, Ken Larner, and Bruce S. Gibson.

Time migration—some ray theoretical aspects.
P. Hubral.

The wave equation applied to migration.
D. Loewenthal, L. Lu, R. Roberson, and J. Sherwood.

Wave-front charts and three dimensional migrations.
Albert W. Musgrave.

Migration by Fourier transform.
R. H. Stolt.

GEOPHYSICS II
TOOLS FOR SEISMIC INTERPRETATION

TABLE OF CONTENTS

GEOPHYSICS III
GEOLOGIC INTERPRETATION OF SEISMIC DATA

Vp/Vs—a potential hydrocarbon indicator.
Robert S. Tatham and Paul L. Stoffa.

AMPLITUDE VARIATION WITH OFFSET

Plane-wave reflection coefficients for gas sands at nonnormal angles of incidence.
W. J. Ostrander.

VERTICAL SEISMIC PROFILING

Vertical seismic profiling—a measurement that transfers geology to geophysics.
B. A. Hardage.

The use of vertical seismic profiles in seismic investigations of the earth.
A. H. Balch, M. W. Lee, J. J. Miller, and Robert T. Ryder.

Prediction of overpressure in Nigeria using vertical seismic profile techniques.
S. Brun, P. Grivelet, and A. Paul.

Offset source VSP surveys and their image reconstruction.
P. B. Dillon and R. C. Thomson.

Vertical seismic profiling technique emerges as a valuable drilling tool.
R. J. Roberts and J. D. Platt.

GRAVITY AND MAGNETIC METHODS

BULLETIN OF THE AMERICAN ASSOCIATION OF PETROLEUM GEOLOGISTS
VOL. 46, NO. 10 (OCTOBER, 1962), PP. 1815-1838, 25 FIGS.

GRAVITY AND MAGNETICS FOR GEOLOGISTS AND SEISMOLOGISTS[1]

L. L. NETTLETON[2]
Houston, Texas

ABSTRACT

Nearly 95 per cent of geophysical expenditures for oil exploration is for reflection seismograph surveys. Most of the other 5 per cent is for gravity and magnetic surveys. To some extent the disproportionately large expenditures for seismograph work probably are from a lack of understanding and appreciation of the usefulness of the other methods. The paper reviews the basic principles of the gravity and magnetic methods, points out their similarities and differences, and gives examples of their application. These examples include samples of basement depth maps from magnetic surveys where independent control on actual depths was available. A comparison with this control indicates that depth determinations from magnetic surveys may have a reliability of approximately ±5 per cent. Regional gravity effects and their removal by graphical and grid calculation schemes are illustrated by examples from the Gulf Coast and California. Methods of quantitative interpretation of gravity are demonstrated by examples from the salt dome area of the Gulf Coast.

INTRODUCTION

An appropriate sub-title for this paper would be "The Other Five Per Cent." This is because, in petroleum exploration, approximately 95 per cent of the dollar expenditures are devoted to the seismic method, and only about 5 per cent to gravity, magnetics, and the occasional other methods that have made their appearance in the field for a short time and usually have not lasted very long.

From its beginning in the early 1920s, geophysical prospecting used three different physical principles in the determination of underground geology. These are (1) the propagation of elastic waves through the earth which is the basis for the reflection and refraction seismic methods, (2) the measurement of small variations in the intensity of the gravitational field, and (3) the measurement of small variations in the magnetic field. In the nearly 40 years since these three principles were first applied, there have been many advances in instrumentation, in field techniques,

and in interpretation, but petroleum exploration still is based almost entirely on these same three physical principles. Many attempts have been made to use other measurements, such as electrical, chemical, and radiation, as well as witch sticks and black boxes, and many of these have had extensive trials in the field in the search for petroleum. In spite of the fact that very great rewards would accrue to any person or company which could devise any successful alternative method of indicating oil or conditions favorable for oil accumulation, none of these other schemes has lasted very long in the field, and we still depend, in oil exploration, on the three basic physical principles.

Therefore, it seems very unlikely that there is to be any great new breakthrough in petroleum exploration and that the advances in the future probably will be along the line of better use of the methods which we now have, especially in interpretations which coordinate the different geophysical indications with each other and with geology.

Let us consider why the seismic method dominates the exploration industry to the extent that it takes 95 per cent of the petroleum geophysical expenditures. To a great many people, the mention of "geophysicists" or "geophysics" has come to mean "seismologists" or the "reflection seismograph." It is even so bad that I have seen a gravity map with the little circles denoting gravity stations indicated in the legend on the corner of the map as "shotpoints" and gravity field parties may be referred to as "shooting" in an area.

In the first place, the seismic method is very effective; also, it is very much more expensive

[1] Lecture delivered on A.A.P.G. Distinguished Lecture Tour, Spring of 1961. Manuscript received, September 5, 1961.

[2] Geophysical Associates International and Gravity Meter Exploration Company, Houston, Texas.
The writer is indebted to many sources for material included in this paper and the many associates in discussions which have led to ideas which are included. Specifically, thanks are extended to *Geophysics* for the use of many illustrations, to Gravity Meter Exploration Company for use of the original data for the Humble dome gravity map and for necessary drafting and other help, to Bureau des Recherches de Petrole for use of the Senegal basement map, and, finally, thanks are due to the American Association of Petroleum Geologists for the privilege of making the very interesting Distinguished Lecture Tour which led to the preparation of this material.

1815

3

than either of the other two methods. In terms of cost per unit of area covered, subject, of course, to many wide individual variations, the relative costs of seismic, gravity, and magnetic surveys are approximately in the ratio of a dollar, a dime, and a penny. There may be reasons other than high cost and greater effectiveness which account for the high proportion of expenditures for the seismic method. The basic principles of the method are simple. An explosion produces a disturbance or noise of some kind which travels downward in the earth to a discontinuity. Reflections from that discontinuity come back to the surface and are detected by suitable instruments, and the time of travel is measured. If one knows the speed of propagation, it is a simple matter to determine the depth at which this reflection occurs. These principles are so simple that, if I may be pardoned for saying so, they can be understood even by a geologist. Furthermore, the immediate result of the reflection seismograph field operation is a map which is essentially geological in nature. The map is readily understood by geologists because it is similar to the familiar subsurface maps they make from well information.

Gravity and magnetic field operations yield maps which are not obviously geological in nature. Another step is required before these maps can be directly related to geology. This step is an interpretive one which often is not readily understood and to some extent has carried an aura of mystery. Geologists often are too much inclined to consider gravity and magnetic maps as directly indicative of structure. This tendency was recognized many years ago by DeGolyer (1928) in a paper entitled "The Seductive Influence of the Closed Contour." This little two-page paper points out that geologists, who have been trained to look for closed contours as indicative of closed structures favorable for the accumulation of oil, are too much inclined to consider that closed contours on gravity and magnetic maps may be directly indicative of closed structure. This is only occasionally true on gravity maps and almost never true on magnetic maps. The much more subtle relations to geology, which are commonly deeply hidden in gravity and magnetic maps, must be brought out by removing regionals, calculating derivatives, or other analytical processes which are not readily understood. Therefore, geologists or management may be inclined to turn to the more easily understood reflection seismo-

graph method more often than they should and to neglect the very substantial exploration help which can be provided by the other two methods.

With these introductory remarks, we will now proceed to a discussion of the fundamental principles with samples of application of, first, the magnetic and, then, the gravity method. The discussion is concerned almost entirely with basic principles and with certain aspects of interpretation and not at all with the instrumentation and field operations.

MAGNETIC METHOD

In Figure 1, the shaded area represents any sort of a magnetic body, meaning simply that it is a volume of material that is more magnetic than its surroundings. Being more magnetic, the lines of magnetic force tend to pass through it more easily and, therefore, to be squeezed together within and next above the area of the body. This squeezing together of the lines of magnetic force is equivalent to an increase in the intensity of the magnetic field.

In Figure 2, we put the same diagram in a geological environment. The more magnetic area now may represent an intrusive body of basic igneous rock with a relatively large content of

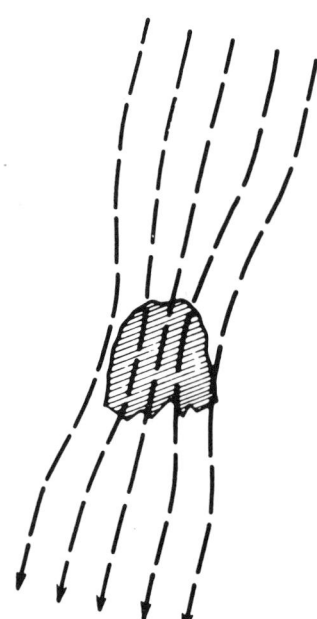

FIG. 1.—Magnetic body and distortion of magnetic field.

4

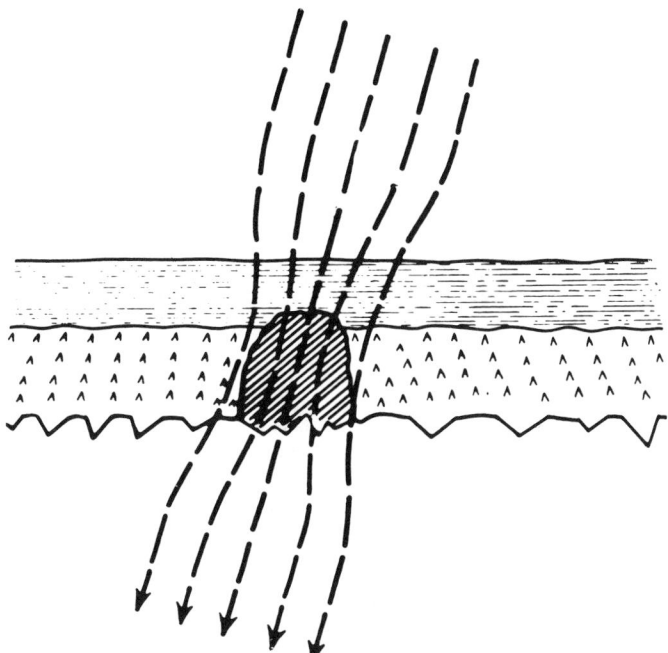

FIG. 2.—Magnetic body in geological environment.

magnetic minerals, largely magnetite. The adjacent area may represent granitic basement rocks with a moderate magnetite content. Overlying both of these is a section of sedimentary rocks with practically no magnetite. Above, and measuring the variatons in the intensity of the magnetic field, is an airborne magnetometer.[3]

The variations in the magnetic field measured by the magnetometer are almost entirely due to variations in the concentration of magnetite in the rocks. Magnetite is not the only magnetic mineral, but it is so much more magnetic and so much more common that it is the dominating cause of magnetic disturbances. The basement rocks nearly everywhere have so much more magnetite in them and, therefore, are so much more magnetic than the sediments that, for practical purposes, we can usually consider that the features measured by the airborne instrument have their origin at or below the surface of the basement or the base of the sedimentary section and that the measurements would be practically the same if the sediments were not present at all. This is a very important feature because, by understanding the nature of the anomalies caused by irregularities in the magnetization of the basement, we can determine the depth of their source and, thereby, the depth to the basement surface and the thickness of the sedimentary section. Also, we can determine major structural features on the basement surface.

In order to make such interpretations, we must understand the general nature of the anomalies. Figure 3 shows examples of the total intensity anomaly as measured by the airborne magnetometer for the same body in different magnetic situations. The body is a sphere, polarized in the direction of the earth's magnetic field. The upper left-hand diagram shows the anomaly when this body is near the magnetic pole where the magnetizing field is vertical. In this case, the anomaly is a simple maximum with a very weak negative zone on each side, so weak that it hardly shows in this diagram. If we move farther south where the magnetic latitude is $67\frac{1}{2}°$, we see the beginning of

---

[3] Most magnetic surveys now are carried out in the air, and therefore this discussion of the magnetic method is concerned almost entirely with its airborne aspect which measures variations in the total intensity of the magnetic field. Certain technical details of anomalies mapped are different from those measured by the formerly common surface magnetometer which measures the vertical component of the field, but the general principles of application and interpretation are the same.

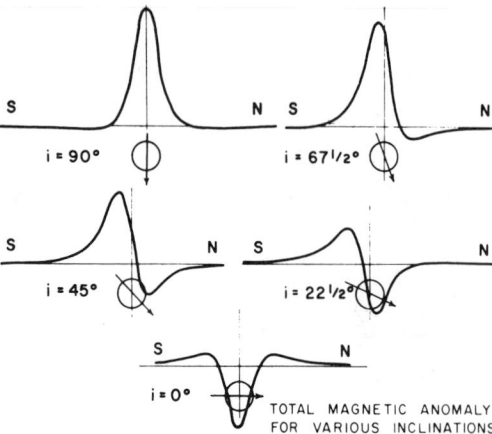

FIG. 3.—Variation in form of anomaly in total magnetic intensity with change in magnetic latitude.

the development of a negative anomaly on the north flank and a shift of the maximum south from the point over the center of the disturbing body. At 45° inclination, the negative anomaly is more strongly developed, and the shift is also greater. At $22\frac{1}{2}°$ inclination, the negative and positive anomalies are of almost the same magnitude with the negative part somewhat sharper. Finally, at the magnetic equator where the polarization is horizontal, the principal part of the anomaly is the minimum centered over the body so that what began at high latitudes as a simple positive anomaly has turned inside out to become a negative anomaly at low magnetic latitudes. (The series would be repeated with a left-to-right change if the diagram were continued to the south magnetic pole.)

Figure 4 shows on the right side a magnetic map in a situation corresponding approximately with the $22\frac{1}{2}°$ inclination as shown on Figure 3. This map is from an area on the west coast of Senegal in West Africa. (The picture is incomplete on the west side because the area is along the coast and the western part would be over the Atlantic Ocean.) The magnetic contour map shows a sharp negative closure on the north side and a positive closure on the south side. The left side of the figure shows the gravity anomaly for the same area, which is a simple maximum. The plus sign at the center of the gravity anomaly appears at the point shown by the plus sign on the magnetic map, and the x's indicating the centers of the negative and positive closures are at the corresponding points shown on the gravity map.

Figure 5 shows profiles across the magnetic and gravity anomalies of Figure 4. The dashed curve shows the magnetic anomaly with its sharp negative part on the north side and a broader positive part on the south side. Circles along this curve show the magnetic effect computed from the spherical body indicated by the dashed circle below. The solid curve shows the profile across the gravity map and indicates a simple maximum corresponding with the nearly circular anomaly shown on the left side of Figure 4. The dots along this curve are calculated gravity effects from the spherical body indicated by the solid circle below. We thus see that the magnetic and gravity anomalies are produced by very similar geological forms and, in fact, the difference between these two forms is not well enough established to be considered as real because of the limitations of the data and also because the disturbing body is only approximated by the simple geometric form. We see that the two very different maps, as shown in Figure 4, actually are two different geophysical aspects of the same geological body which is almost certainly a very large basic intrusive within the basement rocks.

Now let us consider certain features of airborne magnetometer records which are used for the determination of basement depths. Figure 6 shows a set of curves calculated for the body indicated in the lower part of the figure. This is a north-south body, and the calculations are along east-west profiles so the curves are symmetrical. The four curves in the diagram are calculated for four different cases: (1) the case with the bottom

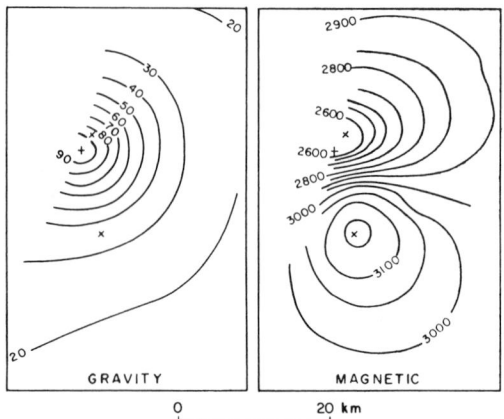

FIG. 4.—Comparison of gravity and magnetic maps in same area.

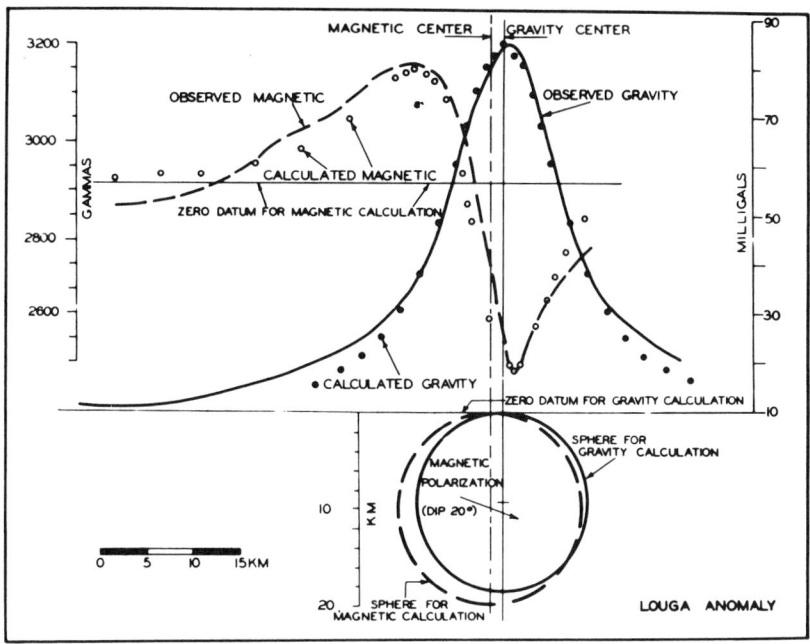

FIG. 5.—Gravity and magnetic profiles from Figure 4 with calculated values and cross sections of anomalous body

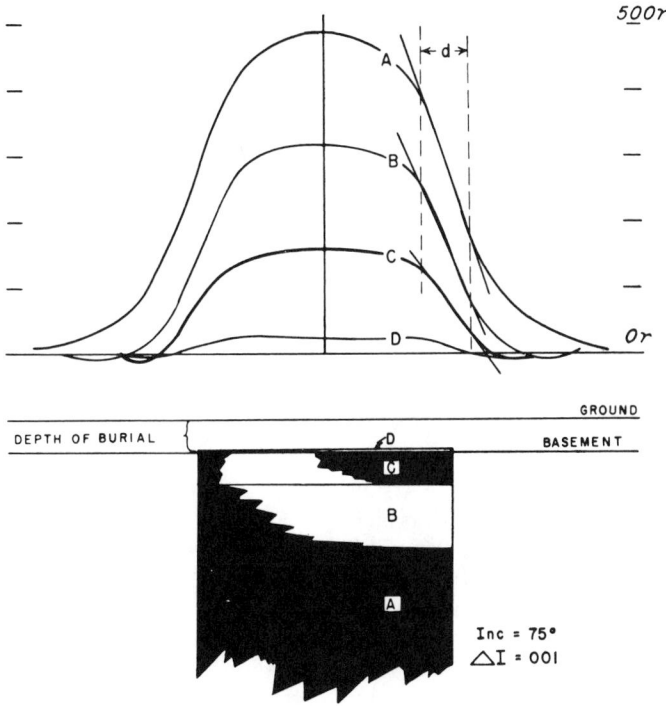

FIG. 6.—Calculated magnetic effects for body with variable depth to bottom; also showing "slope" measurement for determination of depth.

7

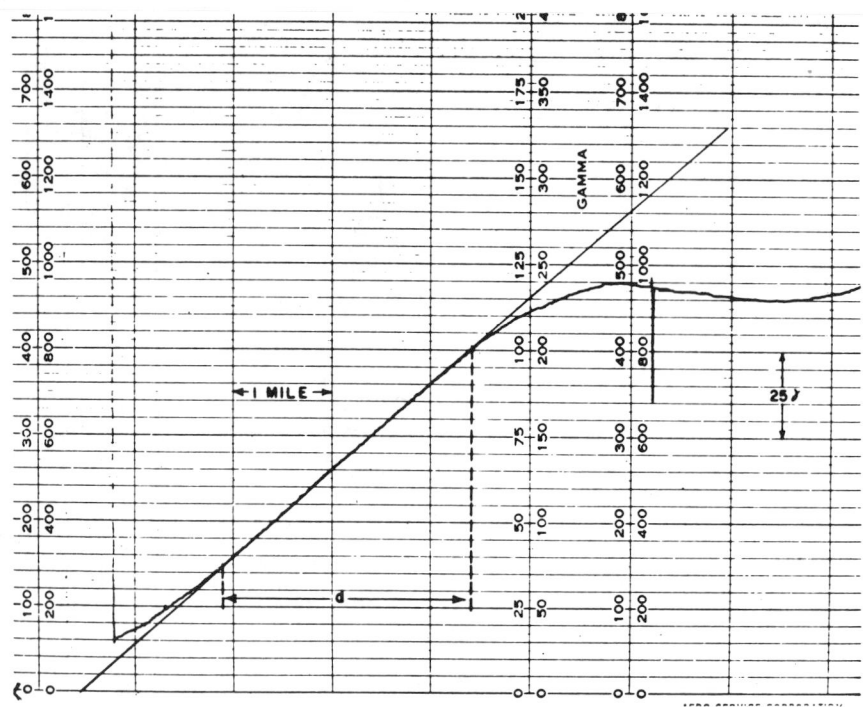

Fig. 7.—Sample aeromagnetic profile showing "slope" measurement.

of the body extending to infinite depth, (2) the case with the bottom extending to four times the depth to the basement surface, (3) with the body extending to twice the depth of the basement surface, and (4) for only the thin slice along the top which has a thickness of about one-tenth of the depth.

One of the most useful of the several parameters which can be used to determine depths from magnetic surveys is the so-called "slope" measurement. This, as indicated on the diagram, consists of measuring the horizontal distance over which the slope on the flank of the anomaly is essentially constant. If we make this slope measurement along all of these curves, we will find that for the first three curves the measurement gives substantially the same horizontal distance, d, as indicated in the diagram. If we have a very thin body, corresponding with the low relief bottom curve, we find that this distance, d, from the slope measurement is substantially smaller. The distance, d, is directly related to the depth of the body below the surface by a factor which is empirically determined by calculation of magnetic fields from suitable models. Since d does not change materially among the three curves, we see that the

depth estimate is not particularly affected by the depth to the bottom of the body, but only by the depth to its top. There are other parameters also which can be measured and used to determine the depth of their origin, but a discussion of these is beyond the scope of this paper. By making appropriate measurements and determining the proper parameters for their use, depths can be calculated at a great many points on magnetic maps and actual contour maps on the basement surface can be drawn. It should be emphasized that the determination of these parameters and making the proper measurements on the magnetic records is often a rather subtle feature in the interpretation and one that is subject to personal judgment qualified by experience.

Figure 7 shows an actual aeromagnetic record on which a slope measurement has been made to give the distance, d, as a depth criterion. The measurement made on the record then must be properly related to the corresponding horizontal distance on the ground. By application of the proper parameter to this distance, the depth to the basement surface can be determined.

Let us now consider an application of these methods. Figure 8 is a geologic map of Senegal in

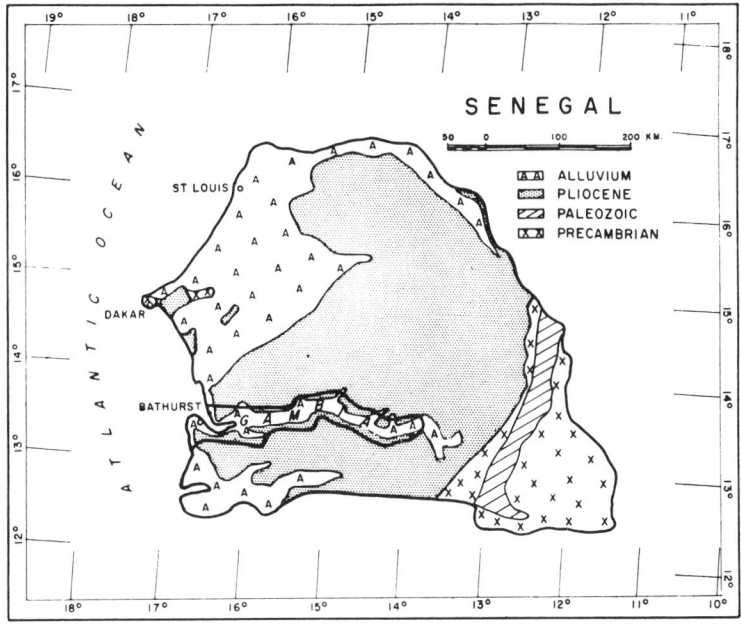

FIG. 8.—Geologic map, Senegal. (Modified from Carte Geologique Internationale de L'Afrique, 1946)

West Africa. Now suppose that you were the exploration manager of a company which had obtained a concession on this area, and let us assume that the concession granted covers 100×100 km. or equivalent, and that it could be placed anywhere in the country, either in one block or divided up and placed anywhere chosen but with the total area limited to 10,000 sq. km. Now, how would you place such a concession on the basis of this surface geological map? The central part of the map shows Quaternary and Pliocene deposits; but there are older rocks and intrusives on the east, making that area look shallow. Also, there is a small area of intrusives around Dakar, making that area also look shallow, but there might be a deep basin in between. With only this much to go on, one probably would be inclined to take the concession area as a number of smaller pieces scattered around the basin or presumed basin in such a way that at least some of them would have a chance to be in a favorable spot.

Let us see what an aeromagnetic survey could contribute to such a problem. An aeromagnetic survey was made over this basin and was interpreted to produce the basement contour map of Figure 9 with a contour interval of 1,000 meters. The northwestern part of the map shows a rather deep basin starting just northeast of Dakar where

the lowest contour indicates 7,000 m., or over 20,000 feet, of sediment in this deepest part of the basin. The basin extends northeastward for nearly 100 miles and has a width of 30–40 miles, so it is a fairly respectable basin. Incidentally, the circular uplift indicated just south of St. Louis is the area of the magnetic and gravity maps of Figure 4, and the intrusive indicated by that analysis is seen here to have caused a very large uplift of the basement surface.

There are other basin areas within the mapped area, including a shallower basin in the eastern part of the survey and another rather deep basin on the south. In the central part of the area there is a broad platform where the depths are only about 2,000 m. With this map before us, we would be in much better position to decide where to put our concession. We might want to put part of it in a deep basin or possibly some of it on the central basin platform or along the steep flank with the idea that there might be pinch-outs there or reefing or some other similar favorable geological condition. We certainly would have a much better basis for choosing concession areas than we would have from the surface map alone.

The general situation is rather similar to the Permian basin in West Texas and New Mexico. If one had been a geologist there some forty-odd

9

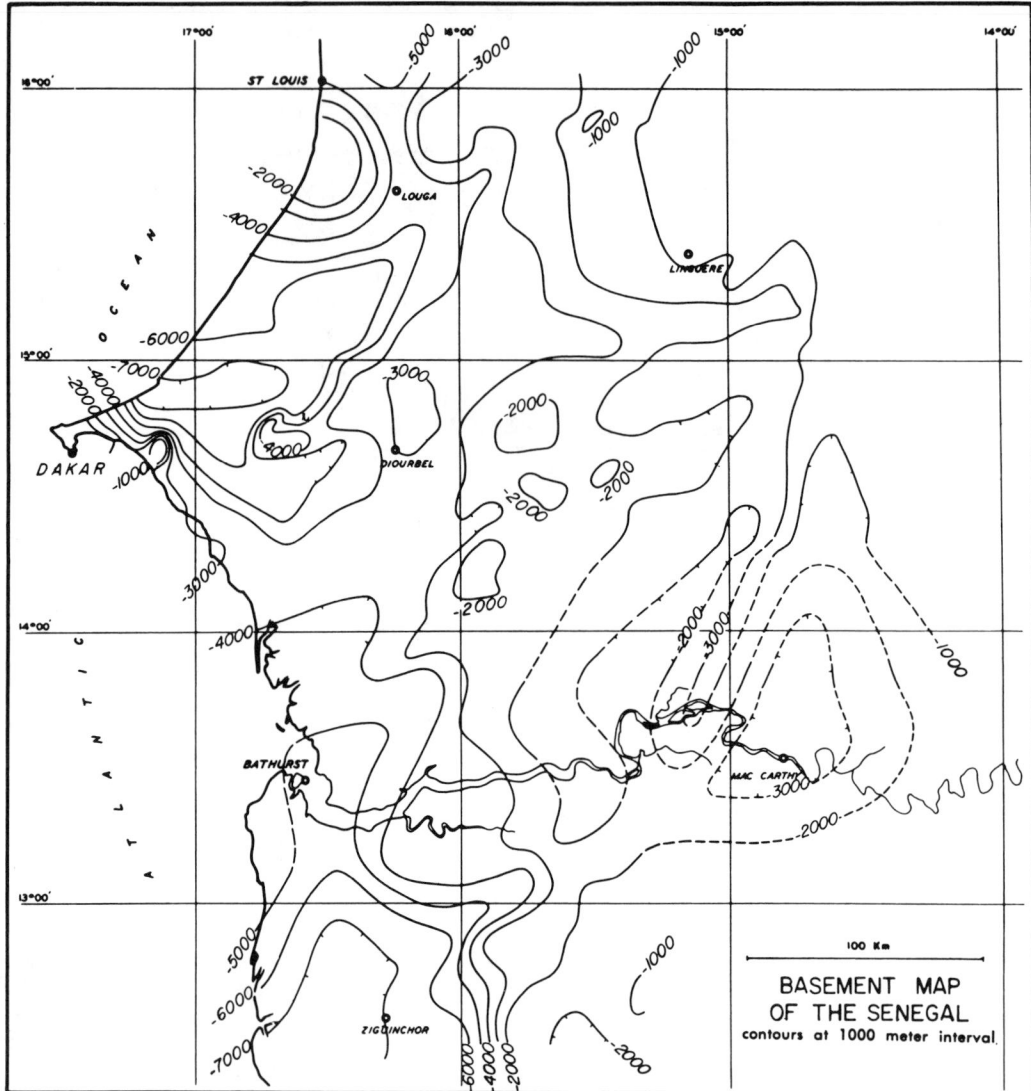

Fig. 9.—Basement depth map, Senegal, calculated from aeromagnetic survey; contour interval, 1,000 m.

years ago, before there had been any drilling, and was considering whether or not the area was favorable for oil accumulation, he would be in the position rather similar to that postulated in connection with surface geologic map of Senegal. If one of the 50,000-acre ranches which were available for leasing in those years had been offered, he might have a hard time deciding whether or not to purchase such a lease. He would not have known of the existence of the Central Basin platform. On the other hand, if the techniques of magnetic surveys and interpretation as here outlined

had been available at that time, the presence of the Central Basin platform would have been clearly revealed. We can say this with assurance because aeromagnetic surveys have been made in this area and their interpretations show clearly and quantitatively the Central Basin platform and its flanks sloping into the much deeper Midland basin toward the east and the Delaware basin toward the west.

One may ask, "How do you know that these basement depth contours are reliable and dependable?" We have some basis to answer this question. A survey made in the Peace River area of northwestern Alberta over ten years ago, which covers about 27,000 sq. mi. was interpreted (Steenland, 1962), and a basement map was constructed. This area now has become an important gas-producing province and a great many more wells have been drilled. Since the gas horizons are in the Lower Paleozoic and near the basement, this subsequent drilling gave many points of control on the basement depth either by actually drilling into the basement or by drilling so deeply into the section that it was possible to estimate the depth to the basement with considerable assurance. There were 169 of these new wells, and at each such well a comparison could be made between the actual depth of the basement as found by the drilling and that which had been predicted by the analysis of the magnetic survey.

Figure 10, from the referenced paper, is a statistical diagram or histogram of the differences between the actual and the magnetic estimates of the basement depth. The ordinates of this diagram are the number of instances having the differences indicated by the horizontal scale where these differences are depth estimate errors expressed as per cent of the total depth. The histogram is somewhat unsymmetrical with more points on the right, or "too shallow" part, than would be expected from a normal or random error distribution. The cause is in certain systematic errors in use of depth parameters which are be-

yond the scope of this paper. (They are covered in detail in the original paper.) These difficulties are now understood much better, and with modern techniques and a magnetic pattern with anomalies reasonably favorable for depth calculations, probable errors of the order of ±5 per cent of the depth below the flight line may be expected.

Another available example (Jacobsen, 1961) to evaluate the dependability of the aeromagnetic basement contour is shown by Figure 11 which shows a contour map on the basement from the analysis of an aeromagnetic survey over the eastern part of the Orinoco basin of eastern Venezuela. The depth contours at 2,000-foot intervals show a generally northward sloping basement with the depth ranging from about 4,000 feet in the southeast part to more than 22,000 feet in the northern part of the area covered. This map is based on the aeromagnetic survey together with the known depth to the basement in the three wells shown in the southern part of the area.

The interpretation as shown in Figure 11 must be qualified somewhat because the survey was flown east and west for certain operational reasons, whereas it would have been much preferable to have had the flight lines north and south, as it is always desirable to have the lines perpendicular to the geologic strike if that is known. The map in Figure 11 shows, in addition to the contours on the basement surface, certain local

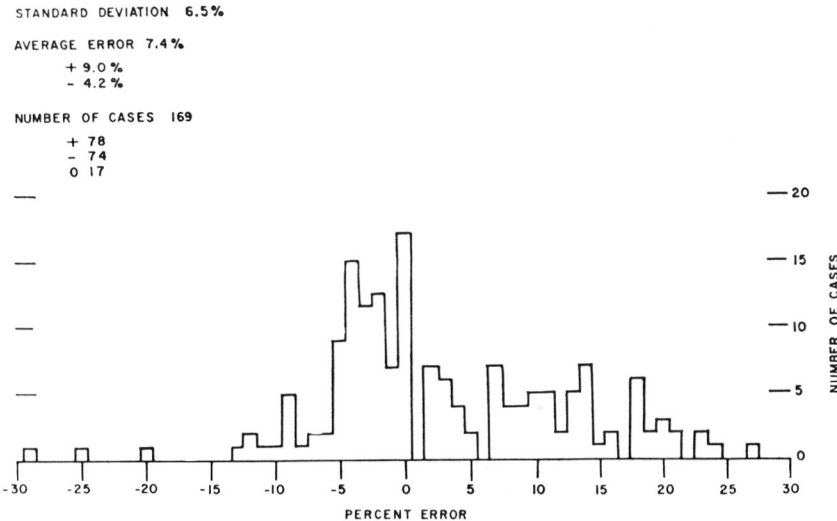

FIG. 10.—Histogram of error in basement depth estimates, Peace River, Alberta, at 169 wells drilled after magnetic survey and interpretation.

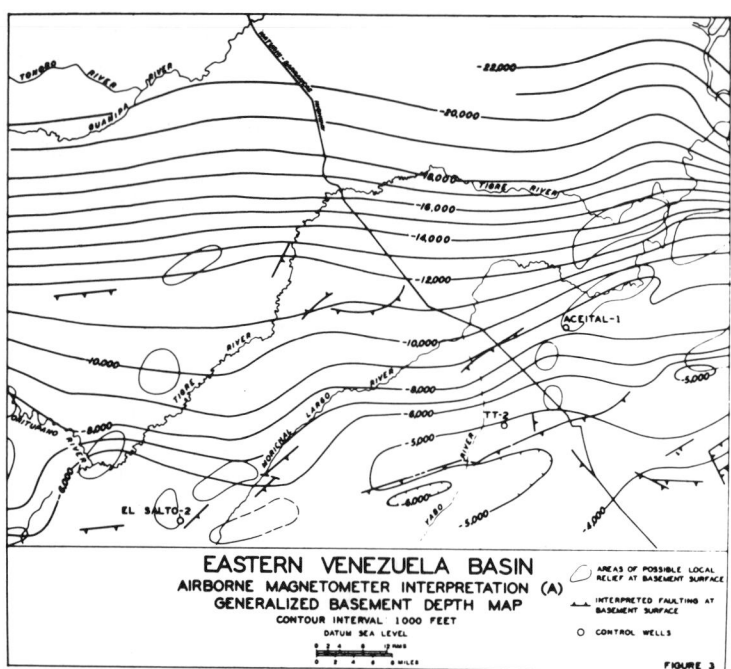

Fig. 11.—Basement map, eastern Venezuela basin, from aeromagnetic interpretation. (Courtesy *Geophysics*)

features on that surface such as faulting and possible areas of local uplift.

The company which made this survey was not interested primarily in the map itself but rather in the evaluation of the method. Since this is an oil-producing area many wells had been drilled there, and much additional control on the basement surface over the southern part of the area was available from this drilling. Furthermore, extensive reflection seismograph work had been carried out, and the reflection identified with the basement surface as determined over the southern part of the area could be carried much farther north so that a basement map as shown in Figure 12 was made from the subsurface and seismograph data. This information was not available to the interpreters of the aeromagnetic survey, and they had basement depths only at the three wells mentioned.

The two maps (Figs. 11 and 12) are not by any means identical; but, quantitatively, the differences are mostly relatively small, and the general picture as given by the aeromagnetic interpretation is very similar to that of the actual basement from the seismograph and subsurface control. Certain of the local features also correspond well. In particular, the fault across the southeastern

part of the magnetic interpretation (just south of well TT2) is very close to the fault shown on Figure 12. The location of the fault where it crosses the kink in the highway is almost exactly the same on the two maps. The strike is almost exactly the same. The direction of displacement on both is down toward the updip side of the basement slope, and the magnitude of displacement as indicated near the western end of the magnetic fault interpretation is about 1,000 feet, and also about 1,000 feet of displacement is indicated by the contours from the subsurface and seismograph data. The cost of the magnetic survey was only a fraction of one per cent of the cost of the information from which the other map was made.

In cases where magnetic conditions are unfavorable or the quality, spacing, or orientation of the flight lines are poor, errors may be larger than have been indicated by the examples shown. However, extensive experience and practice have led to the development of procedures which can handle nearly all types of magnetic conditions so that with careful field work and reasonably favorable general conditions, basement depths can be very reliably determined, and the aeromagnetic surveys offer a dependable means by which knowledge of the basement depth can be obtained. Therefore, in any exploration or geological problem where depth and form of the basement surface are important, the aeromagnetic method offers a means for determination of these factors which is reliable and dependable and which is very much cheaper than any other general exploration process that can be carried out over a large area.

GRAVITY INTERPRETATION

The analysis of gravity surveys is quite different from that of magnetic surveys. The theoretical background is considerably simpler. A body with a specific density contrast will produce the same gravity anomaly no matter where it is on the earth's surface and, therefore, is not subject to variation in its expression with latitude or orientation as is a magnetic anomaly. On the other hand, the gravity interpretation lacks the great geological simplicity which the magnetic method has because of the definite contrast at the surface of the basement. The basement surface may or may not be a surface of density contrast at which gravity anomalies have their origin. For instance, in areas such as parts of West Texas

where heavy limestones lie on the granite, the density of the limestones is about the same as that of the granite, and therefore there is no density contrast at the basement surface. On the other hand, in some of the Tertiary basins where clastic sediments lie on the basement rocks, there is a definite density contrast at the basement surface, and regional or local anomalies may have their origin there.

The origin of gravity anomalies may be anywhere from the grass roots down. There may be disturbances near the surface such as variable thickness of glacial drift or solution sinks which cause local irregularities not related to underlying structure. There may be density contrasts anywhere within the sedimentary section, and, as just mentioned, there may or may not be density contrasts at the surface of the basement. Density contrasts may extend into the basement rocks much more deeply than the magnetic contrasts. In the magnetic case, there is a limit to the depth at which the magnetic contrasts can occur because, with the universal increase of temperature with depth, magnetite loses its magnetization where the temperature reaches the Curie point.

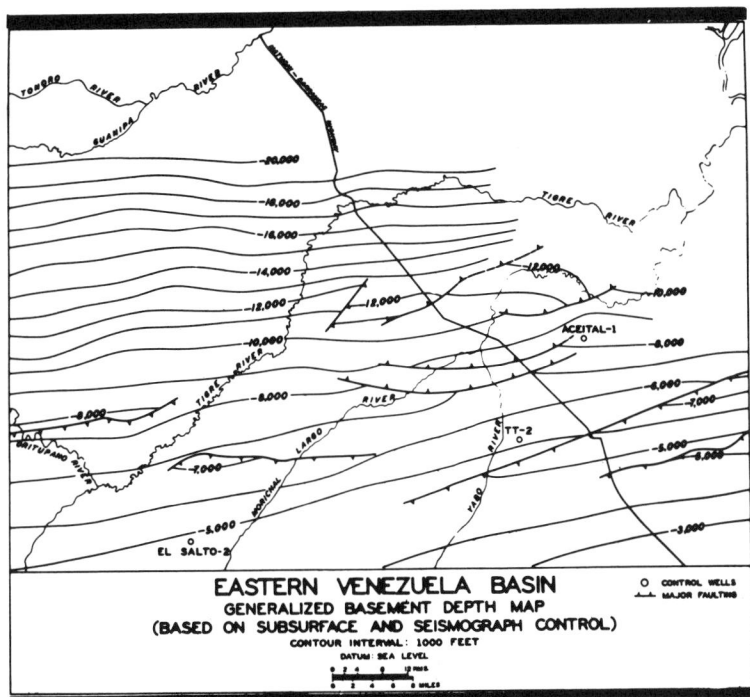

FIG. 12.—Basement map, eastern Venezuela basin, from subsurface and seismograph control. (Courtesy *Geophysics*)

The exact depth of this loss of magnetization is not known but is probably in the range of 15 to 20 miles. There is no such limit on density contrasts and gravity anomalies may have their origin at any depth in the basement extending down at least to the Moho discontinuity. Therefore, unlike magnetic maps, gravity maps can and usually do contain anomalies having their sources over a wide range of depths from near-surface to very deep within the basement rocks. This means that before any quantitative analysis of gravity anomalies and accurate relation of such anomalies to geological disturbances can be carried out, it is first necessary to isolate or separate out the anomaly itself from its background.

Figure 13 illustrates this problem and two different methods of its solution. The upper line of heavy dots represents a hypothetical gravity profile such as might be made by drawing a line across a gravity map and picking off contour crossings to plot the profile. This profile obviously contains two local anomalies and one broad regional anomaly having its apex towards the right-hand side of the diagram. This regional influence must be removed before a quantitative analysis of the local anomalies can be undertaken.

From the earliest days of the analysis of gravity anomalies, extending back into torsion-balance times, a common and very effective approach to this problem has been to draw a graphical regional, as indicated by the smooth dashed curve on the profile. This regional is simply an expression of the form the profile would have if no local anomalies were present. This regional is subtracted from the observed gravity profile to leave the graphical residual as shown by the bottom line of the figure. We can then proceed with appropriate analysis and computation to determine quantitatively the meaning of this graphical residual and relate it to geological possibilities. However, someone else might look at this regional curve and say, "I don't believe you have drawn that curve exactly right." About the only answer I can give is to say that "if you don't like the way I have drawn it, you draw it your own way." The reason for making this statement is that there is no theoretical or analytical basis which will say that one regional curve is right and another is wrong. In the present case, the general form of the regional is simple, and anyone else would draw a curve rather closely similar to the one indicated in the figure. However, there are many cases where the determination of the regional is not nearly so obvious and different people with comparable experience might draw curves which were quite different, which, of course, would lead to differences in the significance attributed to the local anomaly itself.

In an attempt to resolve this problem and provide a more quantitative solution, there has grown up a series of analytical treatments, all of which may be considered together as "grid systems." In all of these systems, a grid of lines is laid down over the map. Values of gravity (or of magnetic intensity) are written at the grid corners, and these values are entered into a systematic numerical operation. Quite commonly the result of such an operation is a "second derivative" map, but other quantities sometimes are calculated. In all cases, a certain group of values

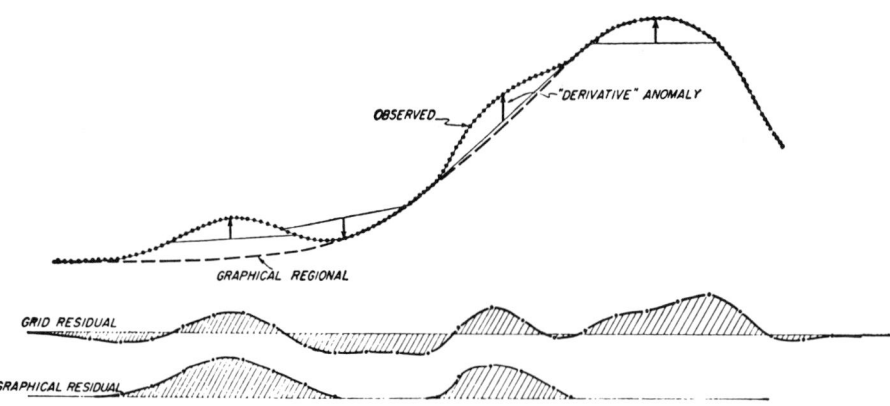

Fig. 13.—Hypothetical gravity profile showing regional, graphical residual, and grid residual (second derivative) anomalies. (Courtesy *Geophysics*)

around the point at which the calculation is made is taken into account, usually by determining an average value around several circles at different distances from the center and multiplying these averages by coefficients which have been determined by rather elaborate mathematical analysis (Nettleton, 1954). The remainder of Figure 13 is drawn to indicate the nature of such a second derivative calculation and to compare its results with those of the graphical residual analysis.

The short bars along the upper curve of Figure 13 represent the diameter of the array taken into account in each grid of second derivative calculation. If, in general, the values in the central part of the array are algebraically more positive than the average of the peripheral values, the calculation, no matter what its details, will lead to a positive number as indicated by the upward pointing arrows under the two anomalies of the diagram. If the values in the central part of the array are algebraically more negative than the peripheral areas, the calculation will lead to a negative number as indicated by the downward pointing arrow. If we carry out such a calculation along the observed gravity profile, we arrive at the "grid residual" curve as shown immediately below the observed gravity curve on Figure 13.

Let us consider the likenesses and differences of the results of these two different methods of removing regional effects and isolating the local anomaly. It will be seen immediately that for the two local anomalies on the profiles, the grid calculation develops an anomaly which centers at the same point as the center of the anomaly determined by the graphical method. On the right end of the diagram there is a substantial difference between the two operations. In the graphical residual, the large regional maximum was considered as entirely regional in origin and, therefore, the regional curve included it completely and no residual anomaly was left when the subtraction was made. On the other hand, the grid calculation has no judgment, and when it comes to the apex of the large regional anomaly it does the best it can and clips off a portion of the top which leads to a derivative anomaly in this area where none resulted from the graphical analysis. If one were to look at the grid residual or second derivative map alone, he would see three anomalies with the one on the right looking rather similar to the other two, as it is of similar amplitude although somewhat broader in lateral extent.

From the derivative map alone, all three anomalies might well be considered as expressing the same general type of geological disturbance. On the other hand, if one would look back at the original observed gravity map, he would see that the source of the grid residual anomaly on the right is entirely different from that of the two local anomalies and that the geological origin must be very different. This illustrates the need to look behind the second derivative or any grid calculation at the observed gravity map from which it came, as such considerations may lead to very different evaluation of certain anomalies.

Another indication of the difference between the two anomalies is the fact that the graphical residual analysis leads to a simple positive anomaly for the two local features on the map, whereas the grid calculation develops negative flanks on these anomalies. Thus, the complete derivative picture of the disturbance includes the positive anomaly in the central part, together with the flanking negative parts, all of which are components of the expression of a single disturbance.

The likenesses and differences of the two approaches to separation of anomalies may be further compared by application to an actual gravity map. Figure 14 shows a gravity map for an area south of Houston which is dominated by a fairly regular northward increase of gravity with generally east-west trends of the gravity contours, but with a distinct negative anomaly indicated by the northward bending of the contours. The problem then is to isolate this anomaly and evaluate its geological significance.

Figure 15 shows the evaluation by the graphical regional method. By a series of profiles across the area, graphical regional curves were drawn, and these were the basis for the smooth background contours at 1.0 mg. interval shown on Figure 15. These contours express the form of the regional gravity as it presumably would be if no local anomaly were present. By subtracting the field represented by these smooth contours from the observed contours as shown on Figure 14, the residual contours of Figure 15 at 0.2 mg. interval were obtained. These show a fairly regular closed minimum which centers over the relatively small South Houston salt dome which is indicated by the contours of heavy dots in the center representing −4,500-foot and −6,000-foot contours on the dome. A calculation of the expected gravity effect of this small dome gives a magnitude of

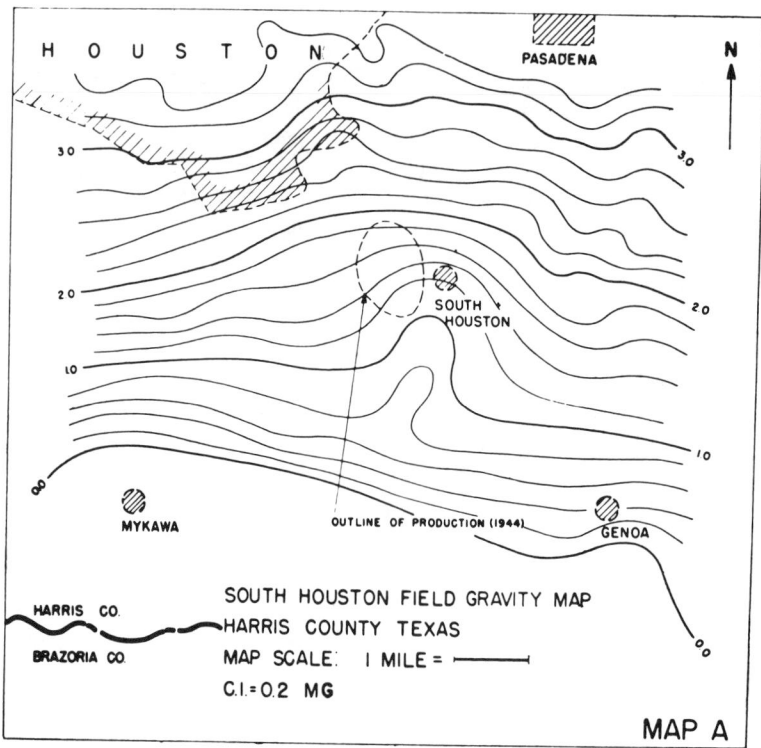

Fig. 14.—Observed gravity map, South Houston area, contour interval 0.2 mg. (Courtesy *Geophysics*)

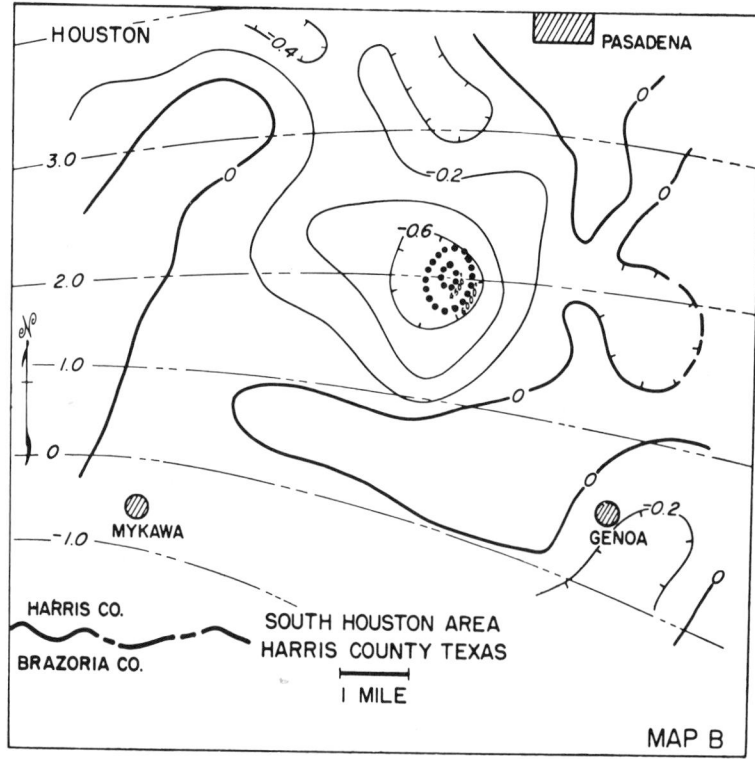

Fig. 15.—Regional contours (at 1.0 mg. interval) and graphical residual contours (at 0.2 mg. interval), South Houston area. (Courtesy *Geophysics*)

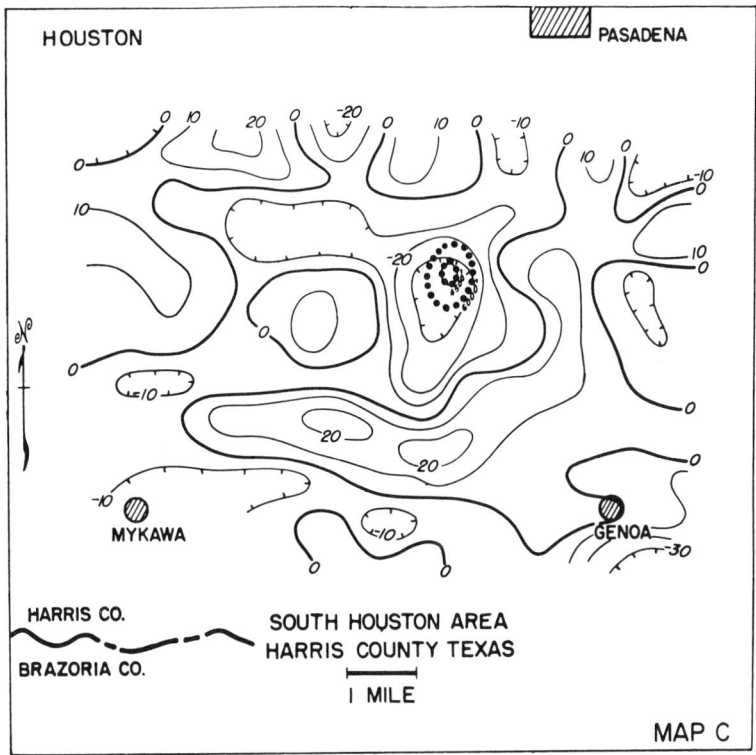

FIG. 16.—Second vertical derivative contours, South Houston area, calculated from grid with ½-mile spacing; contour interval 10×10⁻¹⁵ cgs. units. (Courtesy *Geophysics*)

about 0.7 mg. which is thus in good accord with the magnitude indicated by the residual gravity contours. On the whole, this is a fairly satisfactory quantitative gravity picture of this dome. There are some local irregularities caused by near-surface disturbances or possibly inaccuracies in the survey itself which become fairly apparent on this low relief gravity picture.

Figure 16 shows a second derivative map for the same area calculated from a one-half mile grid. The contours are in 10^{-15} cgs. units which is a common unit for second derivative maps. We note that this calculation develops a rather small, sharp, and somewhat irregular negative anomaly well centered over the dome. If we compare this with the graphical residual anomaly of Figure 15, we see that the second derivative anomaly is considerably smaller in area. This relation corresponds with that shown on the profile analysis of Figure 13 where the positive part of the grid residual anomaly was considerably narrower than the positive anomaly from the graphical analysis. On Figure 16 the central negative anomaly is sur-

rounded by positive anomalies consisting of the arcuate feature on its east and south sides, a circular positive feature on the west, and another on the north. These are all somewhat irregular expressions of the natural reversal around the anomaly and correspond to the negative parts of the grid residual curve on Figure 13; therefore, these positive anomalies do not mean that there is heavier material under these areas as the anomalies really are parts of the expression of the negative anomaly over the dome. Thus, the complete expression of the salt-dome disturbance includes the central negative anomaly and the surrounding positive disturbances which make an area somewhat larger than the graphical residual as shown on Figure 15. This larger area corresponds with the fact that the total disturbance on the profile of Figure 13, as shown by the grid residual calculation including its negative parts, is somewhat wider than the positive anomaly developed by the graphical analysis.

Figure 17 shows a second derivative calculation of the same area made with the same formula and

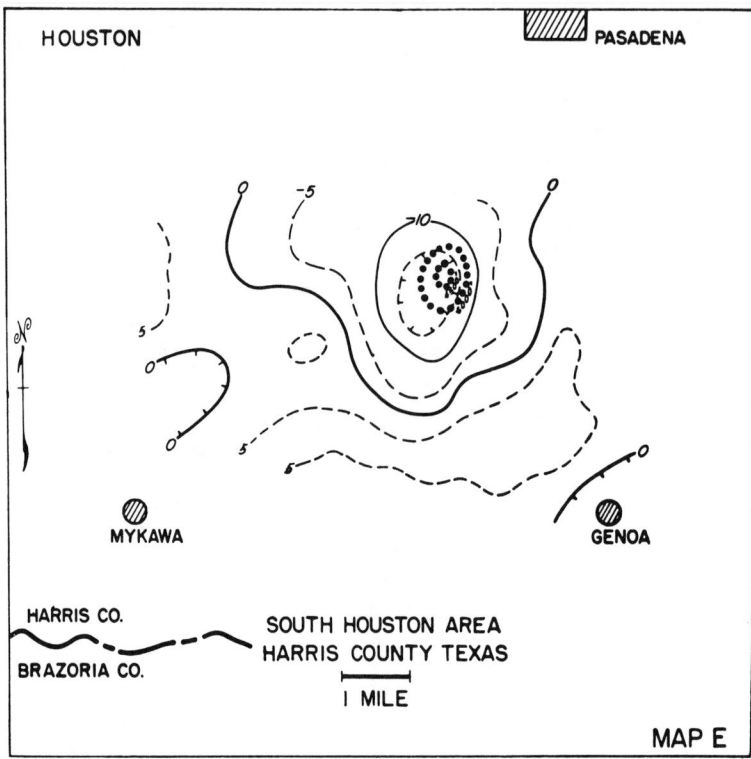

FIG. 17.—Second vertical derivative contours, South Houston area, calculated from grid with 1-mile spacing; contour interval 5×10^{-15} cgs. units. (Courtesy *Geophysics*)

coefficients but with a grid spacing of one mile rather than a half mile. The marked difference in the general appearance of the map and also the quantitative difference in the magnitude of the anomaly show why we do not like to use grid residual or second derivative calculations for quantitative analysis of gravity anomalies. For any such analysis, it is much better, in spite of the uncertainty and lack of any really analytical basis in determination of the regional, to use a graphical residual for quantitative calculations. The second derivative methods serve quite well to locate anomalies but are not suitable for quantitative analysis.

The second derivative calculations are very widely used. This is largely because they are much easier to carry out than the graphical process. Once the formulas and system have been set up, the entire operation can be turned over to relatively inexperienced calculators, and, also, it is readily adaptable to high-speed digital electronic computers. This means that in any office where

there is a large volume of gravity and magnetic data to be analyzed, one is almost certain to find an extensive series of second derivative maps. These can be very useful for determining location of anomalies when proper consideration is given to the source from which the map was derived, but such maps have to be used carefully and usually, if anything more than location of the anomalies is to be determined, further analysis, usually based on the observed gravity maps, is necessary.

An area which illustrates both regional and local gravity anomalies of significance is the Los Angeles basin. Figure 18 is a geologic map of the basin on which the regional gravity at 5 mg. contour interval has been superimposed. The general configuration of the gravity pattern is in accordance with expectations from the general form of the basin itself. The minimum is over the area of deepest sediments, and gravity increases very rapidly southwest toward the basement outcrop in the Palos Verdes hills. The sediments are all

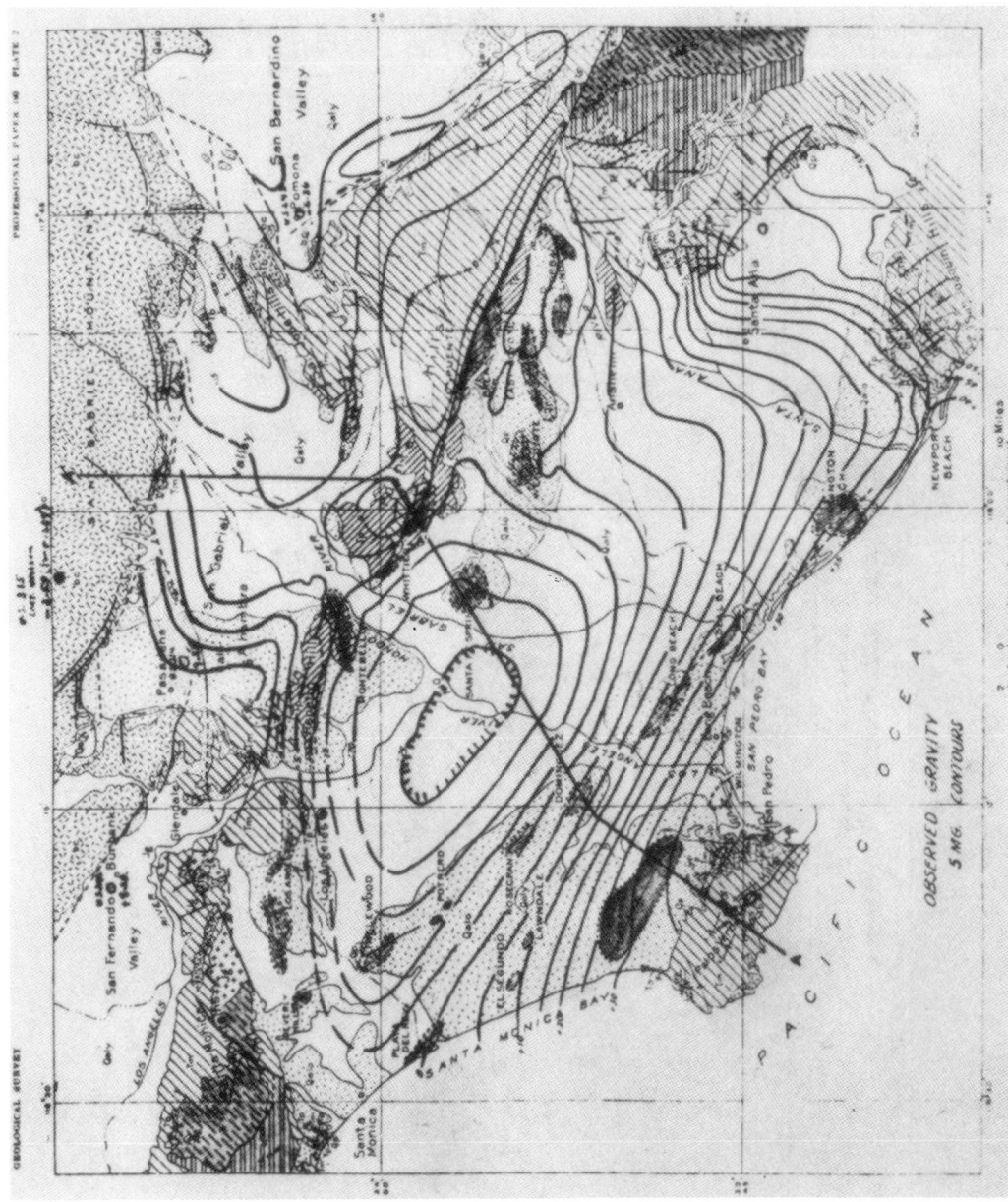

Fig. 18.—Surface geologic map, Los Angeles basin, with gravity contours at 5 mg. interval. (Modified from U.S.G.S. Professional Paper 190)

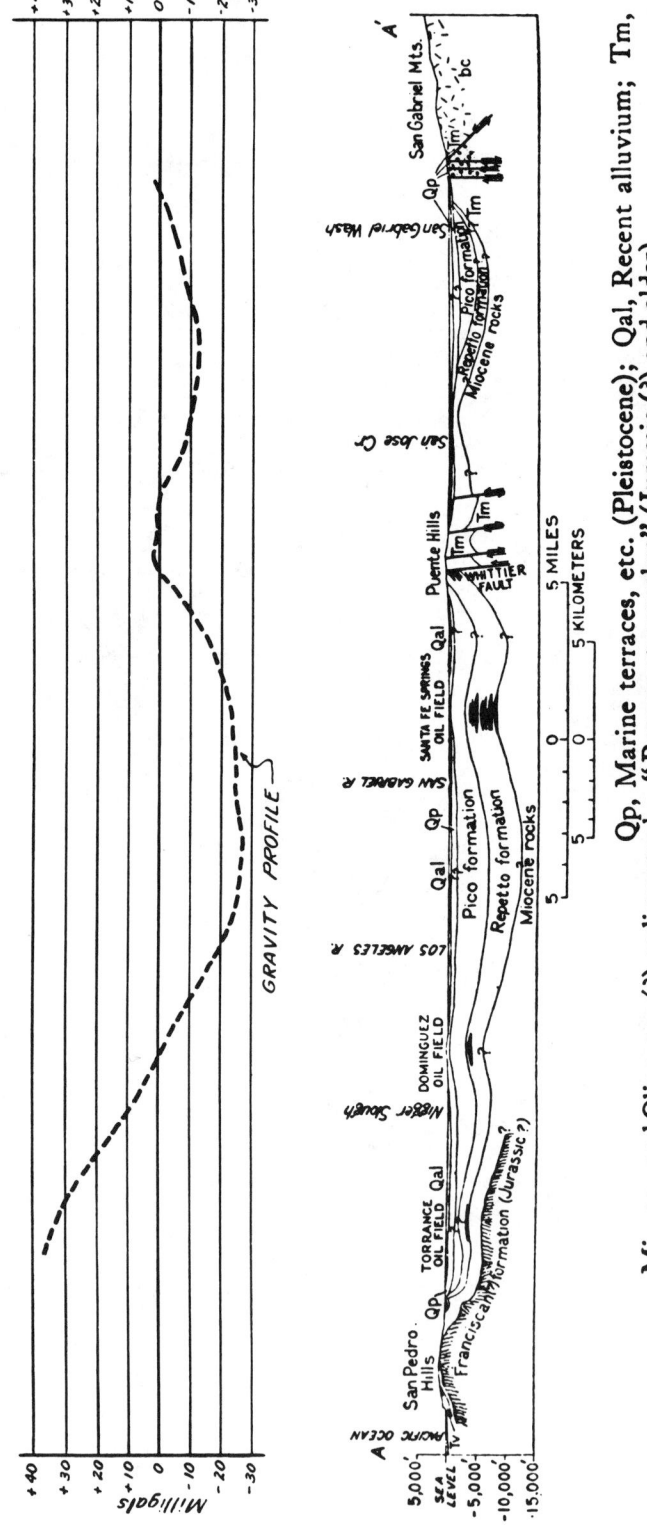

Qp, Marine terraces, etc. (Pleistocene); Qal, Recent alluvium; Tm,
Miocene and Oligocene (?) sediments; bc, "Basement complex," (Jurassic (?) and older)

Fig. 19.—Cross section (line AA', Fig. 18) showing gravity profile. (Modified from Guidebook, XVI International Geological Congress, 1932)

Tertiary clastics and of substantially lower density than the Franciscan schist of the basement and, quantitatively, the 60 mg. southwesterly gravity rise from the center of the minimum to the Palos Verdes Hills is consistent with that which would be expected from reasonable estimates of the total thickness of sediments and of the densities of the sedimentary and basement rocks.

Figure 19 is a profile across the basin along line AA', shown on Figure 18, and shows a geologic section together with the gravity profile from Figure 18. This profile shows the bottom of the

gravity minimum corresponding almost exactly with the deepest sedimentation. The very strong gravity maximum over the Puente Hills structure almost certainly means that the basement rocks are involved in this structure although not so shown on the geologic section. On the southwest side of the basin, we see that the Dominguez structure does not produce a noticeable gravity anomaly.

Figure 20 shows a detailed gravity survey, with ½-mile station spacing, over a part of the southwest flank of the basin which includes the areas of

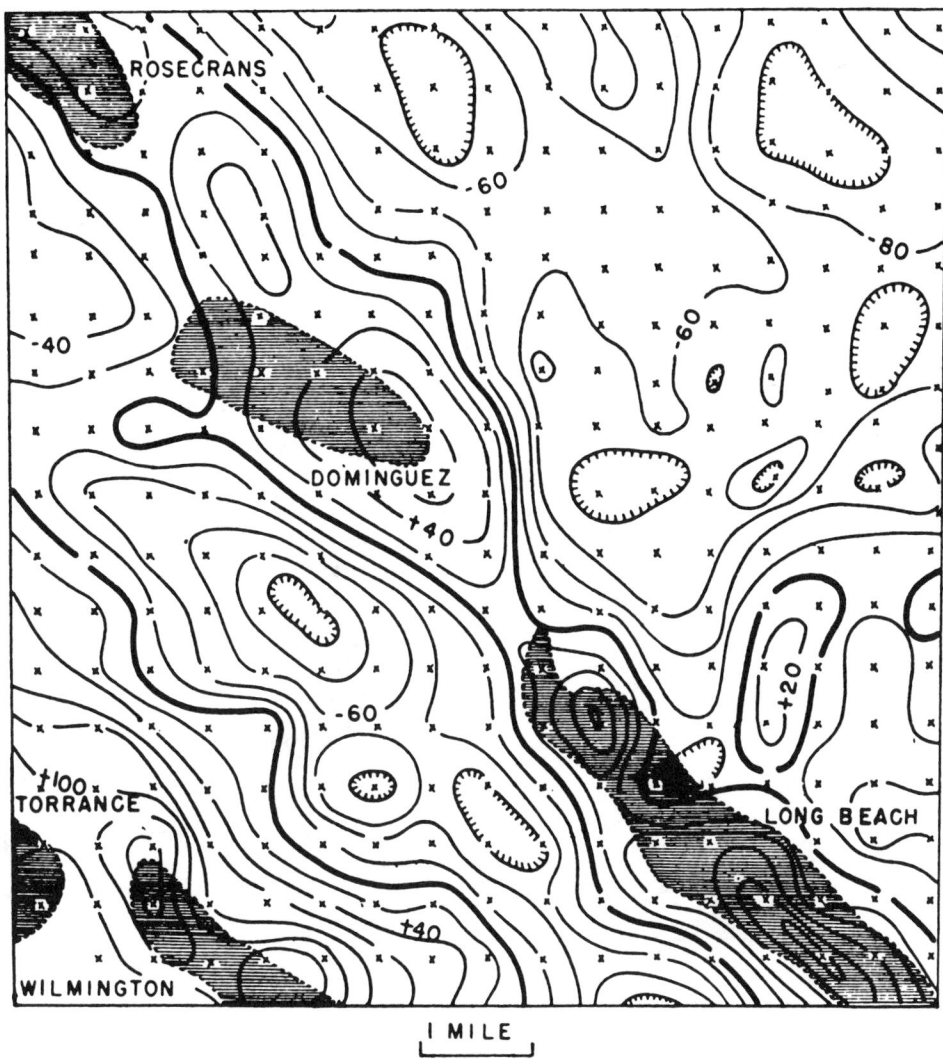

FIG. 20.—Gravity map, part of southwest flank, Los Angeles basin; contour interval 0.2 mg.; station spacing ½ mile.
(Courtesy *Geophysics*)

the Long Beach, Dominguez, parts of the Rosecrans, Wilmington, and Torrance oil fields. On this map, casual inspection would suggest that there is no gravity expression at all of these oilfield structures. However, if one would look carefully, he would notice that, for instance, on the northeast side of the Long Beach field the contours are a little closer together than over the field itself, and, similarly, on the northeast side of the Dominguez field the contours are crowded more closely than they are over the oil-field area. These differences are significant as shown by the

second derivative map for this area (Fig. 21). This map was calculated with a $\frac{1}{2}$-mile grid spacing from the observed gravity map of Figure 20. This second derivative map develops a very strong and definite positive trend which extends through the Long Beach, Dominguez, and Rosecrans fields, and also another positive trend which includes the Wilmington and Torrance fields. The second derivative calculation has brought out these features which were very subtle in the observed gravity map. Of course, they must be present on that map to be developed by the derivative calcu-

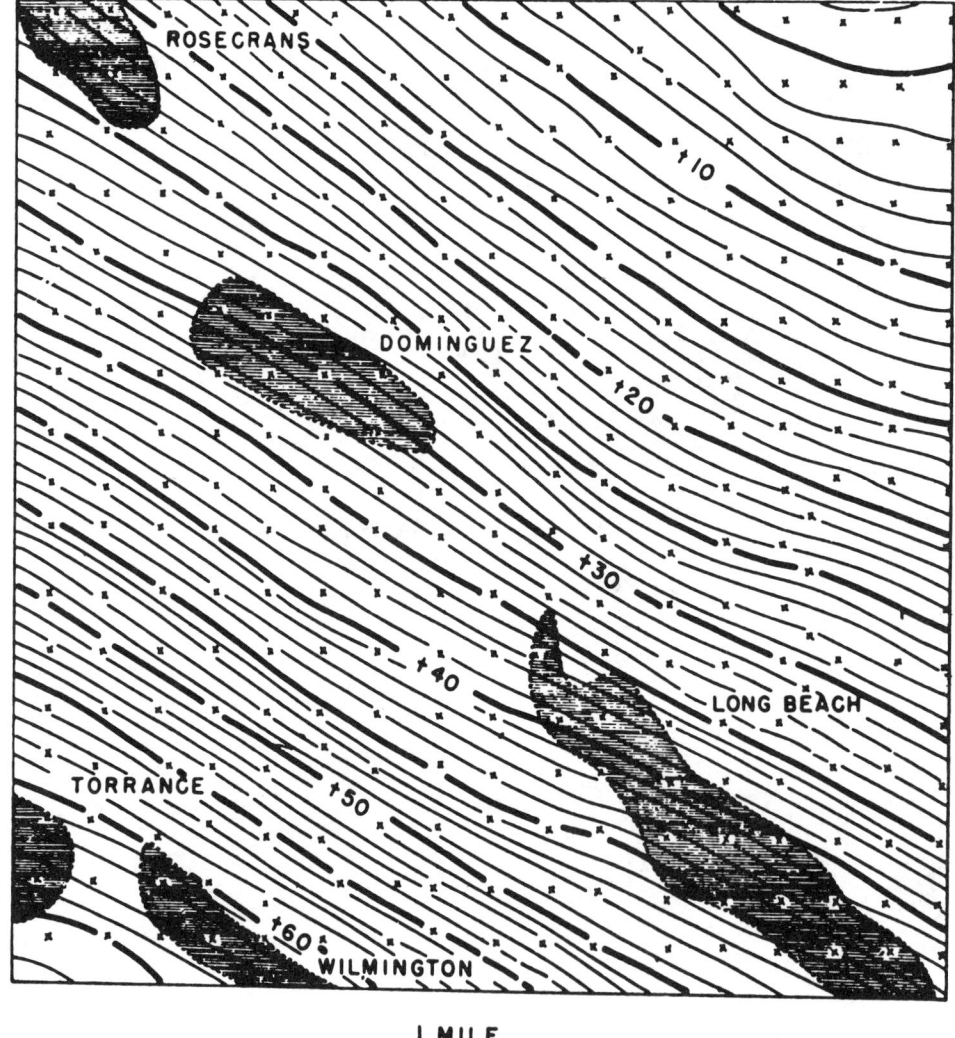

Fig. 21.—Second derivative map, part of southwest flank, Los Angeles basin, calculated from gravity of Figure 20; grid spacing one-half mile; contour interval 20×10^{-15} cgs. units. (Courtesy *Geophysics*)

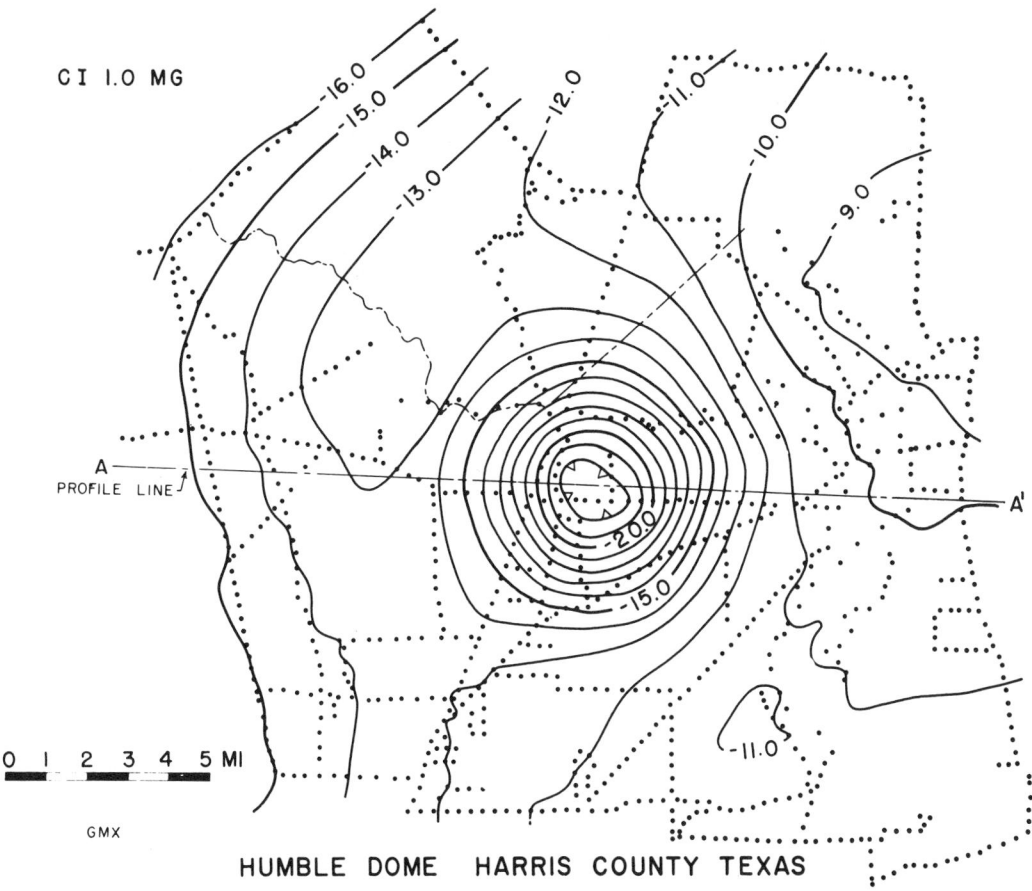

CI 1.0 MG

HUMBLE DOME   HARRIS COUNTY TEXAS

FIG. 22.—Bouguer gravity map, Humble salt dome, Harris County, Texas. Contour interval 1.0 mg. Control shown by dotted lines, each dot representing a gravity station location.

lation, and their presence is indicated by the previously mentioned variations in contour spacing. It should be pointed out also that the negative flanks of the positive trends are parts of the total anomaly in the same way that previously mentioned negative aspects of second derivative maps have been pointed out.

An example of a gravity anomaly subjected to quantitative analysis is shown by Figure 22. This is the gravity contour map, at 1.0 mg. interval, of the area of the Humble dome north of Houston. This dome probably has the largest dome salt mass in the Texas Gulf Coast (there are some larger ones in Louisiana). In this case, the anomaly is so strong and prominent that it dominates the picture to the extent that any regional effect is not very apparent (this is the reverse of the situation on the southwest side of the Los Angeles basin). There is evidence for the regional in the

reversal of the −13.0 contour on the west side of the minimum. The smooth "regional gravity" curve shown at the top of Figure 23 was obtained from a profile along line AA' of Figure 22 together with a number of other profiles across the gravity map. Subtraction of the regional from the observed gravity gives the "residual gravity" profile, and it is now our problem to find a salt dome form which will account for the residual gravity curve.

It is well known that any calculations of gravity effects to determine structures are ambiguous if we do not have information other than the gravity itself (Skeels, 1947; Nettleton, 1940). All such calculations depend on values for density differences. In the Gulf Coast, the densities of salt and of the sediments at depth are well known so that such calculations can be made with considerable confidence. After trials of different forms, the "calculated salt" form as shown on Figure 23 was determined and gives the calculated gravity effects as shown by the points along the residual gravity curve. From the close fit of these points with the curve, one might feel that a high degree of certainty had been achieved in determining the form of the dome. This may or may not be true depending primarily on the certainty with which the density contrasts are known. In this case, we are reasonably sure of the density values, and, furthermore, a limited amount of control by drilling shows that the calculated salt form is in accord with this control.

Figure 24 shows another salt dome calculation which is somewhat more complicated. The observed gravity curve shows a sharp irregularity in its central part. After determination of the regional, as indicated by the dashed line at the top of the diagram, and subtraction from the observed curve, we obtain the "residual" gravity curve as shown. Our problem now is to calculate a dome form which will account for this residual curve.

We note that the central disturbance of the observed gravity curve leads to a flattening and slight reversal in the bottom of the residual gravity curve and that this curve obviously consists of a broad negative part and a sharper central positive part. To one familiar with salt dome behavior and with the density contrasts involved, this immediately indicates that we are dealing with a dome having shallow cap rock. In order to make a quantitative calculation, it is necessary to separate the two components of the gravity picture. This can be done by first finding a salt column which will give calculated effects which fit the outer flank of the residual gravity curve, then carrying this same calculation across the central part of the profile to give the "interpolated negative" curve as shown on the figure. If we then subtract this salt effect curve, including the interpolated negative effect, from the residual curve, we get the "positive residual" which is now the isolated positive effect due to the cap rock. We can now devise a form, depth, and thickness for the cap rock, using the appropriate density contrast for the heavy cap rock minerals, which will give a

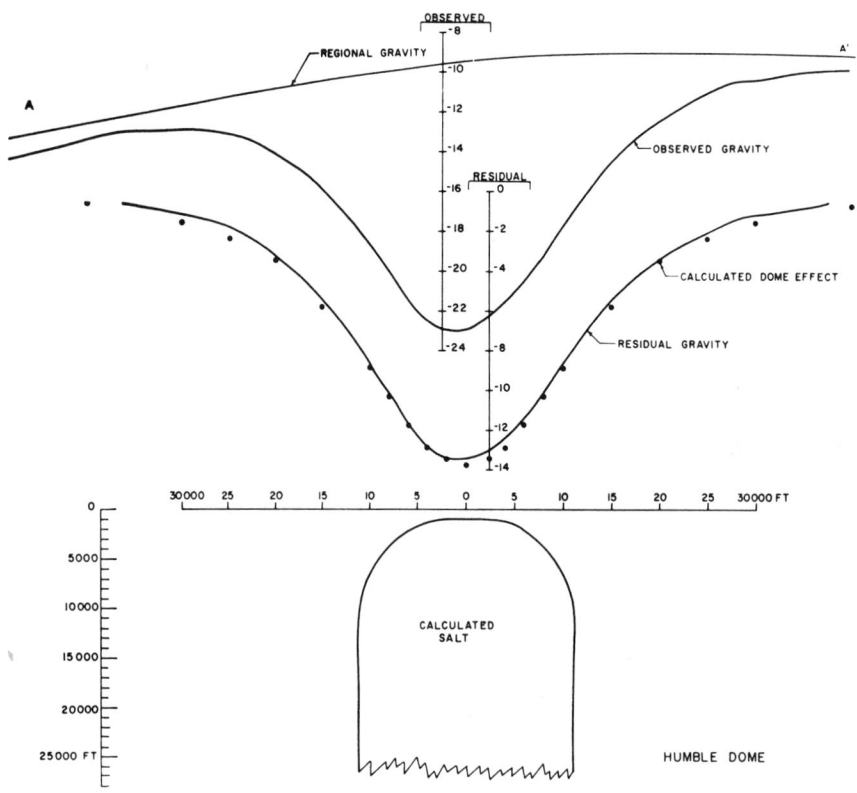

FIG. 23.—Observed, regional, and residual gravity on line AA' of Figure 22, with cross section and calculated gravity effect shown by dots along residual gravity curve.

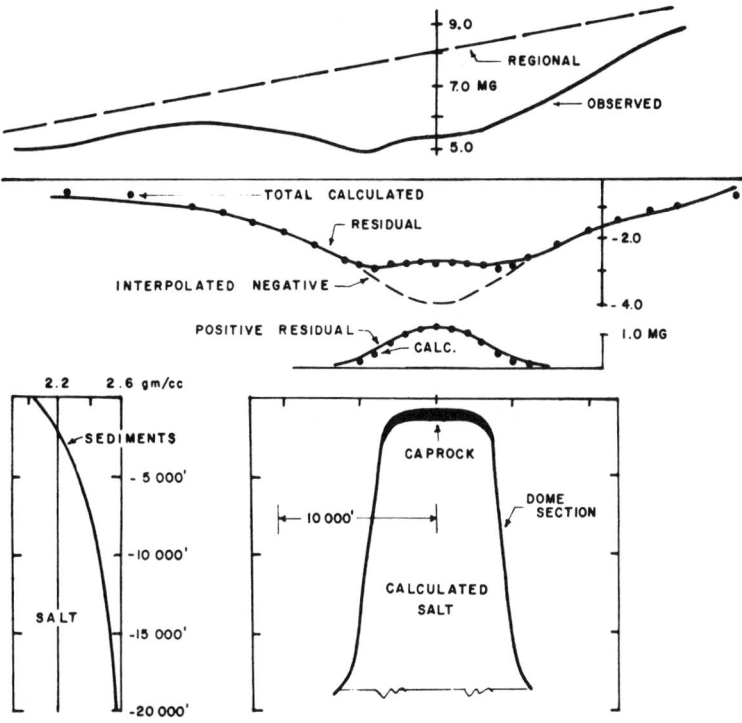

Fig. 24.—Observed, regional, and residual gravity profiles, shallow cap rock dome, with separation into negative and positive anomaly components; also showing dome cross section, calculated gravity effects, and density-depth curve for determination of density contrasts used for calculations.

calculated effect to fit the positive residual as shown by the "calc." points along the curve. We then combine these two calculated effects to give the "total calculated" effect as indicated by the dots along the residual gravity curve and find that the curve has been completely explained.

The diagram in the lower left part of Figure 24 indicates the density contrasts used in the calculation. So far as we know, the salt has a nearly constant density of about 2.2 and is not appreciably compressed or changed in density with depth. On the other hand, the sediments increase in density rapidly from a value of about 2.0 near the surface to a value of about 2.6 at great depth. The depth at which the salt and sediments have the same density is relatively shallow so that, in this geological area, the upper part of the salt contributes very little to the gravity disturbance. In other areas such as the southeastern part of the Louisiana salt dome belt, and especially offshore, the density increases with depth at a much lower rate. This is to be expected from the very rapid deposition of the sediments farther south and, therefore, less geological time for compaction and lithification

at a given depth. For instance, a deep well in the lower Mississippi delta, at a depth of 22,500 feet, was still in middle Miocene, which shows that these sediments have been laid down extremely rapidly. In such areas the slower increase of density with depth due to less compaction means that there may be a substantial thickness of salt in shallow domes which is above the "crossover point" where the densities of salt and sediments are the same and, therefore, that there can be substantial positive gravity effects due to shallow salt without cap rock being involved. This condition is not very common on the onshore domes but is fairly common on the offshore domes. By very careful quantitative considerations and establishment of a correct density-depth curve, it is possible to determine whether the origin of a positive anomaly is from shallow salt or from cap rock and thus distinguish between two interpretations which would give quite different quantitative values for depth and form of the shallow part of the dome.

Figure 25 shows a gravity anomaly which has a quite different solution. Superficially this is quite similar to a salt dome anomaly although it might

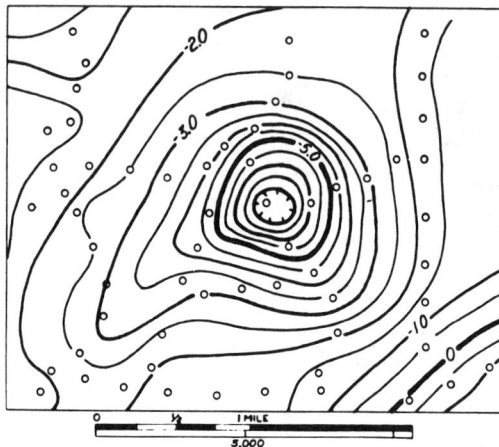

Fig. 25.—Residual gravity map, mud volcano area, Trinidad; contour interval 0.5 mg. Gravity station locations shown by circles. (Courtesy *Geophysics*).

be considered rather sharp if it occurred in a salt dome area. However, this anomaly is from southeastern Trinidad in an area of sharp diapiric structures and of mud volcanoes. Also, there are other mud volcanoes in southern Trinidad which produce negative gravity anomalies. Now, why should a mud volcano produce a negative gravity anomaly? Mud volcanoes are essentially gas seeps with the gas escaping at depth and coming up through several thousand feet of sedimentary section. Apparently water gets into the column somewhere and also the section includes layers of mud, possibly from turbidity current deposits. There is a mud volcano on the south coast of Trinidad which occurs on a slope and from which mud runs off down to the beach. The mud from the volcano comes out in sufficient volume so that vegetation does not grow on it and, in an aerial photograph, it looks like a landslide scar. On the beach, the waves have washed away the mud and have left a collection of rocks and boulders of various sizes which are scattered out and which serve to sample the geologic section through which the volcano has intruded. The geologist familiar with the area can pick out samples here from several thousand feet of geologic section, and, therefore, this mud volcano must have a fairly deep origin. Presumably, the negative gravity anomalies over such volcanoes are caused by mud which has been lightened or leavened by inclusion of a considerable

volume of gas bubbles to give it a low density. In this case, we have no direct information on the densities or density contrasts involved. For the sake of discussion, we might assume that the mud has a density of, say, 1.7 and the adjacent clastic sediments should have a density of around 2.2. This would give a density contrast of 0.5. On this basis, if we calculate the volume of material necessary to cause the anomaly shown in Figure 25, we come up with a hole or cauldron, or vent, through which the mud must arise having an average diameter of the order of $\frac{1}{2}$ mile. This seems a bit large, but until we have more definite density information or other basis to change the calculation, this is the tentative solution to the problem. In this case, our solution is not nearly so definite as is that of the salt domes because we do not have the necessary geological information, particularly as to densities. So far as this writer knows, this little piece of gravity data is the only indication we have as to the underground nature of these mud volcanoes.

It is hoped that by this review of the magnetic and gravity methods, some better appreciation may be gained of their applicability to petroleum exploration so that when geologists or others speak of "geophysics" they may remember that there is something to be considered besides a reflection seismograph. Perhaps in the future when the real possibilities of these more obscure and less well understood methods are realized, we may increase their proportion of the total exploration picture from their present 5 per cent to perhaps 6 per cent or 7 per cent. Also, we may hope to open again the forgotten files of gravity and magnetic surveys for review and correlation with seismograph, subsurface geological, and logging information as well as surface geology. Many new prospects may thus be found.

REFERENCES

DeGolyer, E. L., 1928, The seductive influence of the closed contour: Economic Geology, v. 23, p. 681–682.
Jacobsen, Peter, Jr., 1961, An evaluation of basement depth determinations from airborne magnetometer data: Geophysics, v. 26, no. 3, p. 309–319.
Nettleton, L. L., 1940, Geophysical prospecting for oil: McGraw-Hill, p. 101.
——— 1954, Regionals, residuals, and structures: Geophysics, v. 19, no. 1, p. 1–22.
Skeels, D. C., 1947, Ambiguity in gravity interpretation: Geophysics, v. 12, no. 1, pp. 43–56.
Steenland, N. C., 1962, An evaluation of the Peace River aeromagnetic survey: Geophysics, in press.

GEOPHYSICS, VOL. 40, NO. 2 (APRIL 1975), P. 256–268; 11 FIGS., 4 TABLES

EXPLORING FOR STRATIGRAPHIC TRAPS WITH GRAVITY GRADIENTS

SIGMUND HAMMER* AND RODOLFO ANZOLEAGA‡

The vast and growing literature on the search for stratigraphic traps for petroleum ignores gravity gradients, for which the theory has been available since the heyday of the Eötvös torsion balance decades ago. These are discussed in this paper.

The horizontal and vertical gradients can be measured with available gravimeters and (to a limited extent) with the Eötvös torsion balance. A major advantage of the gradient method is that surveying to determine position and elevation of the station is not required. Both theory and practice have been reported in the geophysical literature, but the important application to stratigraphic traps has not been mentioned. We evaluate here the method for locating "pinchouts", a term which embraces "stratigraphic" and "unconformity" traps in Halbouty's (1972) classification. Both position and depth of the assumed pinchout are determined by the gradient anomaly. The magnitudes of anomalies of horizontal and vertical gradients are about equal. However, pending new instrumentation, only the horizontal gradient is practically useful for field surveys. The gradient method is quantitatively promising and, used in conjunction with other methods, should significantly advance the search for stratigraphic traps for petroleum.

INTRODUCTION

One of the important, but also one of the most difficult, exploration problems confronting the petroleum industry is finding stratigraphic traps. This was emphasized recently in an important volume published jointly by the AAPG and SEG (King, 1972). Completely ignored or forgotten in the vast and growing published literature on the subject is the use of gravity gradients for which the basic theory and practice have been available since the heyday of the Eötvös torsion balance, fifty years ago (Heiland, 1940, 1943; Thyssen-Bornemisza and Stackler, 1956; Thyssen-Bornemisza, 1965). The present paper is intended to fill this void in the literature. The authors believe that total dependence on only one geophysical method is inadvisable and that judicious use of gravity gradients as a complementary approach will significantly advance exploration for stratigraphic traps.

The nature of the gravity problem is illustrated in Figure 1. It is apparent from the diagram that the pinchout is well defined by the horizontal (GX) and vertical (GZ) gradient profiles. The definition is less evident on the gravity curve. In fact, it is difficult to distinguish the gravity anomaly of the pinchout from that of a vertical fault.

The pinchout edge of a stratigraphic trap is sharply expressed in the various gravity gradients (horizontal, vertical, differential curvature) across it. The gradient anomaly depends upon the dip angle and depth of the pinchout and, of course, on the existence of density contrast. (This limitation is not unique to gravity.

Paper presented at the 43rd Annual International SEG Meeting, October 25, 1973 in Mexico City. Manuscript received by the Editor October 4, 1973; revised manuscript received October 24, 1974.
* University of Wisconsin, Madison, Wisc. 53706.
‡ Yacimientos Petroliferos Fiscales Bolivianos, Santa Cruz, Bolivia.

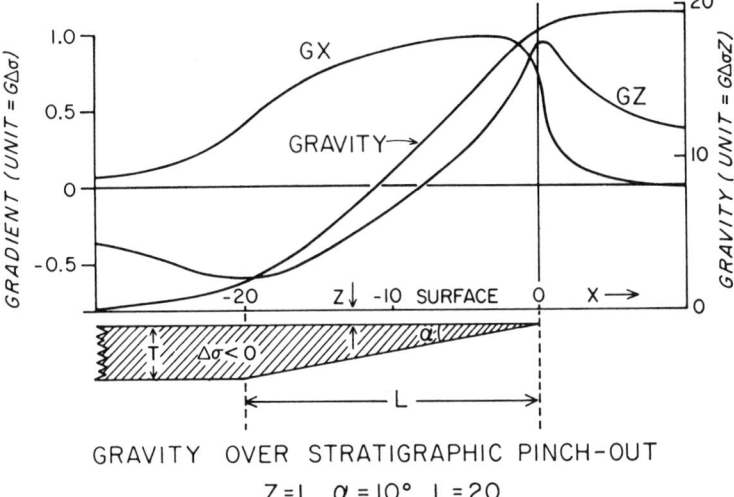

GRAVITY OVER STRATIGRAPHIC PINCH-OUT

Z = 1   α = 10°   L = 20

FIG. 1. Gravity and gradient anomalies on a stratigraphic pinchout. GX and GZ are the horizontal and vertical gradients of gravity, respectively. The ordinates are in dimensionless units. The depth to the top of the pinchout section Z is the unit of distance.

All exploration methods fail unless there are adequate contrasts in the stratigraphy.) For practically important cases, the gradient anomaly is within the range of observation with the original Eötvös torsion balance and with modern gravimeters. An important advantage of a gradient survey is that no connected survey control is needed. Elevation of the station is not required. Location is required only to plot the station on the map. These are not factors in the numerical station data as they are for gravity. Therefore the gradient method has major cost advantages especially in areas of difficult accessibility. All that is required is to be able to occupy satisfactory station sites at adequate spacing. Any type of transportation between stations, including helicopter, may be

used. Another important advantage of a gradient method is its greatly increased sensitivity as compared to the usual gravity survey. Furthermore, the well-known sensitivity to terrain corrections in hilly topography (Thyssen-Bornemisza, 1958) and to local variations in soil density ("gradient noise") is greatly diminished in the present context.

THEORY

A diagram for the gradient theory at a pinchout is illustrated in Figure 2. The horizontal extent of the pinchout, perpendicular to the plane of the figure, is assumed to be infinite. Finite extent can be treated, but that situation is seldom important in pinchout problems.

The gradients for the pinchout in Figures 1

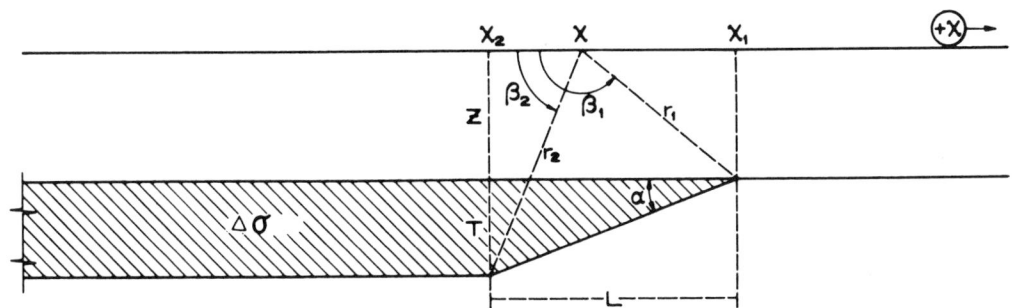

FIG. 2. Diagram for stratigraphic pinchout theory.

and 2 (Heiland, 1940) are

(a) horizontal gradient:

$$\frac{1}{G\Delta\sigma}\frac{dg}{dx} \equiv GX = 2\sin^2\alpha\cdot\ln\left(\frac{r_2}{r_1}\right)$$
$$+ \sin 2\alpha\cdot(\beta_1 - \beta_2), \quad (1)$$

(b) vertical gradient:

$$\frac{1}{G\Delta\sigma}\frac{dg}{dz} \equiv GZ = \sin 2\alpha\cdot\ln\left(\frac{r_2}{r_1}\right)$$
$$- 2\sin^2\alpha\cdot(\beta_1 - \beta_2). \quad (2)$$

Here

$$r_1^2 = (X_1 - X)^2 + Z^2,$$

$$r_2^2 = (X_2 - X)^2 + (Z + T)^2,$$

$$\beta_1 = \pi/2 + \tan^{-1}\left(\frac{X_1 - X}{Z}\right),$$

$$\beta_2 = \pi/2 + \tan^{-1}\left(\frac{X_2 - X}{Z + T}\right),$$

G is the gravitational constant, and $\Delta\sigma$ is the density contrast in the pinchout section. These equations, like all gravity gradient equations, are conveniently nondimensional.

For two-dimensional cases $(dy/dx) = 0$, and the vertical gradient is equal in magnitude and sign to the differential curvature, which is one of the two quantities measured with the Eötvös torsion balance. Therefore, the Eötvös torsion balance could be used for making routine surveys of both the horizontal and vertical gradients across two-dimensional stratigraphic pinchouts. However, modern, sensitive gravimeters are more convenient.

Both horizontal and vertical gradients can be determined with a gravimeter as follows:

(a) *Horizontal gradient* (Heiland, 1943)—. To measure the horizontal gravity gradient at a point, observe gravity differences Δg and elevation differences ΔH at three stations set out to form a small triangle as sketched in Figure 3. Small corrections are made for eleva-

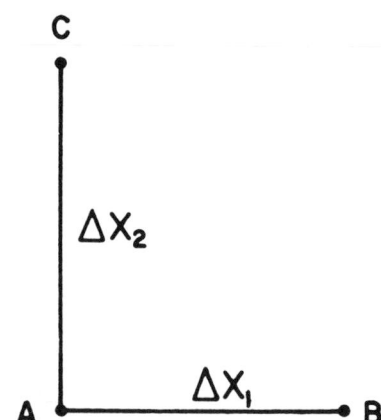

Fig. 3. Field layout for observing the horizontal gradient of gravity with a gravimeter. It is convenient but not necessary that the triangle be a right triangle.

tion differences and normal (global) gravity changes along each leg. The horizontal gradient is then calculated from the corrected "observed" gradient components $(\Delta g/\Delta x)$ along the two legs.

Precision requirements for results of useful accuracy are given in Table 1. The required precision for ΔX is directly proportional to the component of the horizontal gradient along the leg. The tabulated values assume a reasonably high value of 20E°. Since the Δg values represent differences of two determinations, the tolerable error for each individual value is smaller than tabulated in Table 1, and also in Table 2, by the factor $(2)^{1/2}$.

It is evident in Table 1 that modern gravimeters are practical, but that a high-precision gravimeter such as the recent LaCoste and Romberg model D "microgal" meter would be advantageous for faster field progress (i.e., shorter legs and fewer repeat readings for each leg of the triangle).

(b) *Vertical gradient* (Thyssen-Bornemisza and Stackler, 1956, and others)—. The measurement of the vertical gradient with a gravimeter may be done by observing the difference in gravity

Table 1. Precision requirements for measuring the horizontal gradient of gravity to an accuracy of 2 Eötvös units

Length of sides, m	10	20	30	40	50	100
Precision of Δg, mgal	0.002	0.004	0.006	0.008	0.010	0.020
Precision of ΔH, cm	0.9	1.8	2.7	3.6	4.5	8.9
Precision of ΔX, m	1	2	3	4	5	10

Table 2. Precision requirements for measuring the vertical gradient of gravity to an accuracy of 2 Eötvös units

Height of tower, m	2	4	6	8	10
Precision of Δg, mgal	0.0004	0.0008	0.0012	0.0016	0.0020
Precision of ΔH, cm	0.13	0.26	0.39	0.52	0.65

at two elevations, usually at the base and top of a portable, double tower. The inner tower supports the gravimeter and the outer tower the observer. The procedure has been described at length in the technical literature (e.g., Thyssen-Bornemisza and Stackler, 1956). Precision requirements of the observations are given in Table 2.

It is apparent in Table 2 that extreme precision is required for both Δg and ΔH values in order to get reliable measurements of the vertical gradient with a gravimeter. Furthermore, the Eötvös torsion balance does not measure the vertical gradient except in two-dimensional situations where it becomes equal to the differ-

ential curvature. Useful vertical gradient surveying is hardly practicable with presently available instrumentation.

Calculated horizontal gradient profiles, equation (1), across a series of stratigraphic traps with various dip angles α are shown in Figure 4. Similar curves of the vertical gradient, equation (2), are shown in Figure 5. All the pinchout models were calculated with flat top and sloping bottom. More generally, pinchouts with sloping top and bottom are easily found by taking the difference between appropriate curves. The ordinate scales in Figures 4 and 5 are dimensionless. Conversion to practical units is given by multiplying the ordinate value by

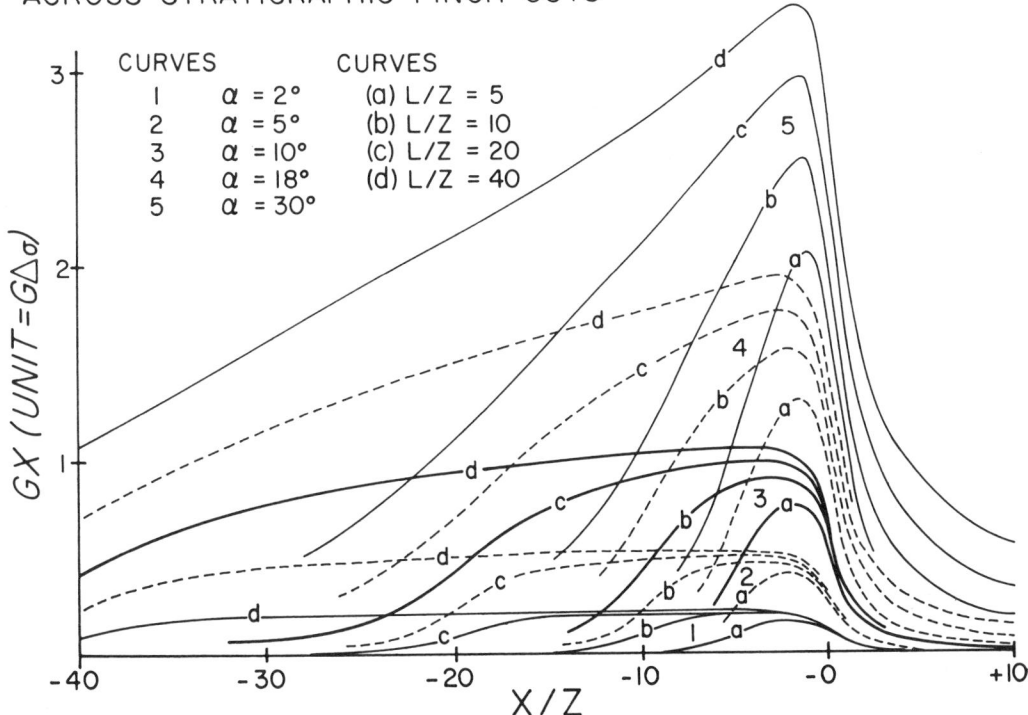

FIG. 4. Theoretical profiles of the horizontal gradient of gravity over stratigraphic traps. Density contrast $\Delta\sigma$ in the pinchout section has been assumed to be negative. The variable parameters are dip angle α and slope distance L.

VERTICAL GRADIENTS OF GRAVITY
ACROSS STRATIGRAPHIC PINCHOUTS

PINCHOUT	α	CURVE	L/Z
1	2°	(a)	5
2	5°	(b)	10
3	10°	(c)	20
4	18°	(d)	40
5	30°		

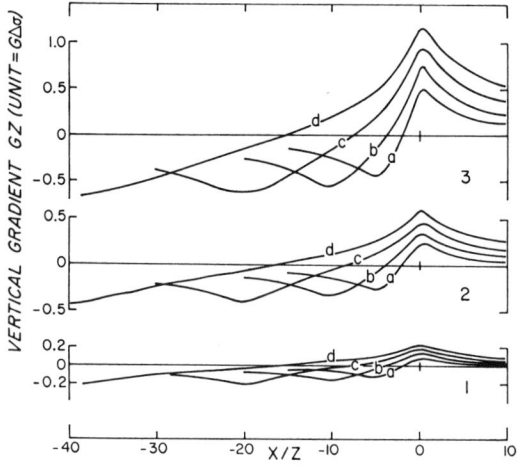
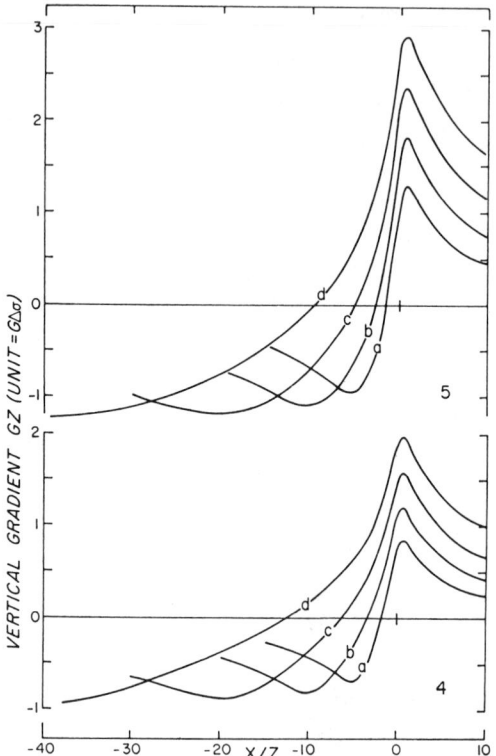

FIG. 5. Theoretical profiles of the vertical gradient of gravity over stratigraphic traps. The models are the same as in Figure 4.

appropriate values of the unit $G\Delta\sigma$, for example,

gradient in mgals/km $= 6.67\Delta\sigma$ times the ordinate value, (3)

gradient in Eötvös units ($E°$) $= 66.7\Delta\sigma$ times the ordinate value.

For a dip angle of 5°43′ (tan $\alpha = 0.1$) and $\Delta\sigma = -0.25$, the magnitudes of the theoretical anomalies are

horizontal gradient $= GX_{max} = 9.5E°$,

vertical gradient $= GZ_{max} = 10.3E°$ (positive portion).

It is clear by comparison of Figures 4 and 5 that the horizontal gradient profile gives a better expression of the up-dip edge of the pinchout than the vertical gradient. For this reason, also, the analysis to follow will be limited to the horizontal gradient method.

Locating the pinchout

It is apparent in Figure 4 that the point of steepest slope of the gradient profile is nearly co-incident with the edge of the pinchout ($x = 0$). In addition, the point of steepest slope is not shifted appreciably by the stratigraphic dip angle α. The position and magnitude of the maximum slope is also nearly independent of the horizontal extension L of the dip slope as illustrated in Figure 6. Consequently, the location of the pinchout, irrespective of its geometry, is well defined by the point of maximum slope on the horizontal gradient profile.[1]

[1] The linear portion of the gravity profile in the center of Figure 1 warrants comment. The associated relation $dg/dx = 2\pi G\Delta\sigma\tan\alpha$, which results from differentiation of the infinite flat plate formula $g = 2\pi G\Delta\sigma t$, has had some use. A more explicit derivation is as follows: At points close to and near the middle of a long sloping plane, we have

$$r_1 \cong r_2, \text{ and } \beta_1 - \beta_2 \cong \pi.$$

This reduces equation (1) in the theory above to

$$dg/dx = \pi G\Delta\sigma\sin 2\alpha.$$

But for small dip angles α we have

$$\sin 2\alpha \cong 2\sin\alpha \cong 2\tan\alpha$$

which completes the derivation. This derivation demonstrates that the linear tangent formula above has restricted validity with respect to dip and, especially, with respect to depth (T. A. Elkins, unpublished).

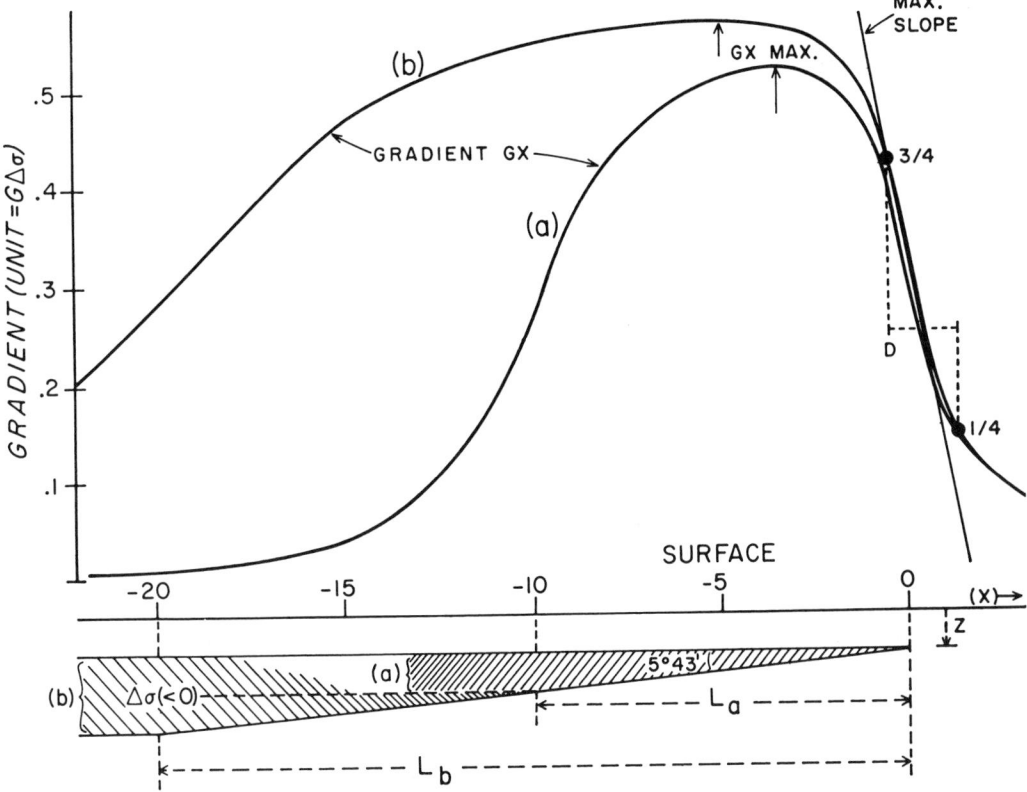

FIG. 6. Horizontal gradient of gravity profiles across pinchouts with different slope lengths L. The factors $1/S$ and D for estimating depth to the pinchout are indicated.

Depth estimation

Since the profiles are plotted in dimensionless units, increasing the depth unit will correspondingly increase the distance unit and will decrease the slope of the curve. Consequently, the slope of the gradient profile is inversely proportional to the depth. The constant of proportionality is evaluated below.

The estimate of depth to the pinchout may be made from the horizontal gradient profile in two ways: (a) from the maximum slope and (b) from the "half-width" D. The factors for the calculation are diagramed in Figure 6. Two cases with different slope lengths L are shown. It is apparent that the right-hand, up-dip portion of the gradient profile is almost independent of L. However, as may be seen in Figure 4, the maximum slope S_{max} across the up-dip edge is appreciably affected by the dip angle α. This is because the magnitude of the gradient anomaly GX_{max} increases with α and also with the slope length L. However, as will

be seen below, the "normalized" inverse maximum slope $(1/S = GX_{max}/S_{max})$ is relatively free of this dependence. The normalizing factor GX_{max} is included automatically in the half-width D.

All the profiles in Figure 4, and several others not shown, were analyzed to obtain values of normalized inverse slope $1/S$ and half-width D. The results are plotted in Figure 7 as a function of the dip angle α. Different curves designate data from different length factors L. It is apparent that for a large range of dip angles ($\alpha < 20°$) the normalized depth factor $1/S$ and, to a more limited extent, the half-width factor D are fairly independent of L and α. This is fortunate because ordinarily α and L are not known. Adopting arbitrary values, we get approximate work formulas for estimating the depth to the pinchout:

$$Z = D/2,$$

and

$$Z = 2.8/S. \tag{4}$$

Of the two formulas, the second is the more reliable because, as seen in Figure 7, it is less sensitive to geometry of the pinchout.

Other interpretation factors

The fact that the D and $1/S$ factors are practically independent of L and α prevents their use for estimating these parameters of the pinchout. An estimate of L can be made from the observation (see Figure 4) that the horizontal gradient drops to about one-half the maximum value at a point above the lower termination of the slope. The dip angle α can then be estimated approximately from the maximum value of the gradient GX_{max} by introducing known or assumed values of the density contrast.

The analysis outlined above is based on the assumption that the gradient anomaly represents a stratigraphic pinchout. Under this assumption, the interpretation becomes unique. More generally, the anomaly might represent a major facies change in which the density of a stratum increases toward the right. In any case, a sharp anomaly such as this warrants evaluation by other data as a normal precaution. Such multisensor procedure should not be limited to gravity prospects.

GRADIENT NOISE

Sensitivity of gravity gradient data to local variations in soil density and adjacent terrain irregularities has been well recognized since the days of torsion balance exploration. In conformity with geophysical usage these extraneous effects in the data can be considered as "gradient noise". To proceed we must specify the term. The noise in gravity data used for conventional petroleum exploration may be the direct objective in mining exploration. Furthermore, noise in mining exploration may be the direct objective for gravity surveys in civil engineering, speleology, and archaeology. This is a new and developing field of application, quite properly called "microgravimetry".

Modern high-precision "microgal" gravimeters have mapped very small anomalies having horizontal dimensions of some decameters and gravity relief of some 100 microgals (0.1 mgal) (Neumann, 1967). The associated gravity gradients reach values as high as 100 E°. For the purpose of the present topic, such tiny anom-

alies are "noise" in the ambient gravity field. Aside from instrumental and other sources of operational error, there are two sources of gradient noise to be considered.

Terrain noise

Practical procedures for evaluating terrain corrections are well developed but are tedious to apply. To simplify the analysis of the magnitude and behavior of terrain effects in the context of the present problem, we simulate a terrain unit by a simple geometric form. A topographic ridge may be simulated by a horizontal cylinder of infinite length placed upon the datum surface, as diagrammed in the inset in Figure 8. The gravity and gradient effects of such a cylinder are (Nettleton, 1940)

$$g = \frac{2Gmz}{x^2 + z^2} = 2\pi G\sigma R^2$$
$$\cdot (R - H)/\{X^2 + (R - H)^2\} \quad (5a)$$

$$\frac{\partial g}{\partial X} = -\frac{4Gmxz}{(x^2 + z^2)^2} = -4\pi G\sigma R^2 (R - H)$$
$$\cdot X/\{X^2 + (R - H)^2\}^2, \quad (6a)$$

where

$$m = \pi\sigma R^2.$$

For convenience in calculation and plotting, we

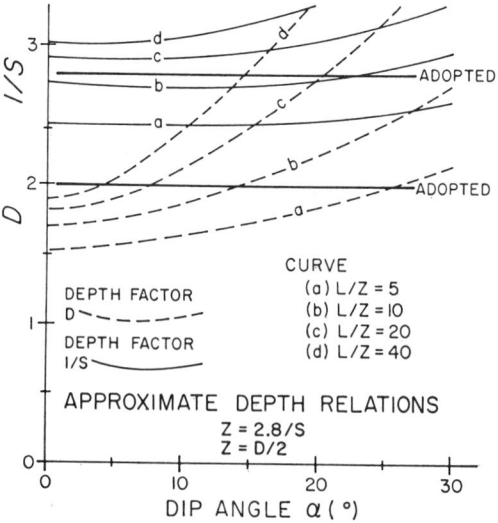

FIG. 7. Dependence on dip angle α and slope length L of the factors for estimating the depth of a stratigraphic pinchout from the profile of the horizontal gradient of gravity.

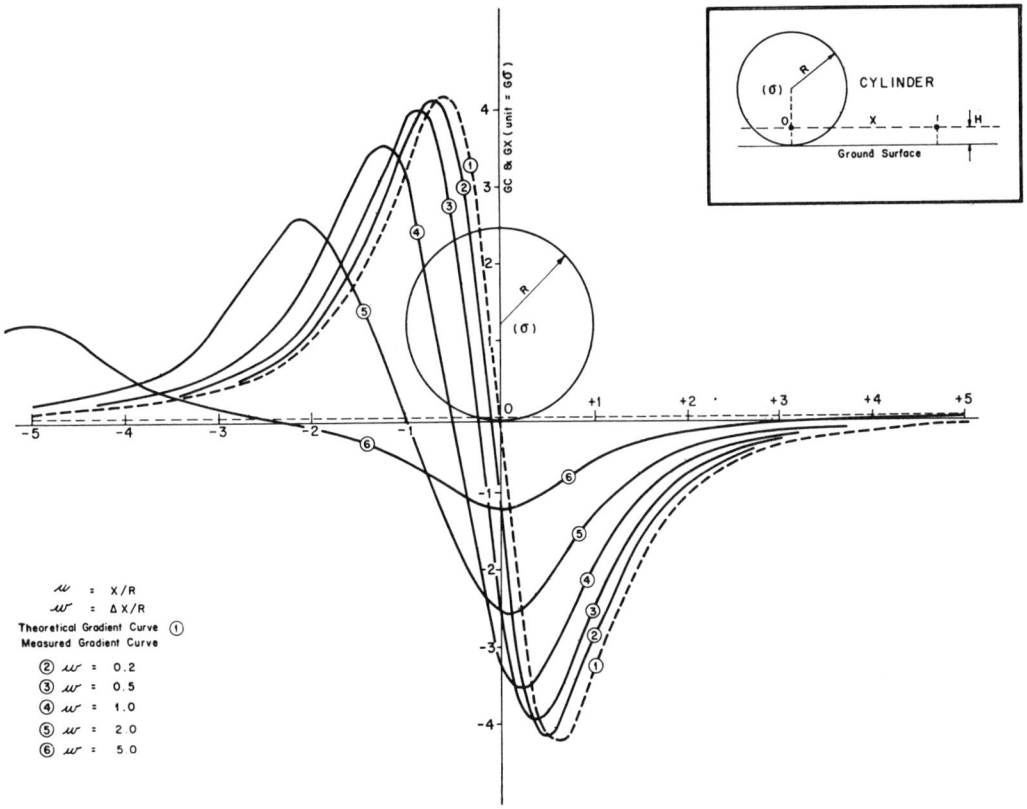

FIG. 8. Terrain gradient effects of a topographic ridge simulated by a horizontal cylinder. H is the height of the instrument; σ is the density in the topographic unit. The variable parameter ($w \equiv \Delta X/R$) is the size of the gradient observation triangle. Note the strong filtering effect for $w > 1$.

rewrite these equations in dimensionless form

$$\mathcal{G} \equiv g/G\sigma R$$

$$= 2\pi(1 - v)/\{u^2 + (1 - v)^2\} \qquad (5b)$$

$$GC \equiv -(1/G\sigma)(\partial g/\partial X)$$

$$= -4\pi u(1 - v)/\{u^2 + (1 - v)^2\}^2, \qquad (6b)$$

where

$u = X/R$ (the horizontal distance for unit R),

$v = H/R$ (the height of the instrument),

σ is the density of the terrain unit.

The gravity gradient measured by the gravimeter method is $\Delta g/\Delta X$, where ΔX is the length of the side of the observation triangle. For convenience the angle of the observation triangle is taken perpendicular to the axis of the cylinder. The "measured" simulated gradient is

$$GX \equiv (1/G\sigma)(\Delta g/\Delta X)$$

$$= \{\mathcal{G}(u + w) - \mathcal{G}(u)\}/w, \qquad (7)$$

where

$$w \equiv \Delta X/R.$$

The calculated gradients GC and GX are plotted in Figure 8 for a range of values of the parameter w. The figure is plotted in dimensionless units and is valid for any desired scale units, whether cm for a molehill or km for a mountain. Conversion of the ordinate scale into practical gradient units is given by simple multiplication as described above, equation (3). The distance scale is given by $X = uR$.

The curves in Figure 8 show two important results: (a) The gradient effect falls off very rapidly with distance; and (b) there is a strong filtering effect in the "observed" gradient values when the dimension ΔX of the gradient triangle

Table 3. Simulated terrain corrections for a topographic ridge. Height = 50 m (165 ft), density = 2.0

w	$(\Delta g/\Delta X)_{max}$	u for $\Delta g/\Delta X = 10$ E°	X for $\Delta g/\Delta X = 10$ E°
GC	559 E°	5.92	148 m
0.2	549 E°	5.29	132 m
0.5	529 E°	5.12	128 m
1.0	469 E°	4.91	123 m
2.0	341 E°	4.48	112 m
5.0	164 E°	3.53	88 m

approaches or exceeds the height of the terrain unit. These results are illustrated in Table 3. It is apparent that what, at first sight, appears to be an entirely unmanageable terrain problem is reduced to easily manageable size by choosing the gradient station at a suitable, relatively short, distance away from the topographic element.

Subsurface noise

A similar simulation for subsurface noise is diagrammed in the inset of Figure 9, in which

the body of abnormal density is assumed to be a sphere. The equations for the calculations are as follows (Nettleton, 1940):

$$g = GMz/(x^2 + z^2)^{3/2}$$
$$= (4/3)\pi G\Delta\sigma R^3(Z + H)/\{X^2 + (Z + H)^2\}^{3/2} \qquad (8a)$$
$$\partial g/gX = -3GMxz/(X^2 + z^2)^{5/2}$$
$$= -4\pi G\Delta\sigma R^3 X$$
$$\cdot (Z + H)/\{X^2 + (Z + H)^2\}^{5/2} \quad (9a)$$

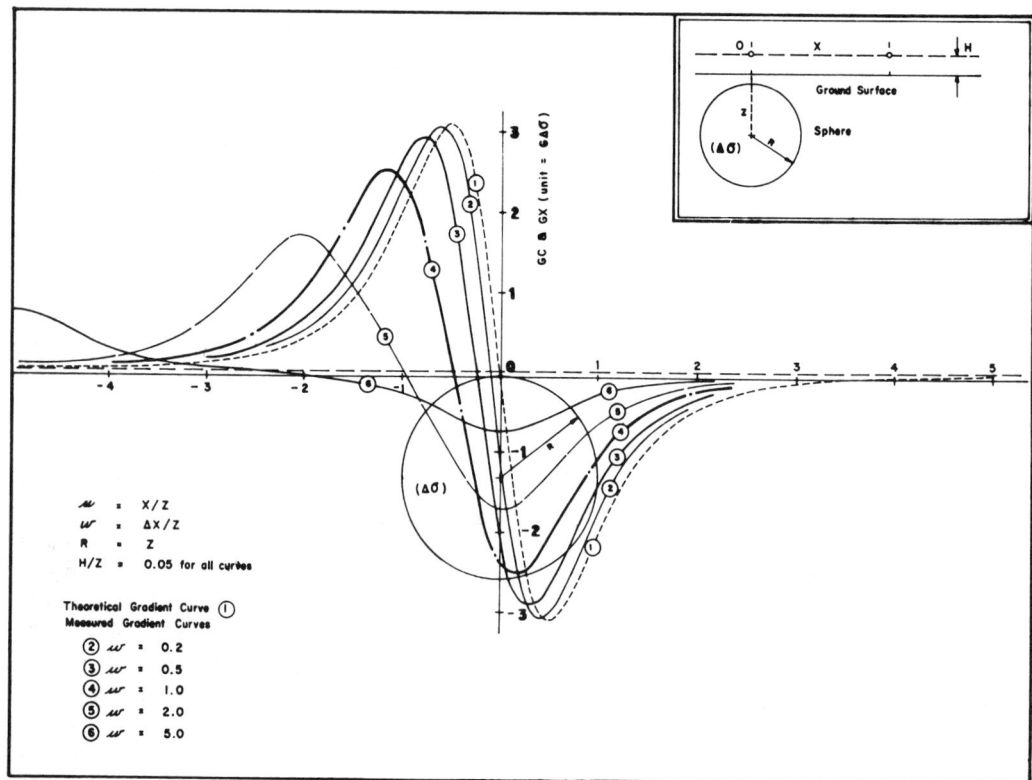

Fig. 9. Gradient "noise" simulated by a subsurface spherical body. For maximum effect, the top of the sphere was taken at the ground surface ($R = Z$). $\Delta\sigma$ is the density contrast in the anomalous body. The variable parameter is the size of the observation triangle ($w = \Delta X/z$). Note the strong filtering effect for $w > 1$.

$$\Delta g / \Delta X = \{ g(X + \Delta X) - g(X) \} / \Delta X. \quad (10a)$$

In dimensionless form these equations become

$$\mathcal{G} \equiv g / G \Delta \sigma Z = (4\pi/3) \mathcal{R}^3$$
$$\cdot (1 + v) / \{ u^2 + (1 + v)^2 \}^{3/2} \quad (8b)$$

$$GC \equiv (1/G\Delta\sigma)(\partial g / \partial X) = -4\pi \mathcal{R}^3$$
$$\cdot (1 + v)u / \{ u^2 + (1 + v)^2 \}^{5/2} \quad (9b)$$

$$GX \equiv (1/G\Delta\sigma)(\Delta g / \Delta X)$$
$$= \{ \mathcal{G}(u + w) - \mathcal{G}(u) \} / w, \quad (10b)$$

where

$$u \equiv X/Z$$
$$v \equiv H/Z$$
$$w \equiv \Delta X/Z$$
$$\mathcal{R} \equiv R/Z.$$

The resultant curves (Figure 9) show behavior quite similar to those in the terrain problem. Note that these curves give maximum effects with the top of the anomalous body at the ground surface ($\mathcal{R} = 1$). For $\mathcal{R} < 1$ the magnitude of the gradient effect decreases rapidly as $\mathcal{R}^3$.

As in Figure 8, the effect drops off very quickly with distance. The results for an assumed radius of the subsurface body of 50 m (165 ft) and density contrast of 0.25 are illustrated in Table 4. The largest gradient effect for this fairly extreme case (gravity anomaly = 0.35 mgal) is 52 E° which is filtered down to 29 E° for a triangle size of 100 m ($w = 2$). The gradient effect decreases to 2 E° at distances of 151 to 82 m.

It is clear from this simulation that the subsurface noise problem in the present context is very local. Therefore, where suspected, it can be identified by observing off-set stations or, as

an alternative, by using two or more triangle sizes at the suspect station. In addition, the noise effect is strongly filtered if $\Delta X/Z \geq 1$. Deeper sources, for which $\Delta X/Z < 1$ (such as prospective geologic structures), are not appreciably attenuated by the filtering effect.

We conclude that the gradient noise problems, from local terrain or shallow variations in soil density, are much less serious in the proposed gradient method than has been generally expected. Where noisy data are encountered, the observation of off-set stations (e.g., "4-stack" stations) will define the extraneous effects to permit their elimination.

FIELD APPLICATIONS

Two examples of the gradient method in stratigraphic pinchout exploration are given below. These examples illustrate two different aspects of the problem, namely, (1) recalculation of conventional gravity data in terms of the horizontal gradient and (2) measurement of the horizontal gradient, by the gravimeter triangle procedure described above, as a new field survey method.

East Texas oil field

A high quality conventional gravity profile across the East Texas oil field has been reported (Nettleton, 1972) and is reproduced in Figure 10.

The East Texas oil field is a typical example of a stratigraphic pinchout and one of the largest oil fields in the world. Nettleton concluded on the basis of his data that discovery of the East Texas oil field by residual analysis of a conventional but high-quality gravity survey is not a realistic possibility. The present authors concur. On the other hand, the gradient data calculated from the difference in gravity values and distance between adjacent stations

Table 4. Gradient noise from a subsurface sphere with top at ground surface. R = 50 m (165 ft), density contrast = 0.25

w	$(\Delta g / \Delta X)_{max}$	u for $\Delta g / \Delta X = 2$ E°	X for $\Delta g / \Delta X = 2$ E°
GC	52 E°	3.02	151 m
0.2	51 E°	2.93	147 m
0.5	48 E°	2.92	146 m
1.0	45 E°	2.60	130 m
2.0	29 E°	2.24	112 m
5.0	12 ½ E°	1.63	82 m

(L. L. Nettleton, personal communication), although quite irregular, demonstrate a definite anomalous tendency. These "observed" gradient results are shown in the center of Figure 10. To evaluate these data, the expected gradient profile was calculated, from equation (1), for the pinchout section shown at the bottom of Figure 10. This theoretical curve was then fitted in amplitude and location to the "observed" gradient profile. In fact, two up-dip pinchouts were inferred, at points A and B. Point A is in good agreement with the up-dip pinchout of the Woodbine sand against the overlying basal Austin chalk and is a little over a mile beyond the up-dip edge of production in the East Texas field. The geologic significance of the inferred gradient anomaly at point B is unknown at present.

There is one difficulty with this analysis. The stratigraphic pinchout angle of the Woodbine sand is so small ($0°23'$) that an unreasonably high density contrast ($+1.0$) was required to

fit the calculated profile to the "observed" gradient anomaly. Unfortunately, density information on the geologic strata involved is not available to the authors to expand on the possible geologic significance of this unexpected result. It may indicate important stratigraphic information at higher or lower levels.

These data and results, although of uncertain geologic significance in some respects, demonstrate that reprocessing available, conventional, high-quality gravity data in terms of horizontal gravity gradients significantly increases the resolution and gives promise of being useful in exploration for stratigraphic pinchouts comparable to the East Texas field. Modern analysis techniques should be advantageous for such applications.

A pinchout interpreted from seismic reflection data

A field test of the gravimeter triangle method of measuring the horizontal gradient of gravity

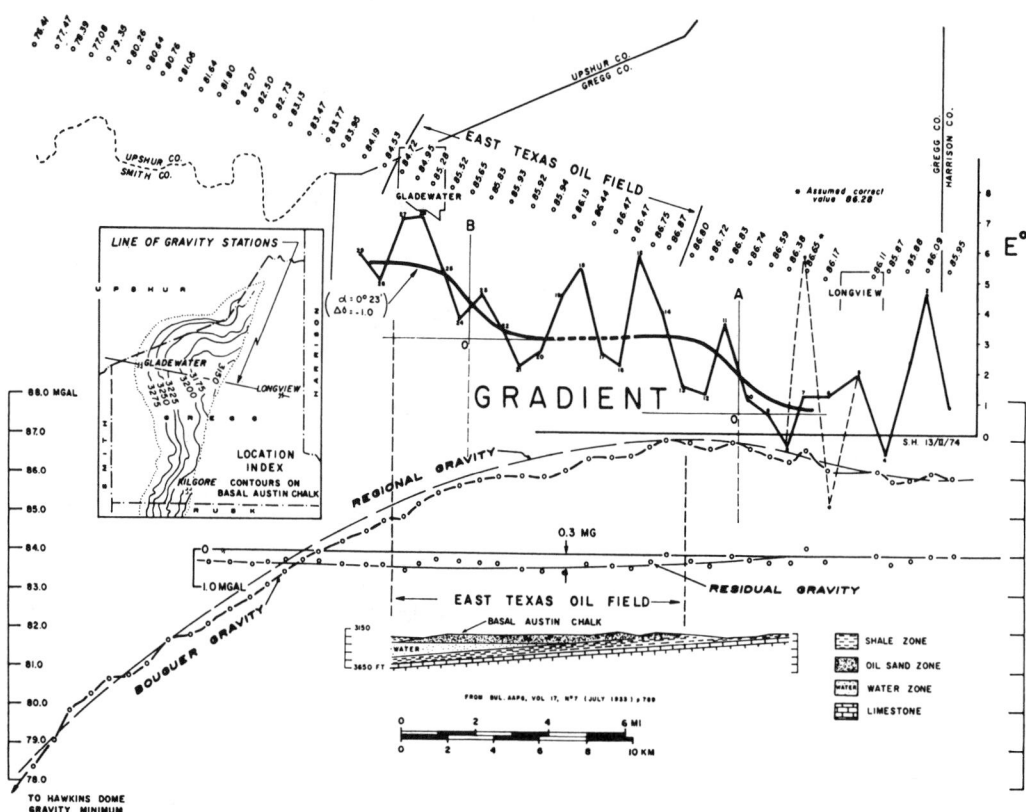

FIG. 10. Gradient analysis of a gravity profile across the East Texas oil field (after Nettleton).

is illustrated in Figure 11. The location is in east-central Bolivia on the southwest flank of the Brazilian Shield, across a probable stratigraphic pinchout interpreted from seismic reflection data. The gradients were determined by observing the gravity and elevation differences along the two sides of right triangles. The length of each leg was 100 m, and each of the three points in the triangle was observed twice with a modern Worden gravimeter. The topography in the area was almost flat. The observed gradient components along the profile are plotted in the upper part of the figure. The measurement accuracy, judged from internal consistency of the data, is ±1E°. A gradient anomaly of about 5E° is indicated.

The theoretical gradient anomaly calculated from the interpreted seismic section, which shows two pinchouts, was adjusted in amplitude and fitted horizontally to the observed gradient data. The locations of the two pinchouts indicated by the calculated gradient curve are displaced 1200 m down-dip from the seismic locations. The adjusted amplitude of the calculated gradient curve yielded a density contrast of −0.3 in the pinchout section, a value which is in good agreement with available data.

We conclude that the gradient profile gives valid support to the seismic interpretation and enhances the pinchout prospect, although the displacement between the two suggests that more data and reinterpretation would be useful for more precise location of the actual pinchout.

SUMMARY AND CONCLUSIONS

(1) Surveying the vertical gradient of gravity (dg/dz) by measurement of gravity at two

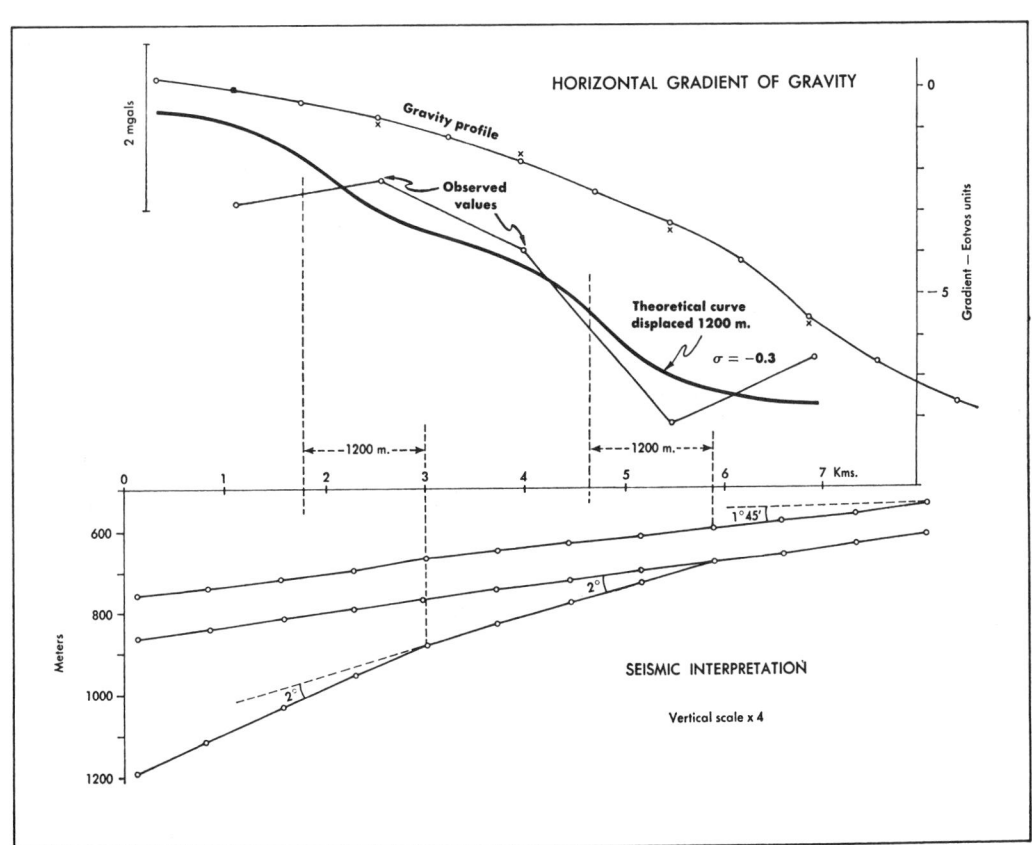

FIG. 11. Observed horizontal gradient profile across a pinchout interpreted from reflection seismic data. The smooth curve shows the gradient profile calculated from the interpreted seismic section and adjusted for amplitude and horizontal location to fit the observed gradient data. Gravity values (x) calculated by integration of the gradients are in agreement with the observed gravity profile.

elevations is not a practically attainable objective with present instrumentation.

(2) Horizontal gradients of gravity (dg/dx) are potentially useful in prospecting for stratigraphic pinchouts, which represent a major source of future petroleum reserves (Lyons and Dobrin, 1972).

(3) There are two aspects to the gradient problem: (a) reprocessing available conventional gravity survey data in terms of the horizontal gradient, and (b) surveying the horizontal gradient of gravity as a field method by observing groups of three stations set out in the form of small triangles.

(4) Important advantages of a gradient survey in comparison with conventional gravity surveys are: (a) greatly increased sensitivity with much sharper resolution, and (b) virtual elimination of position and elevation surveying, which is a major cost factor in gravity surveying.

(5) Anticipated gradient method limitations due to ambient "noise" (terrain effects and shallow variations in soil density) are greatly reduced by cancellation in the gradient triangle method. Gradients from deep events are accurately recorded but diminish in amplitude with depth (Hammer, 1971), and only the stronger anomalies from depth will be mapped.

(6) The geologic principle of multiple hypotheses applies in full force to the increasingly important and extremely difficult search for stratigraphic traps. The gravity gradient method, used for rapid and economical reconnaissance to guide seismic confirmation or after seismic work to confirm the interpretation, is recommended as an effective step forward in exploring for stratigraphic traps of the pinchout type.

ACKNOWLEDGMENTS

The authors are pleased to acknowledge permission to publish this study, graciously granted by Ing. Jaime Oblitas García, Gerente de Exploración; Ing. Rolando Kadima, Director del Centro de Tecnología Petrolera; and Hulusi D. Berilgen, Officer in Charge, United Nations Development Project Yacimientos Petrolíferos Fiscales Bolivianos, Santa Cruz, Bolivia.

REFERENCES

Hammer, Sigmund, 1971, Vertical attenuation of anomalies in airborne gravimetry: Geophysics, v. 36, p. 867–877.

Halbouty, Michel T., 1972, Rationale for deliberate pursuit of stratigraphic unconformity, and paleomagnetic traps, *in* Stratigraphic oil and gas fields, classification, exploration methods and case histories: AAPG-SEG, Tulsa, p. 3–7.

Heiland, C. A., 1940, Geophysical exploration: New York, Prentice-Hall, Inc., equations 7–93C.

—— 1943, A rapid method for measuring the profile components of horizontal and vertical gravity gradients: Geophysics, v. 8, p. 119–133.

King, Robert E., Editor, 1972, Stratigraphic oil and gas fields, classification, exploration methods and case histories: AAPG-SEG, Tulsa, 687 p.

Lyons, Paul L., and Dobrin, M. B., 1972, Seismic exploration for stratigraphic traps, *in* Stratigraphic oil and gas fields, classification, exploration methods and case histories: AAPG-SEG, Tulsa, p. 225–243.

Nettleton, L. L., 1940, Geophysical prospecting for oil: New York, McGraw-Hill Book Co. Inc.

—— 1972, Use of gravity, magnetic, and electrical methods in stratigraphic trap exploration, *in* Stratigraphic oil and gas fields, classification, exploration methods and case histories: AAPG-SEG, Tulsa, p. 244–251.

Neumann, R., 1967, La gravimetrie de haute precision—Application aux recherches de cavities: Geophys. Prosp., v. 15, p. 116–134.

Thyssen-Bornemisza, S., 1958, Vertical gradient of gravity: Geophysics, v. 23, p. 359–360.

—— 1965, A short note on double-track profiling with the gravity meter (horizontal gradients): Geophysics, v. 30, p. 1135–1137.

Thyssen-Bornemisza, S., and Stackler, W. F., 1956, Observation of the vertical gradient of gravity in the field: Geophysics, v. 21, p. 771–779.

REVIEWS OF GEOPHYSICS   VOL. 5, No. 4   NOVEMBER 1967

Measurement of Gravity at Sea and in the Air

LUCIEN J. B. LaCOSTE

LaCoste and Romberg, Inc., Austin, Texas

Abstract. General problems of gravity measurement at sea are discussed. A treatment of the effects of vertical accelerations shows that gravity meter nonlinearities cause errors that ordinarily are proportional to the square of the vertical acceleration. In a treatment of horizontal accelerations, similarities and differences that exist between gimbal supported and stabilized platform gravity meters are pointed out. It is shown that optimization of the parameters of a stabilized reference can reduce gravity meter errors by two or more orders of magnitude at long periods. A method for correcting for inadequate period of a stabilized reference is given. The Schuler-tuned stabilized platform is briefly described. The theory of inherent type cross coupling and examples of imperfection type cross coupling are given.

Theory and significant details of construction are given for the LaCoste and Romberg (L&R) gravity meter and stabilized platform. Tests on L&R air-Sea gravity meters made during the last decade are discussed. The largest error in earlier work is shown to have been due to vertical accelerations. A method of correcting for these errors is described. Tests indicate that presently attainable accuracy of L&R gravity meters is appreciably better than 1 mgal.

INTRODUCTION

Gravity has been measured at sea with underwater gravity meters (operated on the ocean bottom), with Vening-Meinesz pendulums, and with shipboard gravity meters. Underwater gravity meters are well adapted to shallow water operation. They also give greater detail than the other types of instrumentation because they have higher accuracy and because underwater gravity data are taken closer to the anomalies. Underwater gravity meter accuracy falls off as the water depth increases because of increasing errors in measuring water depth and because of uncertainty in the position of the gravity meter relative to the ship. Furthermore, underwater gravity meters can detect detail that is lost in the averaging procedures inherent in instruments operating on moving ships. Until recently nearly all sea gravity work for oil exploration was done with underwater gravity meters. They can be modified for use at any desired water depth, but operation at great depths is slow and expensive.

From the 1920's Vening-Meinesz pendulums have been used in submarines to measure gravity [*Vening Meinesz*, 1929, 1932, 1941; *Browne*, 1937; *Browne and Cooper*, 1950, 1952; *Worzel and Ewing*, 1950; *Worzel et al.*, 1955; *Harrison*, 1960]. For many years they were the only instruments used for measuring gravity in deep water. They have an accuracy of 1–2 mgal, but they are complicated to operate and computation of the data is laborious. They handle very well the small-amplitude, long-period accelerations encountered in a submarine, but they do not handle the larger-amplitude, short-period accelerations present on a surface ship. The preceding difficulties might have been overcome, but the advent of submarine and surface ship gravity meters discouraged work in that direction. At present few if any Vening-Meinesz pendulums are still in use. Consequently, they will not be considered further in this review; they are well described in the references mentioned.

Since the end of World War II several gravity meters have been designed for use in submarines and on surface ships. *Gilbert* [1949] made a vibrating string gravity meter and later B. J. Electronics and *Tsuboi et al.* [1961] made similar instruments. In these gravity meters the vibrating string supported a mass whose changes in weight affected the frequency of vibration. Very recently Worden, Bell Aircraft, and Texas Instruments have designed instruments that employ the 'force-balance' principle that is used in some accelerometers. In this method changes in the gravitational pull on the mass are balanced by electromagnetic or electrostatic forces controlled by a very fast acting servo. The Bell instrument has given good results at sea. The type of shipboard gravity meter that has so far been the most successful is, however, a highly overdamped spring type of gravity meter. The Graf and the LaCoste and Romberg shipboard gravity meters are both of this type.

Originally these highly overdamped gravity meters were designed for operation in submarines. Their accuracy and their ability to withstand accelerations were soon improved, however, to such an extent that they were capable of operating on a surface ship. Recently their accuracy has been further improved so as to make them valuable in oil exploration. Although their accuracy does not equal that of underwater gravity meters in shallow water, it is comparable in deep water because their results are not so much affected by errors in depth measurements. Furthermore, shipboard gravity surveys are much faster and cheaper than underwater gravity surveys. Shipboard gravity meters can also be used on the same ship with magnetometers and sparkers, which reduces costs.

UNDERWATER GRAVITY METERS

Underwater gravity meters are essentially remote controlled land gravity meters, although underwater work involves some substantial additional problems. The electronic requirements for remote operation are well within the state of the art; thus, it is hardly worth while in this review to give electronic details. A recent improvement was to reduce the number of conductors in the controlling cable. This improvement makes it economically feasible to use an armored cable with larger conductors, which is important because cable failures are the greatest source of trouble at sea. Cable conductor breakage near the submersible gravity unit occurs regularly because of continual flexure in this region. The standard operating procedure is to cut off a few feet of cable near the gravity meter every few weeks and reterminate the cable. Broken cable conductors are often very troublesome because electrical conductivity at the break is often restored when the load on the cable is removed. Cable trouble could be reduced by a further reduction in the number of cable conductors, and research is being done to implement this change.

Unlike land gravity meters, underwater gravity meters must withstand very rough treatment. They usually receive a substantial bump when they hit the ocean floor; they are sometimes accidentally dragged on the ocean bottom; and they sometimes hit the side of the ship when they are being raised out of the water. Also, poor hoist operators often bump gravity meters when raising them into their cages on the hoist. The rough treatment makes the errors in sea surveys greater than the errors encountered in a land survey, but the gravity meters are made rugged enough that the errors are only about 0.1 mgal. Rough treatment can also knock gas bubbles in the column of the mercury thermostat which has up to now been used to regulate the temperature of the gravity meters. This trouble has been eliminated in recent gravity meters by replacing the mercury thermostat with a thermistor-transistor circuit.

Seismic motion problems. Seismic motion of the ocean bottom has been a considerable problem in making underwater gravity surveys. This seismic motion is caused by wave action and is particularly troublesome on muddy bottoms at depths less than about 50 feet. The vertical motion in this seismic action is often greater than the motion that can be tolerated in the gravity meter beam without introducing errors caused by mechanical hysteresis in the gravity meter spring. Two methods have been used to overcome this problem.

The first method was to put the gravity meter unit on a servo-controlled elevator to approximately counteract the vertical seismic motion [*LaCoste, 1952a, b*]; a diagram showing this operation is given in Figure 1. Obviously the acceleration of the servo can be controlled so that the gravity meter beam or weight can be held (for a while) at any point on its scale regardless of the seismic motion. This can not generally be done for an indefinitely long time because the elevator will usually reach a limit of its travel. Although the elevator operation described will counteract vertical seismic motion, it will not in itself

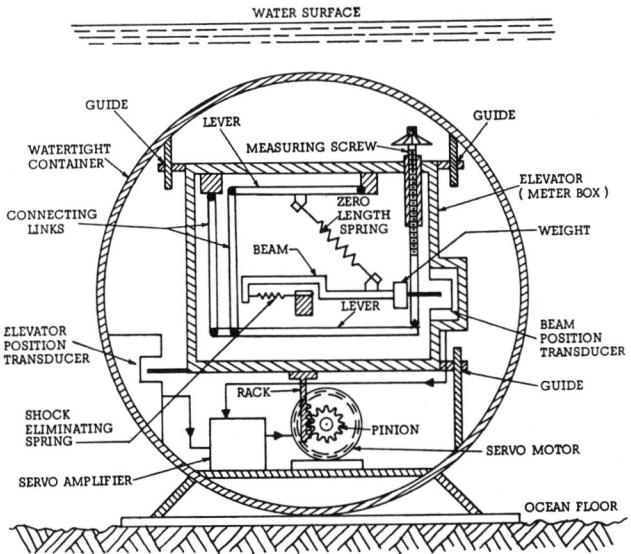

Fig. 1. Elevator for seismic compensation of underwater gravity meter.

permit a measurement of gravity because an unknown acceleration (that of the elevator) has been introduced.

To measure gravity, an additional feature is added that makes the acceleration of the elevator average out to zero. As shown in Figure 1, this is done by feeding to the servo an error signal that is a combination of the beam position and the elevator position. The combination is chosen so as to make the servo null the beam above its center position when the elevator is above its center position, and vice versa. Thus, a stability is given to the elevator operation, as can be seen from the following considerations. As the elevator moves up, the servo nulls the beam closer to the top stop. This beam position reduces the spring tension, which causes the beam to be accelerated down. This downward acceleration is equivalent to a restoring force on the elevator and results in an equilibrum position for the elevator provided that the gravity meter is not too far out of balance.

The elevator system just described would give an undamped oscillatory motion to the elevator. To damp it, some dead space is provided in the servo. More damping can be provided if desired by modifying the electrical circuit. When the servo is first turned on, initial conditions can be troublesome. To damp them out before the elevator reaches a limit of its range, the operator is provided with two controls that apply an electrostatic force to the beam in either an upward or a downward direction. Efficient operation of these controls requires some training.

The elevator system was developed about 1948 and is still in use, but many operators do not take full advantage of its capabilities. It has an advantage over other systems in that the elevator almost completely eliminates vertical accelerations that would otherwise be experienced by the gravity meter unit; this makes it unnecessary for the gravity meter to respond to accelerations linearly.

The second method for overcoming the problem of vertical accelerations of the ocean bottom is to highly overdamp the gravity meter beam or weight. This method is used in most present day shipboard gravity meters. It will be described in detail later, but a few comments about it will be made now. It would be thought that high damping would make a gravity meter very slow to read. This difficulty can be avoided, however, by reading the gravity meter before it has come to rest, which can be accomplished by observing the *velocity* of the beam. If the velocity is zero, the beam has come to rest. If the velocity is not zero, a correction can be made for the observed velocity. By using this method an overdamped gravity meter can be read as fast as an ordinary gravity meter. The overdamped gravity meter is simpler to operate than the elevator type.

Accuracy. The accuracy of underwater gravity meter results depends to a considerable extent on how the gravity meter is handled. The inherent precision of the gravity meter is about 0.01 mgal. In actual sea operations under normal conditions, base station checks indicate an accuracy of about 0.1 mgal [*Beyer et al.*, 1966], although many carefully controlled surveys have been made with an appreciably higher accuracy. An accuracy of 0.1 mgal is generally adequate because uncertainties in water depth and latitude often given larger errors. Water depth is usually measured with pressure gages whose accuracy is not much better than ½%. A 2-foot error in depth corresponds to about 0.1 mgal. Also, a 500-

foot error in the north–south direction corresponds to about 0.1 mgal at a latitude of 30°. An over-all accuracy of about 0.2 mgal is considered good in a survey in water 500 feet deep.

Much gravity exploration for oil has been done in water up to 600 feet deep, but not much has been done at greater depths because of increased costs and lower accuracies of the measurements in deeper water. In exploration for oil, accurate electronic navigation systems such as Raydist and Shoran are used.

Recently some experimental gravity work has been done at depths up to 2900 feet [Beyer et al., 1966]. In this work an over-all accuracy of 0.24 mgal was achieved, but it was necessary to make very precise depth measurements to achieve this accuracy. The depth was measured in several ways. One transponder was mounted on the ship, and it gave good results when the bottom was flat. A second transponder was mounted on the gravity meter, and travel times to the ship were measured. It gave good results when the gravity meter was directly below the ship. This position was found by looking for the shortest travel time. Furthermore, the actual velocity of sound in the water was measured as the gravity meter was lowered, thereby eliminating errors caused by using an incorrect value of sound velocity. The measurements were made with an instrument fixed to the gravity meter. Depth was also measured by a pressure gage (accuracy ¼%), but the depth values obtained in this way were not considered to be as accurate as the other values.

Other uses of underwater gravity meters. Underwater gravity meters have also been used in swamps, on muskegs, and on frozen lakes and ice islands. In such places there is often so much seismic motion that a land gravity meter would not operate satisfactorily. An underwater gravity meter with a light fiberglass container is then used. Either the elevator type gravity meter or the overdamped model can handle the long-period seismic motion and is usually satisfactory. Trouble is sometimes experienced however, by short-period motions caused by the wind blowing the trees in a swamp or by the movement of the operator if he is too close to the gravity meter. In some cases it is desired to hover a helicopter over the gravity meter while taking a reading. Tests indicate that this can be done even in a soft peat bog if the helicopter is more than about 100 feet off the ground. Efforts are now being made to suitably shock-mount the gravity meter so that it will be possible for the helicopter to hover closer to the gravity meter.

SHIPBOARD AND AIRPLANE GRAVITY METERS

Eotvos Effect

Before considering any details of shipboard or airplane gravity meters, it is desirable to discuss some problems that are common to all types of moving instruments. One problem is caused by the motion of the ship or airplane over a curved rotating earth. This motion results in a centripetal acceleration, which must be corrected for. This correction, known as the Eotvos correction, is [Thompson and LaCoste, 1960; Nettleton et al., 1960].

$$E = (R_\varphi + h)(2V_\varphi V_e + V^2)/R_\varphi{}^2 \tag{1}$$

where R_φ is the earth's radius at latitude φ, h is the height above sea level, V_φ is the speed of rotation of the earth's surface at latitude φ, and V_e is the easterly component of V, the total speed of the vehicle. At speeds less than 15 knots the Eotvos correction can be approximated by $7.5 \cos \varphi$ milligals per knot of east–west speed to be added to the observed gravity values for eastward velocities. The Eotvos correction is typically in the range of -50 to $+50$ mgal for measurements at sea and of the order of 1000 mgal for airplane observations.

It is therefore apparent that navigational errors are often the most serious limitation in the accuracy of gravity measurements made on moving platforms. If there is a 1-knot error in the east–west speed of a ship, it will introduce a 7.5-mgal error in the Eotvos correction at the equator and a somewhat smaller error at higher latitudes. Errors in ship velocity of 1 knot can easily occur if a ship is out of range of land and accurate electronic aids to navigation and must depend on astronomical sights. The use of satellite and inertial navigation will reduce navigational errors in the future.

In (1) the term, containing V^2 as a factor would occur even if the earth were not rotating. However, the term containing $V_\varphi V_e$ as a factor is present because of the rotation of the earth. This term is actually the vertical component of the Coriolis force $2\,\mathbf{\Omega}\times\mathbf{V}$, where $\mathbf{\Omega}$ is the vector representing the angular rotation of the earth and $\mathbf{V}$ is the vector representing velocity relative to the earth. The horizontal component of the Coriolis force is directed in the east–west direction; it is produced by north–south motion over the earth. Since the horizontal component is at right angles to gravity, its effect on gravity is generally negligible. It can always be disregarded in ship measurements and is only of the order of a milligal in airplane measurements.

Vertical Accelerations

The biggest problem in the measurement of gravity on a moving platform is caused by vertical accelerations. Since no instrument can distinguish between gravity and acceleration, any gravity measurement made on a moving platform will actually be a meaurement of gravity plus vertical acceleration or $g + z''$. Furthermore, instantaneous vertical accelerations are generally 10,000 to 100,000 times greater than the desired gravity meter accuracy. It is therefore necessary to do some averaging or filtering of the data. If a simple averaging process is used over the time interval T, the data give

$$\frac{1}{T}\int_0^T (g + z'')\,dt = \langle g\rangle + [z']_0^T/T \tag{2}$$

where the angle brackets denote an average value.

Equation 2 shows that averaged gravity measurements must be corrected by an amount equal to the change in vertical velocity during observation divided by the time duration of the observation. The precise method used to make the correction depends on the circumstances of the measurement. In a shipboard measurement the vertical ship accelerations are large, but there can be no long-term change in z', since the vertical accelerations of the ocean surface due to tides are negligible. In this case, therefore, it is merely necessary to filter out the wave frequencies. In submarine measurements filtering can be used also, but an additional correction must be made for long-term changes in depth. In making measurements in aircraft, however, there is no such easy division by period. In this case it is necessary to measure and correct continuously for variations in height, making sure that the corrections are filtered the same way that the gravity data are.

Vertical accelerations also make it necessary for the gravity meter and filter responses to be extremely linear in order to avoid errors in averaged gravity indications [*Harrison and LaCoste*, 1968]. To study the effect of nonlinearities the differential equation for a spring-type gravity meter will be needed. Although the equation to be derived does not apply to any particular spring-type gravity meter, it might be helpful to refer to Figure 2. If horizontal accelerations are negelected, the equation is

$$g + z'' + bB'' + fB' + kB - cS = 0 \tag{3}$$

where z'' is the vertical acceleration of the gravity meter case, B is the displacement of the gravity meter mass relative to a null position in the gravity meter case, and b, f, k, and c are constants if the gravity meter is linear. The first three terms in (3) result from the gravitational and acceleration forces on the mass. If the mass is constrained to move in an approximately straight line, b will be very nearly constant. It will be assumed constant. The term fB' results from damping, and kB is due to the restoring force of the spring. The term cS represents the vertical force per unit mass exerted at the center of mass by the spring acting through the various links in the gravity meter when the mass is nulled. S can be adjusted by moving the point of attachment (A in Figure 2) of the spring on the gravity meter case.

The coefficients f, k, and c will be taken, respectively, as the sum of constants f_0, k_0, and c_0 and variables f_v, k_v, and c_v. Equation 3 then becomes

$$g + z'' + bB'' + (f_0 + f_v)B' + (k_0 + k_v)B - (c_0 + c_v)S = 0 \tag{4}$$

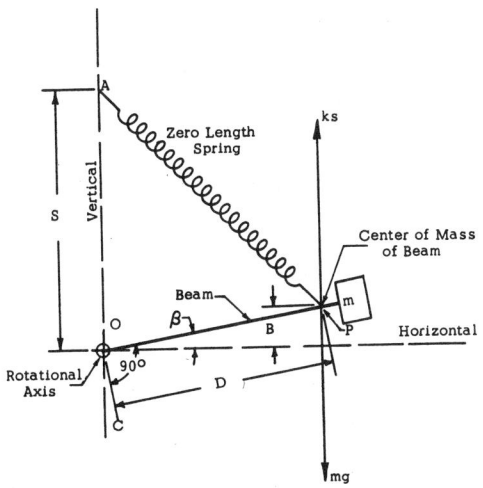

Fig. 2. Diagram of LaCoste and Romberg gravity meter.

Gravity can be determined correctly from (4) provided that z'', S, and B and its derivatives are known; however, if f_v, k_v, and c_v are neglected, an erroneous value g_e of gravity will be obtained. g_e is given by

$$g_e + z'' + bB'' + f_0B' + k_0B - c_0S = 0 \tag{5}$$

Combining (4) and (5) gives the error e in g_e as

$$e = g_e - g = f_vB' + k_vB - c_vS \tag{6}$$

To evaluate (6) the coefficients must be known. They all can have a term proportional to z'' because the forces caused by vertical acceleration can alter the geometry of the gravity meter. The coefficient f can also contain terms proportional to B and B' because the damping can depend on the position of the damping device and its velocity relative to the gravity meter case. k_v can include a term proportional to B but it is not likely to include a term proportional to B', although such a term could easily be included and would be found to have no effect on the average error. c_v can include a term proportional to S. For convenience S will be expressed as the sum of its average value $\langle S \rangle$ plus a variable term S_v, whose average value is zero. Equation 6 then becomes

$$e = (f_1z'' + f_2B + f_3B')B' + (k_1z'' + k_2B)B$$
$$- (c_1z'' + c_2\langle S \rangle + c_2S_v)(\langle S \rangle + S_v) = 0 \tag{7}$$

Some of the terms of (7) can be neglected. For instance, the average value of f_2BB' can be shown to be negligible. Consider B to be expressed as a Fourier series. Then for each frequency B and B' will differ in phase by $90°$, and therefore the average value of the product will be zero. Also the average value of the products of different frequencies will be zero. The average value of $\langle S \rangle z''$ will be zero on a ship because $\langle S \rangle$ is a constant and the average value of z'' must be zero. The average value of $\langle S \rangle S_v$ is zero because $\langle S_v \rangle$ is zero. The term $c_2\langle S^2 \rangle$ will be dropped because it does not involve z'', and therefore it is presumably taken care of in the static calibration of the gravity meter. Equation 7 then gives for the average error

$$\langle e \rangle = \langle (f_1z'' + f_3B')B' \rangle + \langle (k_1z'' + k_2B)B \rangle - \langle (c_1z'' + c_2S_v)S_v \rangle = 0 \tag{8}$$

For particular types of gravity meters, (8) can be simplified as follows. If S is not varied while a reading is taken, the terms involving S_v will of course be zero. If S is varied by means of a servo, S_v will be approximately proportional to z'', or $S_v = S_2 z''$. Also, if the gravity meter is highly damped, B' will be approximately proportional to z'', or $B' = K z''$. The preceding relation is accurately satisfied if the gravity meter has infinite sensitivity; i.e., $k = 0$. If the preceding relations are introduced into (8) and it is noted that $\langle z' z'' \rangle$ is zero because z' and z'' are $90°$ out of phase (as was shown for $\langle B' B \rangle$), then (8) gives

$$\langle e \rangle = (f_1K + f_3K^2 - c_1S_2 - c_2S_2^2)\langle z''^2 \rangle + k_2K^2\langle z'^2 \rangle \tag{9}$$

For a LaCoste and Romberg shipboard gravity meter some further simplifications can be made. In this gravity meter the servo controlling S is so slow that it varies S only a few milligals in response to wave accelerations; therefore, the constant S_2 can be taken as zero. Also, the coefficient k is so close to a constant that the term kB in (3) contributes less than 1 mgal over the entire range of B; therefore, k_2 can be taken as zero. Equation 9 for a LaCoste and Romberg meter then becomes

$$\langle e \rangle = (f_1 K + f_3 K^2)\langle z''^2 \rangle \tag{10}$$

Equation 10 was determined empirically in the laboratory for LaCoste and Romberg gravity meters before it was derived theoretically. It was found to hold accurately. Of course the attempt is made to adjust each gravity meter so that the vertical acceleration error is zero before the gravity meter is put into operation, but (10) shows that it is possible to correct data for vertical acceleration error even if the adjustment has not been made accurately enough. More will be said about such corrections later.

It is probably worth while to consider an example of nonlinearity in order to see how much is required to cause a 1-mgal error at 0.1-g rms vertical acceleration. In this example k will be taken as zero, cS will be taken equal to g, and B'' will be neglected. It is permissible to neglect B'' in actual overdamped gravity meters unless readings are taken in times much shorter than 1 sec. If f is taken equal to $f_0 + f_3 B'$, equations 3 and 7 give, respectively,

$$\langle (f_0 + f_3 B')B' \rangle = \langle -z'' \rangle = 100{,}000 \text{ mgal} \tag{11}$$

and

$$\langle e \rangle = f_3 \langle B'^2 \rangle = 1 \text{ mgal} \tag{12}$$

Combining (11) and (12) gives

$$f_3 \langle B' \rangle / f_0 = 1/99{,}999 \tag{13}$$

This equation states that the damping coefficient can not vary more than one part in 99,999 over a damper velocity range corresponding to a vertical acceleration range of 0 to 0.1 g. This is a very high linearity. Also, for a 0.2-g vertical acceleration, the permissible variation would be only half as much over twice the range; this corresponds to reducing the nonlinearity constant f_3 to one-fourth.

Another effect of large vertical accelerations is that they tend to cause the moving element of the gravity meter to exceed its permissible range, which is determined by mechanical hysteresis errors in the spring. Methods of reducing the motion of the gravity responsive element will be discussed later when describing the different types of gravity meters.

Horizontal Accelerations

Introduction. Since gravity meters operating on moving platforms are subjected to horizontal accelerations as well as to vertical accelerations, their effect must also be taken into account. Two methods have been used in the past: one method is to suspend the gravity meter from a gimbal joint and to correct for the swinging; the second method is to mount the gravity meter on a stabilized platform. The gimbal method was first put into practice by *Vening Meinesz*

[1941]. Until 1965 this method was used on LaCoste and Romberg gravity meters. It now appears that the stabilized platform method is considerably better, but, since so much work has been done and is still being done with gimbal suspensions, it seems worth while to discuss the principle in detail and to make comparisons between it and the stabilized platform method. There are striking similarities and some basic differences.

Gimbal suspension. A gravity meter suspended from a gimbal joint is shown in Figure 3. The gimbal joint is at 0; the center of gravity of the suspended system is at G; and the sensing element of the gravity meter (the mass) is at P. For long-period accelerations it can be seen that the gravity meter measures the vector sum of gravity plus the negative of the horizontal acceleration as indicated in the vector diagram in Figure 3. A solution of the vector triangle gives the gravity meter measurement as

$$g_a \doteq g + g\theta^2/2 \tag{14}$$

where g = gravity. This equation is known as the Browne correction [*Browne, 1937*] or the horizontal acceleration correction, and it has been found to be adequate for the motions encountered in submerged submarines. For motions occurring on a surface ship, *LaCoste and Harrison* [*1961a*] have made a more comprehensive analysis in which simultaneous horizontal and vertical accelerations were considered. The analysis included fourth-order terms, and it showed that (14) was correct up to the fourth order provided that the distance l is the length of a simple pendulum having the same period as the gimbal supported gravity meter. The mathematics is as follows.

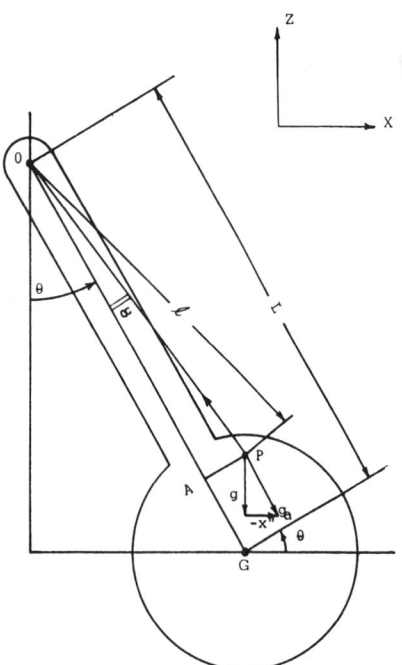

Fig. 3. Gimbal suspended gravity meter.

In Figure 3 we will let $0P = l$, $\angle\ 0AP = 90°$, and the free period of the gimbal system be $T_G = 2\pi/w_G$. The point 0 is subjected to periodic horizontal and vertical accelerations x'' and z''. Then the equation of motion of the gimbal system is

$$\theta'' + F\theta' + w_G{}^2(1 + z''/g)\sin\theta + w_G{}^2(x''/g)\cos\theta = 0 \qquad (15)$$

where F is a damping coefficient. The force per unit mass at P parallel to 0G is

$$g_a = l\theta'^2\cos\alpha - x''\sin\theta + (g + z'')\cos\theta + l\theta''\sin\alpha \qquad (16)$$

and the force per unit mass at P perpendicular to G0 is

$$a_n = l\theta'^2\sin\alpha - x''\cos\theta - (g + z'')\sin\theta - l\theta''\cos\alpha \qquad (17)$$

If (15) is solved for x'' and the result is substituted into (16) and (17), we obtain

$$g_a = l\theta'^2\cos\alpha + l\theta''\sin\alpha$$
$$+ (g/w_G{}^2)\tan\theta[\theta'' + F\theta' + w_G{}^2(1 + z''/g)\sin\theta] + (g + z'')\cos\theta \qquad (18)$$

and

$$a_n = l\theta'^2\sin\alpha - l\theta''\cos\alpha + (g/w_G{}^2)(\theta'' + F\theta') \qquad (19)$$

Expanding (18) in powers of θ, we obtain

$$g_a = l(\theta'^2\cos\alpha + \theta''\sin\alpha) + (g/w_G{}^2)(\theta\theta'' + \theta^3\theta''/3 + F\theta\theta' + F\theta^3\theta'/3)$$
$$+ g(1 + \theta^2/2 + 5\theta^4/24) + z''(1 + \theta^2/2) \qquad (20)$$

neglecting fifth-order terms in θ, its derivatives, and z''.

Since we are interested in obtaining average gravity values over several minutes, we can take θ in (12) as the Fourier series

$$\theta = \sum_i B_{si}\sin w_i t + \sum_i B_{ci}\cos w_i t. \qquad (21)$$

If we then take averages over a length of time that is long compared with the periods in the Fourier series and if we note that $\langle z''\rangle = 0$ on a ship and that $\langle\theta\theta''\rangle = \langle -\theta'^2\rangle$, we obtain from (20)

$$\langle g\rangle = \langle g_a\rangle - \tfrac{1}{2}g\langle\theta^2\rangle - (l\cos\alpha - g/w_G{}^2)\langle\theta'^2\rangle - \tfrac{5}{24}\langle g\theta^4\rangle$$
$$- \tfrac{1}{3}(g/w_G{}^2)\langle\theta^3\theta''\rangle - \tfrac{1}{2}\langle z''\theta^2\rangle \qquad (22)$$

In (22) we note that there is a second-order correction term in $\langle\theta^2\rangle$ and another in $\langle\theta'^2\rangle$. The latter can be eliminated by taking $l\cos\alpha = g/w_G{}^2$, which is the length of a simple pendulum having the same period as the gimbal suspended system. Equation 22 then becomes

$$\langle g\rangle = \langle g_a\rangle - \tfrac{1}{2}\langle g\theta^2\rangle - \tfrac{5}{24}\langle g\theta^4\rangle - \tfrac{1}{3}(g/w_G{}^2)\langle\theta^3\theta''\rangle - \tfrac{1}{2}\langle z''\theta^2\rangle \qquad (23)$$

The second-order correction term is the same as the Browne correction term given in (14). It is the only term that is normally used, but to make this possible, the distance $l\cos\alpha$ is carefully adjusted to equal $g/w_G{}^2$. This adjustment is made

by carefully testing the gimbal-supported gravity meter while it is being subjected to controlled horizontal accelerations. The final results must be independent of $\langle \theta'^2 \rangle$ or independent of the period of the accelerations.

LaCoste and Harrison [1961a] have investigated the effect of the fourth-
order terms is (23) for x'' and z'' equal to 50 gals and found that these terms
contributed about 3 mgal. Therefore, fourth-order terms should be taken into
account even below this acceleration if very high accuracy is desired. The fourth-
order terms plus the need for high accuracy in computing large second-order
terms make analog computers unsatisfactory at high accelerations. Replacing
analog computors with digital computers would solve the problem but would
unduly complicate the system.

It should be noted that (23) is independent of F, the damping in the gimbal
suspension. It will appear later, however, that such damping will cause an error
due to cross coupling between horizontal and vertical accelerations. For this
reason the gimbals are always undamped, and resonance is avoided by the
scheme shown in Figure 4. There the undamped gimbal suspension is shown supported at b by a *damped* gimbal, which is supported on the ship at a. The motion at b is not quite the same as that of the ship at a, but the corrected gravity
reading is independent of any differences. On the other hand, the damping in the
upper support absorbs energy from both pendulums and thereby limits resonant
motion.

An important feature of the gimbal supported gravity meter is that the
suspension can be made so that no forces are exerted on the gravity meter
weight except along the sensitive axis of the gravity meter regardless of the
motion of the gimbal support. This requires that $l \cos \alpha$ be made equal to $g/w_G{}^2$,
which is always done. Then (19) becomes

$$a_n = (g/w_G{}^2)(\theta'^2 \tan \alpha + F\theta') \qquad (24)$$

By making $\alpha = 0$ and $F = 0$, the acceleration a_n normal to the sensitive axis of
the gravity meter becomes identically zero. This parameter choice simplifies the
gravity meter design by eliminating any need to make the meter capable of
withstanding such forces; it is a distinct advantage of the system. A second ad-

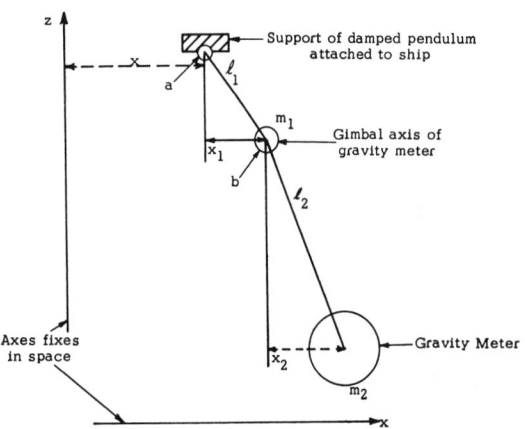

Fig. 4. Damping scheme for gimbal suspended gravity meter.

vantage of making a_n identically equal to zero is that it eliminates cross-coupling errors, which will be considered later. Actually cross-coupling errors can be made negligible by merely making $F = 0$.

To determine the Browne correction given in (14) or the correction given in (23), it is necessary to measure θ; this measurement requires a stabilized reference. The stabilized reference used by *Vening Meinesz* [1941] consisted of an approximately horizontal inertia bar supported for oscillation on knife edges very close to its center of gravity. The inertia bar had a 25- to 30-sec oscillation period, which was long enough to serve as a stabilized reference in a submarine where the accelerations were small. The LaCoste and Romberg gimbal-type gravity meters have used a refined version of this long-period pendulum [*La-Coste* 1960]. In it the knife edges are replaced by a fine wire suspension, as indicated in Figure 5. The restoring forces exerted by the fine wire suspension are approximately counterbalanced by the labilizing spring shown. Also, the box supporting the horizontal bar is rotated by a servo to almost eliminate relative motion between the box and the inertia bar. This further reduces the restoring forces exerted by the wires suspending the bar. The servo support is fixed to the swinging gravity meter, and therefore the angular rotation of the servo is a measure of θ.

Obviously, gyros can also be used for stabilized references. The main reason that they have not been used with gimbal-supported gravity meters is that until a few years ago the estimated lives of good gyros was only 1000 hours. Because good gyros with estimated lives of 1½ years are now available, they should be used with either gimbal or stabilized platform gravity meters because they have more than adequate accuracy, whereas the performance of long-period pendulums is marginal.

The main disadvantages of the long-period pendulums are that vibration or large accelerations can cause interference between moving and fixed parts and

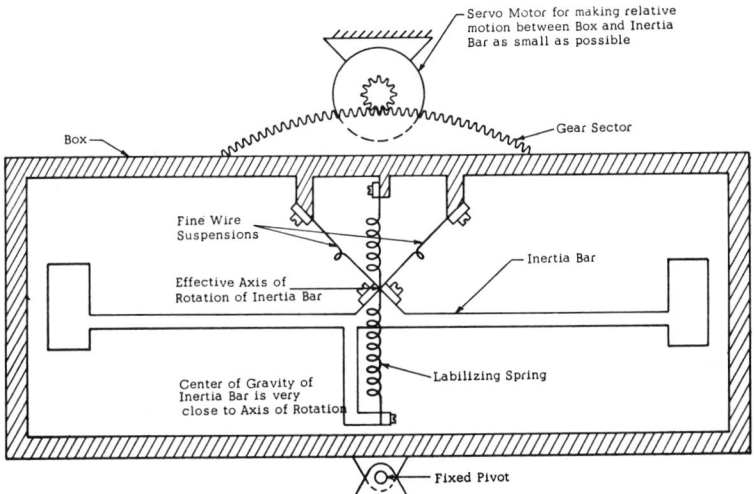

Fig. 5. Long-period inertia bar vertical reference.

that an operator must check them and their servos regularly to keep them in good adjustment. Long-period pendulums are normally operated at periods of 2 min, which appears to be sufficiently long on a ship traveling on a straight line but which is short for airplane work. They also have a slight nonlinearity which causes them to give a slight indication proportional to changes in wave amplitude, but this effect has been made almost negligible up to accelerations of 0.05 g in late models. Their performances can certainly be improved further, but this refinement is not worth while since gyro life is now adequate.

An important element in gravity meter accuracy is the accuracy of the stabilized reference. The effect of an error e in the reference can be determined as follows. Equation 14 gives the Browne correction that must be subtracted from the gravity meter reading as $g\theta^2/2$. If an angle $(\theta - e)$ is substituted into this formula in place of θ, the error in gravity will be

$$e_g = g_a - g(\theta - e)^2/2 - g_a + g\theta^2/2$$
$$= g\theta e - ge^2/2 \tag{25}$$

Equation 25 can be simplified by making use of the differential equation of the gimbal suspended gravity meter:

$$l_0\theta'' + g\theta = -x_G'' \tag{26}$$

where $l_0 = g/w_G^2$ is the length of a simple pendulum having the same period as the gimbal suspension and x''_G is the horizontal acceleration at the gimbal joint. Equation 26 can be written as

$$-g\theta = x_G'' + l_0\theta'' \tag{27}$$

The right side of (27), however, is the horizontal acceleration x'' at the gravity meter, which is always placed a distance l_0 below the gimbal joint. Therefore (25) becomes

$$e_g = -x''e - ge^2/2 \tag{28}$$

It will later be shown that (28) is also the error equation for a stabilized platform.

The last term of (28) refers to steady-state errors or to random errors. For a 1-mgal accuracy in gravity measurement this term requires an accuracy in the stabilized reference of 5'. The first term, however, depends not only on the stabilized reference error but also on the magnitude of the horizontal acceleration and on the correlation between the two. For example we will consider a sinusoidal horizontal acceleration x'' of amplitude 0.05 g. For a 1-mgal accuracy this limits the permissible sinusoidal error in e to an amplitude of 8" in phase with x''. This is not an easy requirement to meet. Laboratory and sea tests indicate, however, that long-period pendulums very nearly satisfy it and the other requirement for accelerations up to about 0.05 g.

Stabilized platform. The second method of handling horizontal accelerations is to mount the gravity meter on a stabilized platform as shown in Figure 6. To determine what accuracy is required in the stabilized platform, it is neces-

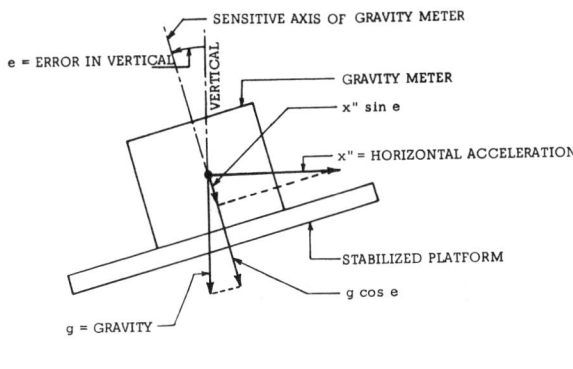

Fig. 6. Gravity meter on stabilized platform.

sary to find out how the gravity reading is affected by errors in verticality [*LaCoste*, 1959]. In Figure 6 the sensitive axis of the gravity meter is shown to be off level by an angle e. The error in measuring gravity is then

$$e_g = g \cos e - x'' \sin e - g \qquad (29)$$

or

$$e_g \doteq -x''e - ge^2/2 \qquad (30)$$

Equation 30 is exactly the same as equation 28, which gives the error for a gimbal suspended gravity meter after the Browne correction has been made. It is therefore apparent that the two systems are mathematically equivalent up to fourth-order terms.

The stabilized platform accuracy requirements given by (30) for a 1-mgal accuracy can easily be satisfied for horizontal accelerations of 0.1 g if good gyros are used and if proper precautions are taken in the design of the stabilized platform. The major precautions will be discussed later. Steady-state and long-period verticality errors have been checked at sea by comparison with oil filled levels* when the ship was traveling in a straight line, and these errors were found to be less than 1′, which is considerably less than the 5′ requirement for 1-mgal accuracy. It is difficult to check directly the component of stabilized platform error in phase with the horizontal acceleration, because this check would require an extremely accurate additional vertical reference. The accuracy that has been achieved in laboratory and sea tests shows, however, that the requirements have been satisfied.

The operation of the stabilized platform is indicated in the block diagram of Figure 7. This figure shows only one of the two required vertical erecting units. Each of the two units can operate from a single-axis gyro, or the two units can both operate from a two-axis gyro. In the unit shown the gyro controls a servo amplifier and motor that makes the platform follow the gyro. This feedback loop is not sufficient, however, to ensure verticality of a reference line on the platform because: (1) the reference line might not have been vertical to begin with, (2) gyros have some drift, and (3) the earth rotates and gyros tend to remain fixed in space. To attain verticality it is necessary to mount accelerom-

---

Author's note, 1989: Oil filled levels are not dependable in dynamic measurements. They often have large cross coupling errors.

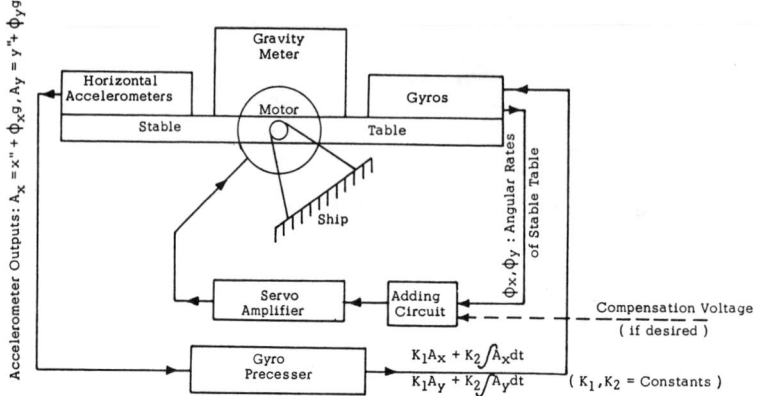

Fig. 7. Block diagram of stabilized platform.

eters (or levels) on the platform as shown and to use them in the second feedback loop. If the accelerometers do not indicate level, they put out error signals. A function of each error signal is fed to the corresponding gyro to precess it gradually to bring the reference line on the platform to vertical.

The simplest function of the accelerometer error signal to feed to the gyro is a constant times the error. This gives the gyro precession rate in space as

$$\phi' = -k_a(\phi + x''/g) \tag{31}$$

where $(\phi + x''/g)$ is the accelerometer error signal and k_a is a constant. Since ϕ represents the angular position of the stabilized platform, (31) is the differential equation for the assumed accelerometer feedback for the stabilized platform. An inspection of (31) will show that it does not represent very good stabilized platform behavior. For one thing ϕ will have to match the corresponding component of the earth's rotation in order for the platform to remain fixed relative to the earth. This will mean that the error signal $(\phi + x''/g)$ must differ from zero and there will be an error in platform verticality. Although this error can be adjusted out, the adjustment has to be changed whenever the platform is rotated about the vertical. A similar error is produced by gyro drift. Furthermore, it can be shown that long-period wave motions cause greater errors with this simple type of accelerometer feedback than with some other types of feedback.

A better type of signal to feed to the gyro is a constant times the accelerometer output plus a constant times its integral. This expression gives for the equation of motion of the platform

$$\phi' + 2fw_0\left(\phi + \frac{x''}{g}\right) + w_0{}^2 \int \left(\phi + \frac{x''}{g}\right) dt = 0 \tag{32}$$

or

$$\phi'' + 2fw_0(\phi' + x'''/g) + w_0{}^2(\phi + x''/g) = 0 \tag{33}$$

It should be noted that the integral term in (32) eliminates errors in ϕ due to rotation of the earth. Also, an inspection of (33) will show that it is the differential equation for a pendulum. The period is determined by the amount of integral feedback, and the damping is determined by the amount of ordinary feedback. It is interesting that there is mathematical equivalence between this type of stabilized platform and the long-period pendulums used with gimbal suspended gravity meters.

We will now determine permissible accelerometer feedback constants for the stabilized platform by means of the stabilized platform equation (33) and the error equation (30). We will consider a sinusoidal x'' of circular frequency w and will examine the steady-state solution of (33). We can then take $\phi = \phi_0\, e^{iwt}$. Substituting this expression into (33) gives

$$\phi = -\frac{w_0{}^2 + i2fw_0w}{w_0{}^2 - w^2 + i2fw_0w}\left(\frac{x''}{g}\right) \tag{34}$$

For perfect operation ϕ should be identically equal to zero; therefore, the value of ϕ given in (34) represents the error e in (30), or $\phi = e$. The average value of the second term in (30) then becomes

$$
\begin{aligned}
-\frac{g\langle e^2\rangle}{2} &= -\left|\frac{w_0{}^2 + i2fw_0w}{w_0{}^2 - w^2 + i2fw_0w}\right|^2 \left(\frac{\langle x''^2\rangle}{2g}\right) \\
&= -\frac{w_0{}^4 + 4f^2w_0{}^2w^2}{w_0{}^4 + 2w_0{}^2w^2(2f^2 - 1) + w^4}\left(\frac{\langle x''^2\rangle}{2g}\right)
\end{aligned}
\tag{35}
$$

In the first term of (30) only the component of e (or ϕ) that is in phase with x'' is of significance. This component is the real part of the coefficient of x'' in (34). The average value of the first term therefore becomes

$$-\langle x''e\rangle = \frac{w_0{}^4 + w_0{}^2w^2(4f^2 - 1)}{w_0{}^4 + 2w_0{}^2w^2(2f^2 - 1) + w^4}\left(\frac{\langle x''^2\rangle}{g}\right) \tag{36}$$

Adding (35) and (36) gives

$$\langle e_s\rangle = \frac{(w_0/w)^4 + 2(2f^2 - 1)(w_0/w)^2}{(w_0/w)^4 + 2(2f^2 - 1)(w_0/w)^2 + 1}\,\langle x''^2/2g\rangle \tag{37}$$

Equation 37 shows that the error is small only when w_0/w is small, or when the wave frequency is considerably greater than the natural frequency of the stabilized platform. The equation also shows that in such cases the error can be decreased by making $2f^2 - 1 = 0$, which corresponds to making the damping $1/\sqrt{2}$ times critical. This has been done for several years on all gimbal supported gravity meters and on LaCoste and Romberg stabilized platform gravity meters. Equation 37 then becomes

$$\langle e_s\rangle = \frac{(w_0/w)^4}{(w_0/w)^4 + 1}\,\langle x''^2/2g\rangle \tag{38}$$

To show how (38) limits the stabilized platform parameter w_0, (38) is plotted in Figure 8 for three different values of w_0 ($= 2\pi/$ natural period of the stabilized platform) and for a horizontal acceleration $x'' = 0.1\,g\sin wt$. The curves show

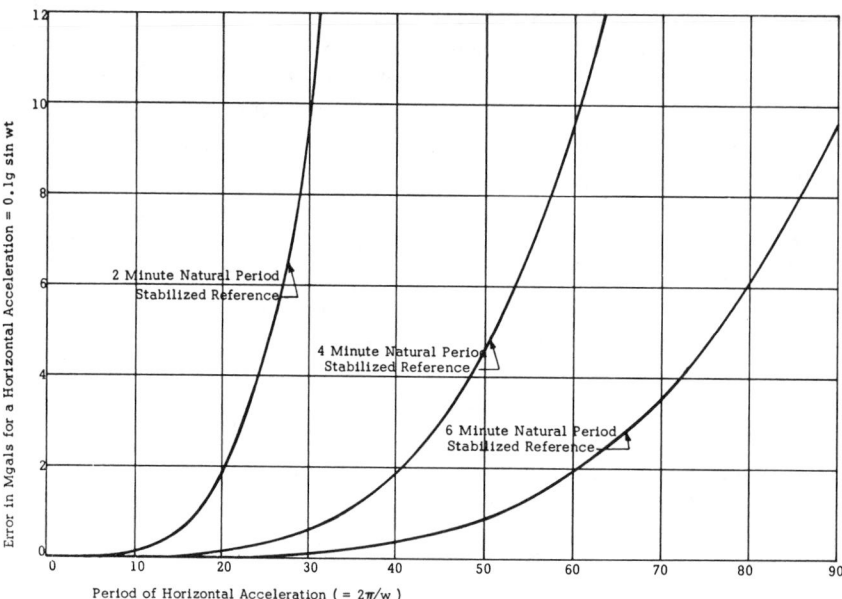

Fig. 8. Variation of gravity meter error with period of horizontal accelerations and period of stabilized reference for $1/\sqrt{2}$ times critical damping of stabilized reference.

that for a stabilized platform with a 2-min period the error is less than 1 mgal for period shorter than 17 sec. Ocean waves have appreciable amplitudes at periods longer than 17 sec, but they do not often have amplitudes as high as 0.1 g at such periods; therefore, a 2-min period stabilized reference can be used. For a horizontal acceleration of 0.05-g amplitude, a 2-min platform would operate up to a wave period of 24 sec with only a 1-mgal error, since the error varies with the square of the acceleration. Another indication that a 2-min period stabilized reference can be used is that gimbal supported gravity meters gave good results with stabilized references of that period.

In order to have some factor of safety the stabilized platforms used with LaCoste and Romberg gravity meters were designed with 4- and 6-min periods and with a selector switch for setting the period. The selector switch makes it possible to change from one period to the other during a run to determine whether the period has an effect on the gravity reading. In actual tests at sea no difference was found; therefore, either period appears to be adequate. The curves in Figure 8 show that for a 4-min stabilized reference the error is within 1 mgal for horizontal accelerations of periods less than 35 sec and amplitudes of 0.1 g. It is very unlikely that ocean waves have amplitudes as high as this at periods as long as 35 sec.

The preceding discussion has dealt with the effects of horizontal accelerations produced by ocean waves. However, horizontal accelerations are also produced when the ship (or plane) does not travel in a straight line at constant speed. To make such accelerations negligible, it is general practice to use an automatic pilot and in all cases the ship (or plane) is run in as straight a line as is practical and as close to constant speed as is feasible. There are, of course, deviations in both course and speed, and such deviations can have periods longer than the periods of ocean waves. To estimate the magnitude of the errors produced by deviations of the ship's track from a straight line (fishtailing), we will assume that the ship travels along a sinusoidal path whose deviation from a straight line is

$$y = D \sin 2\pi t/T \tag{39}$$

where T = the period. The resulting horizontal acceleration is

$$y'' = -(4\pi^2 D/T^2) \sin 2\pi t/T \tag{40}$$

Substituting (40) into (38) gives

$$D^2 = g\langle e_g \rangle (T^4 + T_0^4)/4\pi^4 \tag{41}$$

Equation 41 is plotted in Figure 9 for a 1-mgal error and for stabilized platform periods of 2, 4, and 6 min. The 2-min curve at short periods shows a permissible lateral ship motion of ±24 feet. For a 4-mgal error the permissible lateral amplitude is twice as great. Since accuracies of 1–4 mgal have been obtained with gimbal supported gravity meters using 2-min stabilized references, it appears that lateral ship motion resulting from imperfect steering can be kept under ±48 feet and probably under ±24 feet. The 4-min curve in Figure 9 shows a permissible lateral amplitude of ship motion of ±95 feet, which can easily be achieved, and the 6-min curve shows a permissible amplitude of over ±200 feet.

Good accuracy has been obtained in airplane gravity work with 1- and 2-min stabilized references [*Thompson and LaCoste*, 1960; *Nettleton, LaCoste, and Harrison*, 1960; *Nettleton, La Coste, and Glicken*, 1962] provided that the

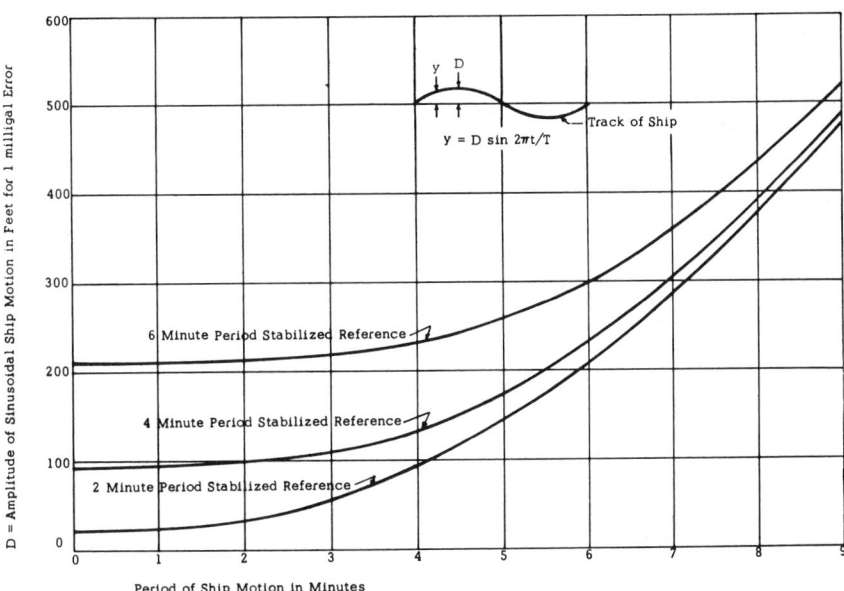

Fig. 9. Gravity meter errors for various sinusoidal ship tracks.

automatic pilot was carefully adjusted, provided that the air was relatively smooth, and provided that course changes were not made. In the tests described in the first two of the above references the automatic pilot had a hunting frequency of about 1 min when the plane flew north, which resulted in large errors. The problem with the automatic pilot was overcome before making the tests described in the third reference, but errors then occurred when the air became somewhat turbulent and when course changes of approximately one degree were made. Such errors could of course have been made negligible by using 4- or 6-min period stabilized references, but it was also found possible to correct for them by the following method.

The correction is based on (38), which gives the error for any frequency. To use this equation it is necessary to have approximate (or accurate) data on the horizontal accelerations. Such data are normally obtained and recorded. Their accuracy, of course, is limited by the accuracy of the stabilized references and therefore is somewhat in error at the long periods being considered, but they are accurate enough to make a good first-order correction. In the case being considered the computations required by (38) were performed by an analog computer. The analog computer consisted of the actual vertical references used in the airplane tests. The horizontal accelerations were applied to the vertical references as tilts rather than as accelerations. It was possible to estimate the accuracy of the corrections by comparing corrected gravity values obtained during course changes with gravity values obtained just before and just after such times. In all cases the gravity data were smooth, and the corrected gravity data were found to be nearly as accurate as those obtained in smooth air when no course changes were made.

There are also other ways in which (38) can be used to make corrections for too short a period in the vertical reference. For instance, data on horizontal accelerations might be obtained from inertial navigation, Doppler radar, or photography if such data are sufficiently accurate. Of course, it is preferable to use adequate periods in the vertical references, but even in this case (38) can be used to check on the adequacy of the periods.

Errors due to inadequate leveling of the gravity meter can also be caused by such things as dead space, hysteresis or too slow a response in the accelerometers, gyros, or servo motors controlling the stabilized platform. Errors can also occur if the sensitive axis of the gravity meter is affected by horizontal or vertical accelerations rather than being in a fixed direction on the gravity meter, as was assumed in the derivation of (30). Evidence that horizontal accelerations or forces can affect the sensitive axis of a gravity meter was given by Alan Goodacre in a private communication. Goodacre's experiments showed slight discrepancies between tilt-table calibrations of gravity meters and other types of calibration. It will also be shown later that the so-called inherent type of cross-coupling effect can be considered as a special case of off- leveling error resulting from shifting of the sensitive axis of the gravity meter when it is subjected to vertical accelerations.

To compensate for errors of the type mentioned in the preceding paragraph, it is possible to add a slight tilt to the gravity meter relative to the stabilized platform. One method of doing this is indicated in Figure 7. A correction volt-age is added to the gyro output, and the sum is fed to the corresponding servo motor. In L&R (LaCoste and Romberg) gravity meters the magnitude of the compensating voltage and its possible dependence on period are determined from tests made on a testing machine that subjects the complete gravity meter system to horizontal accelerations of various amplitudes and periods.

In the previous discussion it has been intimated that it is adequate to use platforms that are stabilized in roll and pitch but not stabilized about the vertical axis. Actual tests indicate that such two-axis stabilization is sufficient for approximately straight line motion of the ship or airplane at constant speed. There are, however, some advantages to be gained by stabilization also about the vertical, as can be seen from the following considerations. In a two-axis system, it is necessary for the sensitive axes of the two gyros to be accurately horizontal; otherwise, rotation about a vertical axis (ship yaw) will affect them in the same way that roll or pitch does. This effect will produce errors in the verticality of the platform. Furthermore, such verticality errors will be in phase with yaw and therefore probably well correlated with horizontal accelerations. Such correlation can produce substantial errors, as shown in (30); therefore, accuracy depends to a considerable extent on the accuracy with which the sensitive axes of the gyros are set.

Since it is impossible to set the sensitive axes of the gyros perfectly, it might be advantageous to stabilize the platform also about the vertical. Furthermore, to initially set the sensitive axes of the gyros accurately, it is necessary to make tests by rotating the entire stabilized platform about the vertical to determine the effect of the rotation. Since a turntable for making this test is not normally available at sea, there is a serious problem in changing gyros at sea unless their sensitive axes are accurately adjusted for uniform orientation in the gyro cases. In view of these considerations a stabilized vertical axis is being provided for LaCoste and Romberg stabilized platform gravity meters. This design has not yet been used at sea and might prove to be more trouble than it is worth, but a test should be made. Long-term vertical stabilization to a definite compass direction is not required; therefore, the LaCoste and Romberg device will be slowly precessed to a fixed direction on the ship.

Until now there has been no need to make gravity observations when travel-ing in any way other than a straight line at constant velocity. Therefore, there has been no need to use a stabilized platform any more sophisticated than the ones previously described. If, however, the need should ever arise, it is possible to measure gravity when moving in a curved path at nonconstant speed by using what is known as a Schuler tuned stabilized platform. Such a platform is used in inertial guidance; it has the property of remaining vertical regardless of the motion of the ship or airplane. Since there might be reason to use such a plat-form at some time, a brief description of its principles might be worth while.

In Figure 10 we will assume that the Schuler platform is at A and that it moves to B over the surface of the earth. We will let the angle ϕ denote the direction in space of a reference line on the platform. If the platform is stabilized as the previously described stabilized platforms were, its differential equation

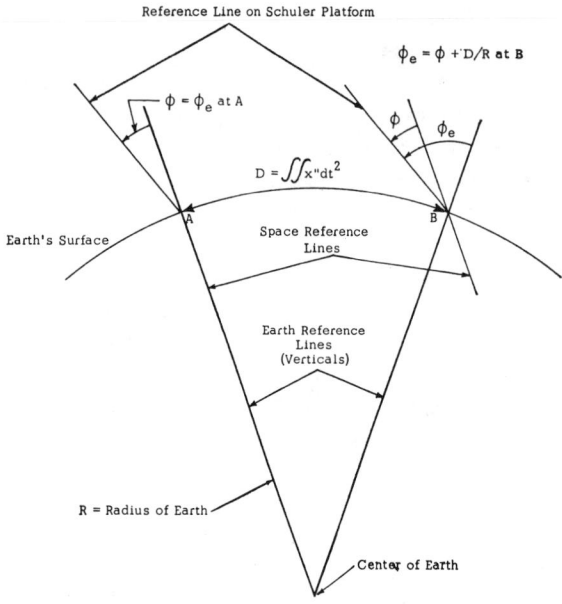

Fig. 10. Diagram for derivation of Schuler stabilized platform equations.

will be given by (33) for an earth of infinite radius. To take into account the finite radius of the earth, (33) must be changed to

$$\phi'' + 2fw_0\left(\phi' + \frac{1}{R}\int x''\,dt + \frac{x'''}{g}\right) + w_0{}^2\left(\phi + \frac{1}{R}\iint x''\,dt^2 + \frac{x''}{g}\right) = 0 \qquad (42)$$

where R is the radius of the earth.

We wish to use a direction normal to the surface of the earth (vertical) as a reference rather than a direction fixed in space. If we designate angles referred to such a vertical reference by ϕ_e, then ϕ_e will be given by

$$\phi_e = \phi + \frac{1}{R}\iint x''\,dt^2 \qquad (43)$$

Substituting (43) into (42) gives

$$\phi_e'' - x''/R + 2fw_0(\phi_e' + x'''/g) + w_0{}^2(\phi_e + x''/g) = 0 \qquad (44)$$

It is desired to make (44) independent of x'', which can be done by taking $f = 0$ and

$$w_0 = \sqrt{g/R} \qquad (45)$$

Equation 44 then becomes

$$\phi_e'' + (g/R)\phi_e = 0 \qquad (46)$$

A stabilized platform satisfying these conditions is known as a Schuler tuned stabilized platform.

The characteristics of the Schuler platform follow from the preceding equations. Equation 46 shows the behavior of ϕ_e, which is the angle between the stabilized platform and the earth's vertical at the point at which the platform happens to be. We note that, if ϕ_e and ϕ_e' are initially zero, ϕ_e will remain identically zero regardless of any horizontal accelerations. Therefore, a Schuler platform will operate satisfactorily for any ship or airplane motion.

There are difficulties in making and operating a Schuler stabilized platform, and therefore it should not be used unless it is needed. Some of the difficulties are as follows: The period of the Schuler platform is given by (45); it is $2\pi/w_0$ $= 2\pi (R/g)^{1/2}$, which is about 84 min. To attain this very long period, it is necessary to use very accurate gyros and accelerometers and to adjust the system very accurately. Furthermore, (46) shows that the system is undamped, and therefore its initial conditions must be carefully controlled to prevent oscillation. Oscillations are also caused by gyro drift and accelerometer imperfections. A Schuler platform requires accurate stabilization about the vertical axis as well as about the two horizontal axes. For these reasons a Schuler tuned stabilized platform has both a high initial cost and a high maintenance cost.

It has been mentioned that in the L&R stabilized platform the damping is adjusted to $1/\sqrt{2}$ times critical. The importance of this adjustment can be demonstrated by comparing the performance of such a stabilized platform with one in which the precessing signal to the gyro is equal to a constant times a filtered signal from the accelerometer. For a single-stage filter the precession rate is given by

$$s\phi = -k_a(\phi + x''/g)w_1/(s + w_1) \tag{47}$$

where s is the Laplace operator. To evaluate the platform (47) can be treated similarly to the way in which (32) was treated. This will determine the resulting error in gravity reading as a function of the period of the horizontal accelerations.

Differentiating (47) gives

$$s^2\phi + sw_1\phi + k_aw_1(\phi + x''/g) = 0 \tag{48}$$

Equation 48 is a second-order equation that is fairly similar in form to the equation for the L&R stabilized platform (and to the equation for the long-period pendulum references). It differs in having no term in x'''. Writing (48) in the standard form of (33) gives

$$\phi'' + 2fw_0\phi' + w_0^2(\phi + x''/g) = 0 \tag{49}$$

If the same analysis is applied to (49) that was applied to (33), the resulting average error becomes

$$\langle e_s'\rangle = \frac{(w_0/w)^4 - 2(w_0/w)^2}{(w_0/w)^4 + 2(w_0/w)^2(2f^2 - 1) + 1} \langle x''^2/2g\rangle \tag{50}$$

For frequencies considerably higher than the natural frequency, (50) becomes approximately

$$\langle e_s'\rangle \doteq -2(w_0/w)^2\langle x''^2/2g\rangle \tag{51}$$

which is the approximate expression for (37) for the L&R stabilized platform for the case of no damping. If, however, the damping constant f in (37) is taken equal to the optimum value $1/\sqrt{2}$, $\langle e_g \rangle$ is given by (38), which reduces approximately to

$$\langle e_g \rangle = (w_0/w)^4 \langle x''^2/2g \rangle \qquad (\text{for} \quad w \gg w_0) \qquad (52)$$

which is much smaller than $\langle e_g' \rangle$. For example, for $w_0/w = 1/10$, $\langle e_g \rangle$ for the L&R platform is 200 times smaller than $\langle e_g' \rangle$ for the other platform.

Another requirement of stabilized platforms is that they must not produce or transmit any vibrations that can cause resonance in the gravity meter itself. Most, if not all, gravity meters have certain resonant frequencies that must be avoided if they are to give accurate results. The resonant frequencies are generally higher than 10 per sec and therefore it is possible to avoid them or to shock-mount to eliminate them. A possible source of objectionable vibration is the hunting of the erection servo motors controlling the stabilized platform. Good servo design will of course eliminate such vibration.

Ship or airplane vibrations can easily be made negligible by shock-mounting a gimbal supported gravity meter, but shock-mounting a stabilized platform introduces some difficulties because an ordinary shock-mounted frame does not provide a firm base against which the servomotors can react. This makes it difficult to avoid hunting unless the servo gain is kept low or unless carefully designed lead networks are used. It is possible to provide a firm base against which the servos can react by making the shock-mounted supporting frame heavy compared with a gravity meter or by using parallel linkages between the supporting frame and the fixed base so as to permit translation but not rotation. The first L&R stabilized platforms [*LaCoste, et al.*, 1967] used parallel linkages, but the linkages were discarded when the servos were improved enough to make them unnecessary. This was a welcome simplification because the parallel linkages had to be well made and carefully adjusted to avoid transmitting high-frequency angular oscillations of the ship. The angular oscillations are transmitted unless the centers of gravity of the supporting frame and of the gimbal ring are on the apropriate gimbal axis.

There is another reason for preventing vibrations from reaching the stabilized platform, which was pointed out by J. J. Jarosh of Hughes Research Laboratories in a proposal to the Naval Oceanographic Office. Jarosh called attention to the well-known fact that torques on a stabilized platform are produced by horizontal and vertical accelerations when the platform is not perfectly balanced, and these torques must be counterbalanced by the servos. He also noted that the servo gain falls off at high frequencies(limited servo bandwidth), and therefore appreciable verticality errors can be produced by horizontal accelerations at these frequencies. Since these verticality errors are in phase with the horizontal accelerations, significant errors in gravity can occur. Jarosh made some computations with reasonable values of servo gain and bandwidth and with reasonable amounts of platform unbalance and found errors of the order of a milligal. He also showed that suitable shockmounting reduced them greatly.

Cross-Coupling Effects

General. Another source of error in shipboard gravity meters is cross coupling between horizontal and vertical accelerations. There is an inherent type of cross-coupling effect in certain kinds of gravity meters. It can be avoided by making the gravity meter symmetrical about a vertical axis or by accurately nulling the meter continuously, or the effect can be corrected for. There is also an imperfection type of cross coupling that is due to imperfections in the gravity meter. It can be made negligible by careful design and construction, or it can also be corrected for. The so-called 'inherent' type was described by *LaCoste and Harrison* [1961]. The equations for it can be obtained as follows.

Inherent cross coupling. The differential equation for a spring type of gravity meter is given in (3) for the case in which horizontal accelerations are absent or neglected. This equation will be modified to take care of horizontal accelerations for a gravity meter with a beam hinged about a horizontal axis as shown in Figure 2. If all terms in (3) are multiplied by $MD \cos \beta$, they will represent torques. The torque $-MDx'' \sin \beta$ due to horizontal acceleration can then be added. The resulting equation is

$$g + z'' - x'' \tan \beta + bB'' + fB' + kB - cS = 0 \tag{53}$$

This equation gives the correct value of g. If the horizontal acceleration term is omitted, an incorrect value g_e is given for gravity; the equation is

$$g_e + z'' + bB'' + fB' + kB - cS = 0 \tag{54}$$

The error in g_e is then

$$\begin{aligned} e = g - g_e &= x'' \tan \beta \\ &\doteq x'' \beta \end{aligned} \tag{55}$$

If we take

$$\beta = \beta_0 + \beta_1 \sin (w_3 t + \psi) \tag{56}$$

and

$$x'' = x_1'' \sin w_1 t \tag{57}$$

where $\beta_0, \beta_1, x''_1, \psi, w_1$, and w_3 are constants, then (55) becomes

$$e = x_1'' \beta_0 \sin w_1 t + x_1'' \beta_1 \sin w_1 t \sin (w_3 t + \psi) \tag{58}$$

The average value of e is zero except when $w_1 = w_3$, in which case it becomes

$$\langle e \rangle = \tfrac{1}{2} x_1'' \beta_1 \cos \psi \tag{59}$$

Equation 59 is the expression for the 'inherent' type of cross coupling. Since the beam is driven by vertical accelerations, β will have the same period as the vertical accelerations. Therefore, inherent cross coupling can exist if components of horizontal and vertical acceleration have the same period. This condition generally exists to some extent at sea. One reason is that water particles undergo approximately circular motion in waves. This results in a phase difference of $\pi/2$.

Also, if the gravity meter is mounted off the ship's roll axis in both the horizontal and the vertical directions, the gravity meter will be subjected to a ramp motion when the ship rolls. This motion will give zero phase difference between horizontal and vertical accelerations.

For a highly overdamped gravity meter the damping coefficient f in (3) is very large, and therefore B' (and β') are approximately proportional to z''. There will therefore be a phase difference of about $\pi/2$ between β and z''. If x'' and z'' also have a phase difference of $\pi/2$, then $\cos \psi = \pm 1$ in (59), and $\langle e \rangle$ can be large. This condition exists if the ship follows the water particle motion in the waves. This is roughly what takes place.

It is of interest to estimate the magnitude of the inherent cross-coupling effect for a recent LaCoste and Romberg gravity meter. In such a gravity meter the damping allows an angular beam motion of ± 0.001 radian for a vertical acceleration of $\pm 0.1\ g$ at a period of 3.5 sec. Therefore, for a horizontal acceleration of $\pm 0.1\ g$ and $\psi = 0$, (59) gives a cross-coupling effect of

$$e = 49 \text{ mgal} \tag{60}$$

Sea data have given cross-coupling effects as high as 20 mgal for LaCoste and Romberg gravity meters. Cross coupling has also been reported by *Bower* [1966], *Talwani* [1966], *Talwani et al.* [1966], *Wall et al.* [1966].

As mentioned in the stabilized platform section, inherent cross-coupling errors can be considered as a special case of off-leveling errors caused by shifting of the sensitive axis of the gravity meter when subjected to vertical accelerations. It will be recalled that in the derivation of the off-leveling equations (29 and 30) the sensitive axis of the gravity meter was assumed fixed on the gravity meter. However, as previously mentioned, the direction of the sensitive axis can sometimes be affected by accelerations. A study of Figure 2 will show how this effect can occur in a beam-type gravity meter. The sensitive axis is the axis along which there is a maximum effect of gravity or acceleration and perpendicular to which there is no effect. The sensitive axis is therefore in the direction OC perpendicular to the beam OP. It is apparent that this axis shifts as the beam moves in response to vertical accelerations.

If the gravity meter is tilted relative to the stabilized platform through the required angle to keep its sensitive axis vertical, horizontal accelerations will have no effect on the gravity meter and there will be no inherent cross-coupling errors. One method of accomplishing this tilting is to add a suitable compensating voltage to the gyro output that controls the corresponding servo motor as shown in Figure 7. In this case the compensating voltage should be proportional to the displacement of the gravity meter beam from horizontal.

The preceding method of handling cross coupling by tilting the gravity meter relative to the stabilized platform is simple, but in the case of L&R gravity meters it introduces a small second-order error because of tilting of the gravity meter relative to vertical. An analysis will show that the tilting results in changes of gravity meter sensitivity to the pull of gravity which introduce an error of $g\ \phi^2/2$ where ϕ is the angle of tilt. This error is entirely negligible for the tilting previously described in the stabilized platform section but might be an appreciable fraction of a milligal if cross coupling were compensated by tilting. For this reason tilting has not been used in operation at sea to compensate L&R gravity meters for inherent cross coupling, although it has been used to compensate them for other errors. Corrections for cross-coupling errors in L&R meters have been made by using a simple analog computer to evaluate (55), which is an accurate equation. Means for compensating for cross-coupling effects are also described by *Talwani* [1966] and *Jacoby and Schulze* [1967].

Cross-coupling effects can be eliminated in gimbal supported gravity meters by correctly placing the gravity meter relative to the gimbal as was explained in the discussion of (24). Equation 24 applies when $l \cos \alpha$ in Figure 3 is taken equal to g/w_G^2; this relation determines the distance the gravity meter weight is below the gimbal. If the angle α and the gimbal damping F are also taken equal to zero, (24) gives the acceleration a_n normal to the sensitive axis of the gravity meter as zero. Since a_n corresponds to the acceleration x'' in Figure 2 and equation 55, it can be seen that the cross-coupling effect e will be zero in this case. *LaCoste and Harrison* [1961] have considered the cross-coupling effect for the case in which $F = 0$ but $a \neq 0$ and have found that cross coupling is negligible even in this case.

Imperfection cross coupling. The imperfection type of cross-coupling effect was described by *LaCoste et al.* [1967]. This effect is due to imperfections in the design or construction of the gravity meter, and there are many ways in which it can occur so that determining its major cause is often difficult. It can be comparable to or larger than the inherent type of cross coupling.

Imperfection type cross coupling can occur if the damping coefficient in (3) is affected by horizontal accelerations or if horizontal accelerations cause vertical forces to vary. Two examples will be given to illustrate. Figure 11 shows a gravity meter with a hinged beam moving up and down a ramp. There will always be some elasticity in the beam and in the wires providing a hinge for it. This elasticity will allow the center of gravity of the beam to move horizontally in response to the horizontal accelerations, thereby changing the moment arm of the beam. The moment arm on the right of the figure will be denoted by $L + \Delta L$ and that on the left by $L - \Delta L$. Also, because of vertical acceleration the vertical force per unit mass on the right will be $g + z''$, where z'' is the vertical acceleration necessary to reverse the motion. Similarly, on the left the force per unit mass will be $g - z''$. The average of the two moments will be $Lg + z''\Delta L$, as shown. However, the average would have been Lg if there had been no motion. Therefore, the term $z''\Delta L$ is a cross-coupling effect.

The previous considerations show how important it is to restrict the gravity meter beam to only one degree of freedom. Even if this is done, however, the center of gravity of the spring can still shift in response to horizontal accelerations, and such a shift can have the same effect as a shift in the center of gravity of the beam.

Another example of imperfection type of cross coupling is shown in Figure 12, which is a top view of the gravity meter. The spring supporting the beam is shown at the right, and wires providing a hinge are shown at the left. The hinge axis is labeled. To restrict horizontal translation along the axis of rotation, the two wires

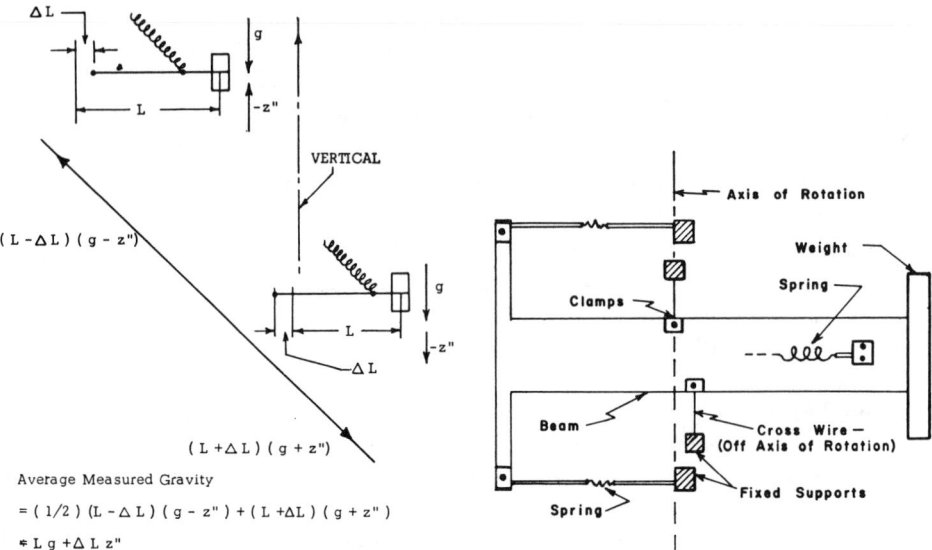

Average Measured Gravity

$$= (1/2)(L - \Delta L)(g - z'') + (L + \Delta L)(g + z'')$$

$$\doteqdot L g + \Delta L z''$$

Fig. 11. Example of imperfection coupling. (Reprinted from *Geophysics, 32*, p. 107, 1967.)

Fig. 12. Example of imperfection cross coupling. (Reprinted from *Geophysics, 32*, p. 108, 1967.)

shown are used. The wire at the top of the figure is shown in the axis of rotation and gives no trouble with cross coupling. The wire at the bottom, however, is shown to the right of the axis of rotation and will give a cross-coupling effect, as can be seen from the following considerations.

The effect of vertical accelerations will be considered first. It will be assumed that the beam will yield to some extent in response to such accelerations. The point of attachment of the lower (off axis) wire to the beam will then move up and down in space (in and out of the plane of the figure). When it moves down, the wire will tend to raise the right end of the beam and vice versa. Since the upward and downward accelerations will average out to zero, there will be no net effect.

If horizontal accelerations are present as well as vertical, however, there can be a cross-coupling effect. The horizontal accelerations will have the effect of tightening or loosening the wire, and, if the wire is always tight when it is pulling up and always loose when it is pulling down, there will be a net upward pull on the weight, which is a cross-coupling effect.

To make imperfection types of cross coupling negligible, it is necessary to make tests with various types of motion in addition to designing and adjusting carefully. Imperfection cross coupling can of course be corrected for with an analog or digital computer if its effects have been calibrated in tests.

The imperfection types of cross coupling just described can not be considered as special cases of off-leveling errors, because they do not involve a shift in the direction of the sensitive axis of the gravity meter. The mathematical expression for them ($x''z''$ times a constant) is, however, very similar to the more important first term $-x''e$ in the off-leveling equation (30). It therefore is possible to very nearly correct for imperfection types of cross coupling by intentionally introducing an equal and opposite off-leveling error. For example, in Figure 7 the compensating voltage can be made proportional to z''. This will give an off-leveling error proportional to $-x''z'' - gz''^2/2$. With a proper choice of the proportionality factor the first term will balance out the imperfection cross coupling and the second term will generally be negligible because it is of the second order in z''.

The LaCoste and Romberg Shipboard Gravity Meter

Figure 13 is a photograph of an L&R gravity meter on a stabilized platform. A diagram showing the operation of the gravity meter is shown in Figure 2 [*La-Coste*, 1961]. The beam is pivoted about the horizontal rotational axis O and is supported at its center of gravity P by a spring, whose unstretched length is zero (a 'zero length spring'). Actual mechanical details are shown in Figures 14 and 15. In these figures the horizontal rotational axis is shown to be provided by the fine wires h. The beam B is very highly damped by the air dampers D and D1. The two movable dampers on the beam and the two fixed dampers on the frame consist of several concentric cylinders. The cylinders of the fixed and movable dampers interweave each other so as to provide high resistance to the flow of air

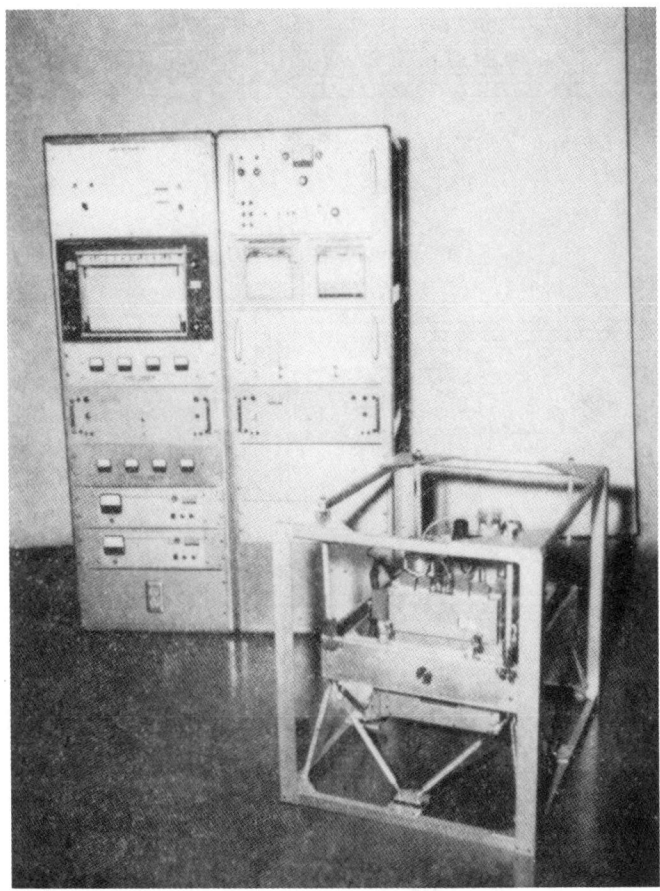

Fig. 13. Photograph of LaCoste and Romberg stabilized platform air-sea gravity meter.

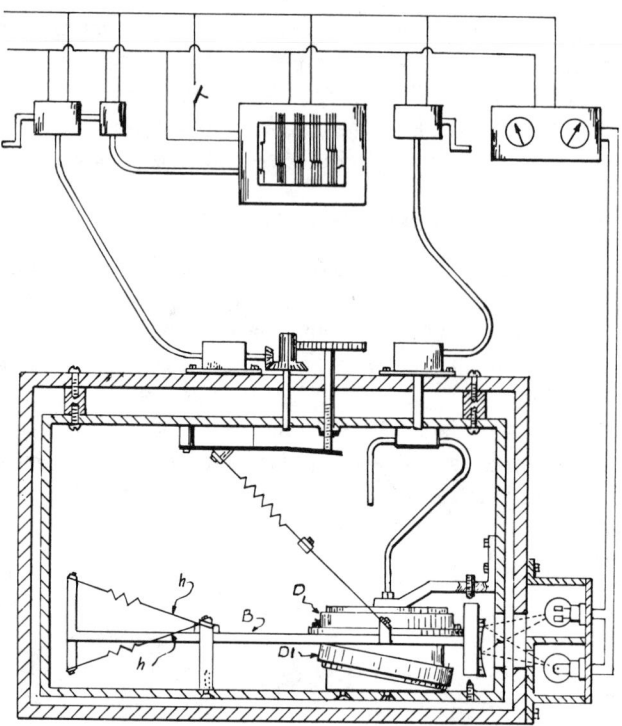

Fig. 14.　LaCoste and Romberg air-sea gravity meter.

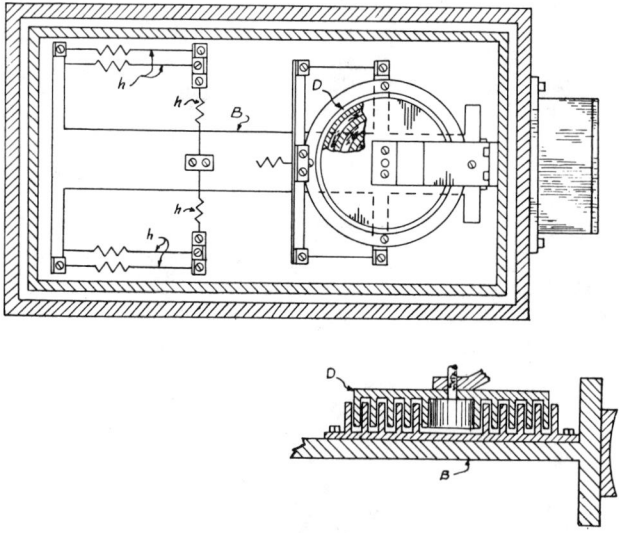

Fig. 15.　LaCoste and Romberg air-sea gravity meter.

into and out of the dampers. The upper end of the spring, as is shown in Figure 14, can be adjusted to take care of variations in gravity. A more sophisticated means of adjusting the spring is actually used [*LaCoste and Romberg*, 1945].

The differential equation can be derived from Figure 2 as follows. The spring PA is a zero length spring; i.e., its unstretched length is zero. If K denotes the spring constant, the force exerted by the spring at P is K times the distance PA, according to Hooke's law. The force can therefore be represented by the vector *PA*. The force exerted by the beam at P is in the direction OP, and its horizontal component equals the horizontal component of the force exerted by the spring if no horizontal accelerations are present. The force exerted by the beam can therefore be represented by the vector *OP*. The vector sum of the forces exerted by the spring and beam at the point P can therefore be represented by the vector sum *OA*. If the distance OA is taken as S, then the vector *OA* represents a force KS where K is the spring constant.

The total static vertical force acting at P will then be $KS - mg$. The gravity meter is operated with the beam approximately horizontal and therefore its center of gravity P has very little horizontal motion. If this horizontal motion is neglected, the differential equation becomes

$$B'' + FB' = KS - m(g + z'') \qquad (61)$$

This equation is the same as (3) with $K = 0$. The time constant is equal to the time required for B to reach 0.6 of its steady-state velocity when an input step function is applied to it. From the equation it can be seen to be $1/F$, which shows that the time constant is reduced by increasing the damping.

A typical value of the time constant is 4×10^{-4} sec. Since this time constant is so much smaller than any time over which gravity readings are averaged, there is no need to consider the transient response of (61) and the equation can be replaced by

$$g + z'' \doteq (K/m)S - (F/m)B' \qquad (62)$$

or, if it is desired to include cross coupling, by

$$g + z'' = (K/m)S - (F/m)B' + \beta x'' \qquad (63)$$

(See equation 53.)

Equation 63 shows one of the features of the L&R gravity meter. Since the gravity reading does not depend on B, it is not adversely affected by drift in the device used to measure B as long as the drift is not fast enough to give an appreciable contribution to the B' term. This feature helps to give long-term stability to the gravity meter.

Even though the gravity meter accuracy is relatively unaffected by drift in the photoelectric device used to measure beam position, a chopper is used in it. Its operation is as follows [*Clarkson and LaCoste*, 1957]. A beam of light is reflected by a mirror on the gravity meter beam. When the beam is nulled, the light falls equally on two slits; when the beam moves up, more light goes through the upper slit and vice versa. A chopping disk lets light through the two slits alternately. The light going through the two slits falls on a single photoelectric cell.

From this description it can be seen that, when the beam is nulled, the photocell output is ideally dc; there is no ac component. As the beam moves from null, an ac output appears and is proportional to the displacement of the beam from null. The direction of the displacement determines the phase of the ac. The ac is rectified by a phase sensitive detector. The advantage of using a single photocell is that drift of the photocell does not affect the reading when the beam is nulled. Actually the light from the two slits is even focused on the same part of the photocell.

Another feature of the L&R spring suspension was pointed out by *Harrison* [1960]. He considered the effects of spring vibrations on the average pull exerted between the ends of a zero length spring. He first pointed out that longitudinal vibrations had no effect because they increased and decreased the force equally. He then analyzed the effects of transverse vibrations and showed that their effects also were negligible for a zero length spring. The following considerations will show why transverse vibrations have no effect.

Referring to Figure 16, consider the spring to be made up of weightless zero length springs with identical masses between spring segments and attached to the fixed points A and B, which are in line vertically. For simplicity the effect of gravity on the masses will be ignored. Since the spring segments are zero length springs, the forces each spring exerts on the adjacent masses are proportional to the spring length. Also, the vertical components of the forces are proportional to the vertical component of the spring length. Therefore, if the masses are equally spaced vertically, the vertical component of force exerted on each mass will be zero regardless of its horizontal position or horizontal motion. Also, the vertical force on A and B will be independent of any horizontal motions.

Until 1965 LaCoste and Romberg shipboard gravity meters were operated suspended from gimbals and were placed at the proper distance below the gimbal joint to make negligible all acceleration forces normal to the sensitive axis of the gravity meter. The method of doing this has already been described in the gimbal suspension section. In these gravity meters it was not necessary to make the beam and the springs in the fine wires providing a hinge for the beam very stiff. Therefore, they were not made stiff, although there actually was an advantage in making them stiff that was not realized at the time. Stiffness makes it easier to achieve

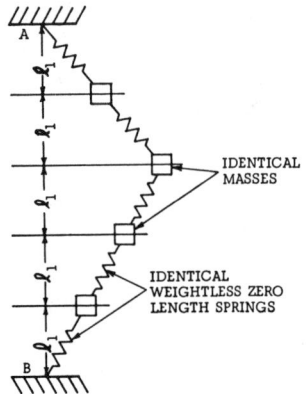

Fig. 16.   Lateral vibrations in a zero length spring.

high gravity meter linearity, which in turn determines the magnitude of vertical acceleration errors. At the time, the linearity was considered adequate for operation at vertical accelerations of ±50 gals, which was the limit of operation for horizontal accelerations. Later it was realized that operation at greater vertical accelerations would have been a distinct advantage because vertical accelerations are often several times larger than horizontal accelerations.

When it was decided to mount L&R gravity meters on stabilized platforms, it was realized that the previously made gravity meters might not be satisfactory, but even so it was considered worth while to make tests with them. The tests showed the need for modification, and the gravity meter was accordingly redesigned to more nearly restrict the beam motion to one degree of freedom. Stiffness with respect to unwanted modes of motion was increased by a factor of 40 without noticeably affecting stiffness in the desired mode.

These modifications not only made the imperfection type of cross coupling negligible but also made it easy to attain sufficient gravity meter linearity to limit vertical acceleration errors to less than 1 mgal at vertical accelerations of ±100 gals. Another modification was to approximately double the damping so as to reduce the inherent type of cross-coupling effect. The increased damping permits operation at ±1g at a period of 3.5 sec without having interference between the beam and the case of the gravity meter.

Both the inertia bar vertical references used with gimbal supported L&R gravity meters and the stabilized platforms used after 1965 are described in the section on horizontal accelerations. The stabilized platforms use inertial guidance quality gyros controlled by accelerometers to give 4- or 6-min periods (or longer if desired) and damping of 0.7 times critical. The platform has a range of ±30° and is powered by fast acting torque motors. It has never failed to operate because of roughness of the sea. To compensate for small platform errors of undetermined origin, a compensating voltage is added to the gyro output as shown in Figure 7 and as explained in the stabilized platform section.

The cross-coupling computer is a simple analog computer that multiplies beam position by horizontal acceleration along the beam. A servo follows the beam position, and the servo drives a potentiometer that forms two adjacent arms of a Wheatstone bridge. The voltage across the bridge is made proportional to the horizontal acceleration, and therefore the bridge output is proportional to the cross coupling.

The block diagram of Figure 17 shows the operation of the gravity meter as well as the operation of the stabilized platform. A simple analog computer referred to as a beam nuller uses the beam position as a signal to a slow servo to approximately null the beam. The servo is made slow so that it will not have to follow the wave accelerations, which normally amount to thousands of milligals. It performs the approximate nulling by controlling the 'spring tension,' S in (63).

A second analog computer referred to as the automatic reader computes an average value of gravity from (63). The computation is accomplished by filtering S, filtering and differentiating B, filtering the cross coupling, and adding them. The filtering is the same for each variable. Obviously this second function can be performed by a digital computer.

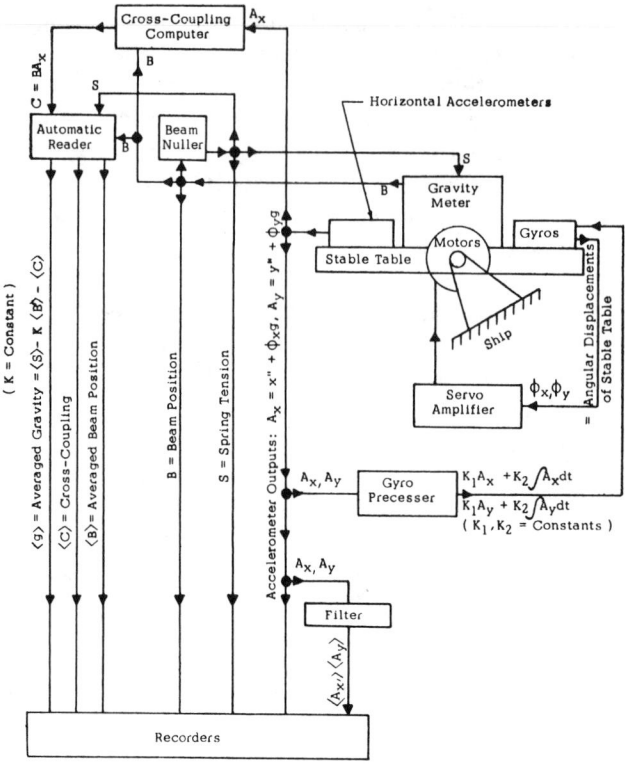

Fig. 17. Block diagram for LaCoste and Romberg air-sea
gravity meter.

The value of gravity computed by the reader is recorded on a strip chart recorder and on magnetic or punched tape. Spring tension and averaged beam position are also recorded so that gravity can be computed even though there is a malfunction in the reader. Instantaneous beam position and filtered and unfiltered horizontal accelerations are also recorded for monitoring purposes. The filtered horizontal accelerations are useful in checking gyro performance. In late models a voltage proportional to vertical acceleration is provided by a tachometer mounted on a servo that follows beam position.

The Graf Askania Sea Gravity Meter

The Graf Askania sea gravity meter is also a highly damped type of gravity meter, but it uses magnetic damping and torsion springs. Because the author has not had experience with it, details of its construction and operation will not be given here. However, many of the fundamental principles previously discussed also apply to it. It has been well described in the literature [*Graf and Schulze*, 1961; *Schulze*, 1962; *Schulze et al.*, 1964; *Hayes et al.*, 1962].

Accuracy at Sea

Worzel [1959] made the first extensive gravity measurements on a surface ship in 1957, using a Graf Askania sea gravity meter on a stabilized platform on the U.S.S. *Compass Island*. Since that time parallel developments of the Graf Askania and the LaCoste and Romberg sea gravity meters have taken place. Since the author was not involved in the development of the former gravity meter but actively participated in the development of the latter, this article will trace through only the development of the L&R instrument. For information on the development of the Graf Askania sea gravity meter the reader is referred to the original articles in the literature [*Worzel*, 1959; *Graf and Schulze*, 1961; *Allan et al.*, 1962; *Fleischer*, 1963; *Wall et al.*, 1964; *Loncarevic*, 1964; *Bower and Loncarevic*, 1967; *Graham and Hales*, 1965; *Bower*, 1966; *Wall et al.*, 1966; *Loncarevic*, 1966].

The accuracy of gravity meters at sea depends greatly on the accelerations to which they are subjected. The previously discussed gravity meter theory indicates that most errors can be expected to be proportional to the squares of the three components of acceleration and to cross-coupling effects, which are products of velocity and acceleration components or products of two different components of acceleration. The first L&R gravity meter that was operated on a surface ship [*LaCoste*, 1959] could tolerate vertical accelerations slightly greater than ±50 gals but required a linearity correction for vertical accelerations even below ±50 gals [*Harrison*, 1959]; Harrison made these corrections. (After the tests the gravity meter linearity was adjusted in the laboratory to give errors within 2 mgal at a vertical acceleration of ±50 gals, which was the standard at that time.) Harrison estimated an accuracy of ±5 mgal in his tests.

In 1960 the damping in the L&R gravity meter was increased, but the meter was designed and adjusted for operation at horizontal and vertical accelerations of only ±50 gals. The horizontal acceleration limitation of ±50 gals was made in order to retain adequate accuracy in the analog computer for the Browne correction, which is required on a gimbal supported gravity meter. The use of a gyro stabilized platform instead of a gimbal was rejected at that time because of the short advertised lives of gyros. The design limitation for vertical accelerations was set at ±50 gals because of the erroneous impression that the vertical accelerations would not greatly exceed the horizontal accelerations. This conclusion was based on the assumption that a ship undergoes approximately the same accelerations as the water particles in waves; it neglects the effects of the response of the ship. The fact that vertical accelerations of a ship can be several times larger than horizontal accelerations when going into the waves was not unknown at the time; therefore, a more thorough investigation would have avoided a serious mistake.

The 1960 L&R gravity meters were capable of withstanding vertical accelerations several times greater than the design limit of ±50 gals, but their linearity was not sufficiently well adjusted to make the errors small at accelerations greater than this limit. In a way it was a disadvantage that the meters were capable of being operated above their design limit because operators often did so. On the other hand, it was also an advantage because, when the error was finally located, it was possible to compute corrections for previous data.

Another cause of trouble in the 1960 L&R gravity meters, which was not known at the time, was that the linearity adjustments in some of them changed during use. It is still not known whether the change was due to rough handling in shipment or to a gradual aging effect. It is known, however, that the change was due to balancing large nonlinearities against each other rather than reducing each nonlinearity to a small amount before balancing them. Balancing the large nonlinearities required extremely critical adjustments and therefore made it likely that the adjustments would be adversely affected by time or rough handling. As soon as it was determined that linearity adjustments changed, design modifications were introduced to make linearity very insensitive to the parameters affecting it. In the meantime, of course, the linearity changes that occurred resulted in increased errors, but fortunately these errors can be corrected even in old data, as has been mentioned.

Because of the acceleration design limitations of ±50 gals on the 1960 L&R gravity meters, they were not capable of operating under very adverse conditions. This situation continued until 1965, when a redesigned L&R gravity meter was introduced that operated on a stabilized platform. Some of the main tests made on L&R gravity meters in the period from 1960 to 1965 will now be discussed.

The largest errors observed with L&R gravity meters during that time were reported by *Allan et al.* [1962] and *Dehlinger and Yungul* [1962]. Allan's test was made on the 3000-ton ship *Aragonese* in the Mediterranean Sea with L&R meter S9. It gave errors of only about ±2 mgal in calm seas, but the errors increased to 10 to 25 mgal when the ship was headed into a moderate sea. Dehlinger's tests were also made with S9 on the 250-ton ship *Hidalgo* in the Gulf of Mexico. He too found large errors when going into the sea, even as high as 49 mgal. It now appears to be almost certain that the large errors were due to operating the gravity meter at vertical accelerations greater than the design limits of the meter, but this was not realized at the time, nor were any measurements of the vertical accelerations made either by the users or by the manufacturers of the gravity meters.

One reason for not having measured vertical accelerations was that they were not expected to exceed horizontal accelerations, as was previously mentioned. Another reason was that the long-period pendulum horizontal references of the gravity meter were almost universally thought to be the weakest link in the system and to be the main source of error. This belief was strengthened when *Dehlinger and Yungul* [1962] noted correlation between error and horizontal accelerations. The correlation very likely existed because there is also some correlation between horizontal and vertical accelerations. Nevertheless, Dehlinger was able to work out an empirical correction based on horizontal acceleration that reduced the errors to about ±10 mgal.

In later tests with S9 *Dehlinger* [1964] was able to establish a criterion, other than excessive vertical accelerations, for rejecting poor data. He noted that the gravity meter beam record (which indicates the integral of gravity when the spring tension is constant) was smooth on good readings but irregular on bad readings. These irregularities can probably be explained as variations of the error when the vertical acceleration varies. The acceleration varies because waves generally come in packets and because the amplitude of the ship motion changes slowly as the ship resonates with the waves (beat notes). Dehlinger also stated that errors were often present when there were large variations in the amplitudes of the accelerations. This statement is reasonable because large variations occur in resonance and resonance in heave is generally responsible for the large vertical accelerations that occur when going into the waves.

Although these criteria were not as relevant as excessive vertical accelerations, Dehlinger was able to use them to improve greatly the accuracy of the results obtained with S9. In his article [*Dehlinger*, 1964] he gives errors of 3 mgal at Browne corrections up to 300 mgal, 5 mgal at corrections of 300 to 400 mgal, and 8 mgal at corrections of 400 to 500 mgal.

The next significant tests made with S9 were made in 1963 by Harrison [*Harrison and LaCoste*, 1968]. In the three years preceding these tests the performance of the long-period pendulums had been greatly improved at periods shorter than 2 sec and at periods longer than 12 sec. In view of this improvement, better performance was expected from the gravity meter, since it was still not realized that vertical accelerations were the main source of error. No criteria were used to reject bad data except that the horizontal accelerations should not exceed ±50 gals. The errors were still large, particularly when going into the sea. The improvements in the long-period pendulums appeared to have made no significant improvement in accuracy.

It was not until tests were made on an experimental L&R stabilized platform gravity meter in 1964 [*Harrison and LaCoste*, 1968] that large vertical accelerations were recognized as the major source of error in L&R gravity meters. The tests were made in very rough weather, which probably helped to show the source of the error. Vertical accelerations were measured in this test and were found to have far exceeded the design limits during much of the test. Vertical acceleration corrections were made from data previously obtained in the laboratory and were found to account for the larger part of the errors. The remainder of the errors appeared to be due to cross-coupling effects.

After it was determined that large vertical accelerations accounted for most of the errors in L&R gravity meters, Harrison [*Harrison and LaCoste*, 1968] applied the corrections to the 1963 data he had obtained with S9. Before making the correction the mean error was 8.4 mgal and the rms errors had ranged from 5 to 20 mgal on lines made with horizontal accelerations ranging from ±20 to ±45 gals. After making the corrections the mean error was reduced to 1.1 mgal and rms error to 3.6 mgal. LaCoste made a similar correction to data obtained with S11 over a test range and obtained a similar improvement. LaCoste's corrections were sent to various users of L&R gravity meters with data for making similar corrections for the individual meters, but these corrections were never published.

To make vertical acceleration corrections, *Dehlinger et al.* [1966] worked with several years of data obtained with S9. In making the corrections, however, the vertical accelerations were inadvertently taken proportional to B/T^2 rather than proportional to B/T, where B is the amplitude of the beam motion and T is the period. The former expression applies to a long-period gravity meter with normal damping, in which case the acceleration is proportional to the second derivative of the beam displacement. The latter expression applies to a highly damped gravity meter, in which case the acceleration is proportional to the first derivative of the beam displacement as previously explained. The error in Dehlinger's computations made it impossible to use the numerical constant determined

in the L&R laboratory for vertical acceleration corrections, and therefore Dehlinger et al. determined a constant for a best fit of the data. The best fit constant
turned out to give the same corrections as the L&R formula for a 7-sec period,
and therefore it gave too small a correction at periods longer than 7 sec and too
large a correction at periods shorter than 7 sec. Even so, Dehlinger's corrections
did improve the accuracy considerably. He estimated his gravity meter errors
to vary from 1.5 mgal at low Browne corrections to ±4 mgal at Browne corrections of 500 mgal.

It has been mentioned that the linearity adjustments of some L&R gravity
meters has changed with time, before certain design modifications were made.
Gravity meter S9 has been particularly bad in that respect. According to the
L&R records its error at vertical accelerations of ±50 gals was within 1 mgal in
1960, when it was completed. In 1961, when it was tested again, the error was
found to be 6 mgal at the null position. In view of what is now known, it should
certainly have been readjusted then, but it was not. It was not tested again
until 1964, when the error was found to be 16 mgal. It has of course been readjusted.

In the period from 1960 to 1964 a considerable amount of good data has
been obtained with other L&R gimbal type gravity meters. One comprehensive
and well controlled test was made by *Bower and Loncarevic* [1967] [*Loncarevic*,
1964, 1966] on the C.S.S. *Baffin* in October 1963 over the Halifax gravity range.
L&R gravity meter S8 and a Graf meter were both aboard the ship. Two precisely
located tracks were traversed a total of 108 times. Decca was used for navigation and was considered to have more than the required accuracy. Gravity values
on the range had previously been obtained with an underwater gravity meter
[*Goodacre*, 1964]. A wide variety of weather conditions was experienced. Vertical accelerations ranged from 2 to 78 gals rms or about ±3 to ±110 gals.

About the same number of usable records were obtained from the two instruments, although the Graf operated in rougher weather. Loncarevic reported
useful readings with the L&R meter with vertical accelerations up to about ±42
gals and with the Graf up to ±71 gals. The mean error was +0.6 mgal for the
L&R meter and −0.4 mgal for the Graf. The standard deviation was 3.9 mgal
for the L&R and 2.7 mgal for the Graf. Drift for the entire test was negligible
for the L&R and irregular over a 7.0 mgal range for the Graf. D. R. Bower
(personal communication) noted a correlation between vertical acceleration and
error in the L&R gravity meter but did not make corrections for it.

In 1959 *Harrison and Spiess* [1963] made tests in the Gulf of California on
the research vessel *Horizon*. Sea conditions were better than average. Shipboard
gravity readings were compared with readings taken with an underwater gravity
meter at 27 stations. The mean difference at the 27 stations was 2.7 mgal, and
the rms departure was 1.5 mgal. Gravity meter drift was 1.6 mgal in 7 weeks.
Discrepancies at intersections were not given.

Caputo et al. [1963] describe gravity measurements made in the continental
borderland off southern California. They obtained 231 track intersections made
on different ships. Most of the data was obtained with L&R gravity meter S3
in various stages of development, but some data were obtained with S5. The

average discrepancy at all 231 crossings was 6.8 mgal, and two-thirds of the discrepancies were within 7 mgal. These values included navigational errors, and at 40 track intersections there were depth discrepancies of more than 90 meters. If these 40 intersections are disregarded, the mean discrepancy is only 5 mgal and two-thirds of the intersections are within this value. The 5-mgal discrepancy at intersections implies a mean observational error of 3.5 mgal. Data from the later cruises were more self-consistent than data from the earlier cruises, which indicates that the gravity meter had been improved somewhat. All the data considered by Caputo et al. were obtained before it was realized that the greatest source of error in L&R gravity meters was due to large vertical accelerations; no corrections have yet been made for these vertical accelerations.

L&R gravity meter S8 has been used in oil explorations. The first tests for this purpose, made in 1961, were used to make a gravity contour map of an area where gravity data were known only to the oil company sponsoring the test. When the oil company data were later made available, the map was found to be accurate to about 2 mgal and showed several salt domes that were present. The following year S8 was used in actual oil exploration by a major oil company. Track intersections and occasional data obtained where gravity had been previously measured with underwater gravity meters indicated that anomalies of 2 mgal could be detected.

Although cross-coupling effects were described in the literature in 1960 [LaCoste and Harrison, 1960] and were known several years earlier, they have not been considered as a serious source of error until fairly recently. Early estimates of their magnitude were given as small [Wall et al., 1964]. Later measurements [Wall et al., 1966; Bower, 1966] showed however, that they were often of the magnitude of 10 mgal or even 30 mgal in very rough weather. It is a simple matter to compute these errors with an analog computer; thus, corrections will certainly be made where they are not already being made.

As previously mentioned LaCoste and Romberg substantially redesigned their gravity meter in 1965 for use on a stabilized platform [LaCoste et al., 1967]. The first of the new models, S20, was tested against the last of the old gimbal models, S19, in the Gulf of Mexico over an area where gravity had previously been accurately surveyed with an underwater gravity meter. The results of the test are given by LaFehr and Nettleton [1967].

The test was made over the San Luis Pass salt dome about twenty miles southeast of Freeport, Texas. A 100-foot boat was used in sea conditions considered average for shallow offshore areas, running about 2 to 6 feet. Navigation was done by means of Raydist, which gave adequate accuracy. The comparison underwater gravity data had a probable error of about 0.1 mgal. Shipboard gravity data were taken on ten lines run in six different directions across the salt dome, giving twenty-five line intersections. The sea was rough enough to prevent the gimbal meter from operating 15–20% of the time, but the stabilized platform meter gave good data at all times. In six months of subsequent operation the stabilized platform gravity meter has never failed to operate because of roughness of the sea.

For the stabilized platform gravity meter the average observed difference between lines for all twenty-five intersections is 0.90 mgal. Of these differences 64% are within 1.0 mgal. The maximum observed difference, 2.5 mgal, occurred on line A-1 on whose run there was a short power failure on the ship. These statistics compare with an average difference for the gimbal gravity meter of 4.6 mgal and a maximum difference of 19.5 mgal. Actually, the 19.5-mgal discrepancy should have been rejected because the horizontal acceleration exceeded the ±50-gal design limit on one of the lines concerned (personal communication).

If line A-1 on which a power failure occurred is discarded, the remaining eighteen line intersections give the following results for the stabilized platform meter: the average observed difference is 0.68, and 78% of the differences are 1.0 mgal or less; the maximum observed difference is 1.5 mgal.

LaFehr and Nettleton point out that the preceding statistics do not represent the true capability of the instrument for measuring gravity anomalies either in the relative or the absolute sense. They note that shipboard gravity data usually have systematic errors that are fairly constant along each line and these systematic errors can be adjusted out. They have written a computer program to make this adjustment and to automatically make contour maps.

In this computer program the systematic error of each line is determined from its discrepancies at line intersections. (See Tables 1 and 2.) Table 1 shows the observed intersection differences. It can be seen that line C is running about 0.8 mgal higher than the lines intersecting it; this average line difference is listed in column 4 of Table 2. The average adjustment for each line is given in column 5 of Table 2 and indicates a systematic error considerably less than 1 mgal.

To compare the shipboard gravity data with the reference data obtained with the underwater gravity meter, it is necessary to refer them to the same datum. An absolute datum was available for the shipboard data but not for the underwater data; therefore, a best fit datum shift was determined for 158 samples. This datum shift was applied to the original data to give column 8 and to the computer adjusted data to give column 9, which shows smaller errors than column 8.

Errors determined at 158 points of the known gravity field were determined for the computer adjusted shipboard data, after making the single datum change

TABLE 1. Observed Differences at Intersections in Milligals
Sign is Plus if the Row Line is High

	A−1	B	C	F	A−2	C−1	D−1	D	E
A − 1		+2.1	+1.2	−0.5	+1.6	+0.7	−2.5	+1.5	
B	−2.1		−1.5	−0.8	−0.9	−1.0	+0.9	−0.3	
C	−1.2	+1.5			+0.6	+0.2	+1.5	+1.2	
F	+0.5	+0.8			−0.1				+0.4
A − 2	−1.6	+0.9	−0.6	+0.1		−0.3	−0.3	0.0	−0.6
C − 1	−0.7	+1.0	−0.2		+0.3		+0.3	+1.4	
D − 1	+2.5	−0.9	−1.5		+0.3	−0.3			
D	−1.5	+0.3	−1.2		0.0	−1.4			
E				−0.4	+0.6				

TABLE 2. Computer and 'Best Fit' Adjustments and Known Field Comparisons

| | | Based on Intersections Alone | | | Based on Known Field Comparisons | | | |
Line	Heading	Number of Inter- sections	Statistical Line Mean Difference,[a] mgal	Average Computer Adjustment,[b] mgal	Absolute Datum Difference,[c] mgal	Relative Datum Difference,[d] mgal	Known Field Minus 'Best Fit',[e] mgal	Known Field Minus Adjustment,[f] mgal
$A-1$	NE	7	+0.5	0.9	−6.7	+0.5	1.1	0.4
B	W	7	−0.7	0.8	−8.7	−1.5	0.5	0.4
C	SE	6	+0.8	0.5	−5.5	+1.7	0.5	0.3
F	S	4	+0.2	0.3	Insufficient known gravity			
$A-2$	NE	8	−0.1	0.3	−6.7	+0.5	0.4	0.4
$C-1$	NW	6	+0.4	0.2	−6.8	+0.4	0.4	0.5
$D-1$	N	5	−0.5	0.8	−8.3	−1.1	0.3	1.1
D	S	5	−0.4	0.7	−7.4	−0.2	0.9	0.0
E	W	2	+0.2	0.2	Insufficient known gravity			

[a] Computer determined mean difference for line intersections based on entire network; negative of first computer adjustment.
[b] Average difference between observed data and adjusted data.
[c] Datum for each line arbitrarily determined by 'best fit' technique.
[d] Difference between mean datum shift and line datum shift: a measure of systematic error; should agree qualitatively with column 4.
[e] Average difference between known field and line data after constant shift of column 6 using 'best fit' technique on 158 samples.
[f] Average difference between known field and computer-adjusted intersections using 25 intersections.

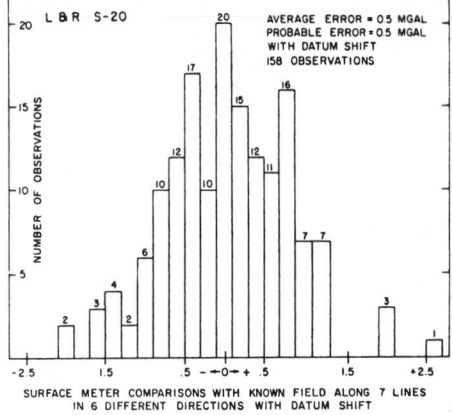

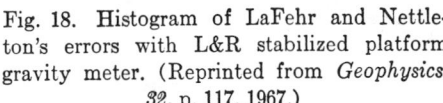

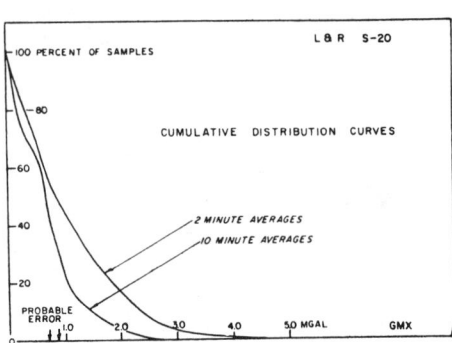

Fig. 18. Histogram of LaFehr and Nettleton's errors with L&R stabilized platform
gravity meter. (Reprinted from *Geophysics*,
32, p. 117, 1967.)

Fig. 19. Cumulative errors in LaFehr and
Nettleton's test of L&R stabilized platform
gravity meter. (Reprinted from *Geophysics*,
32, p. 117, 1967.)

previously mentioned. The results are shown in the histogram of Figure 18; the
probable error is 0.5 mgal. Since data were recorded only to ±0.5 mgal, it appears that results might be improved by recording data more accurately. These
more accurate recordings are now being made.

A comparison was also made of the relative accuracy of 2- and 10-min
averaging of unadjusted data. The results are shown in the cumulative error
curve of Figure 19. Here it is seen that the probable error is smaller for the 10-
min average. Even for 2-minute unadjusted data, however, the probable error is
less than 1 mgal.

It has been mentioned that LaFehr and Nettleton's article is based on
only 25 intersections and about 53 miles of continuous data. Since writing the
article, however, they (personal communication) have obtained data at thousands of intersections and have found that their results have been borne out by
the new data. They also have evidence that navigational errors are appreciably
greater than the gravity meter errors even when the best means of navigation
are used. They have found that they can improve their accuracy by better control of the ship motion, so that it more nearly travels in a straight line at con-
sant speed, thereby reducing changes in the Eotvos correction.

The existence of some systematic error in LaFehr and Nettleton's results
is not surprising, because errors can always be found if they are searched for
diligently enough. Although LaFehr and Nettleton's computer program substantially reduces these errors, work is under way to locate their causes and
then to correct for them.

The great improvement in the new model L&R stabilized platform gravity
meter over the old model has made it practicable to use it in detailed exploration
for oil, which is now being done on a considerable scale. There is no reason to
believe that the accuracy cannot be improved further, because the L&R shipboard gravity meter is essentially the same as the land gravity meter except for

high damping and therefore the same instrumental accuracy can be approached by tracking down the various errors still present. Stationary readings are accurate to ±0.01 mgal, as can be seen in the earth-tide record shown in Figure 20 which was obtained with shipboard meter S22. Data for this record were obtained overnight when the earth tide happened to be particularly small. The record shows that the integral of gravity and, therefore, its slope represent gravity. Reference slopes corresponding to ±0.01 mgal are shown.

An interesting way of looking at systematic errors is as follows. Since these errors are approximately constant along any line, they depend on quantities whose average values are also approximately constant along each line. The quantities are certainly functions of the accelerations and velocities; therefore, it appears that a good expression for the systematic error e_s will be given by a power series of the accelerations and velocities. Furthermore, there will be no first-order terms in this power series because the average value of all first-order acceleration terms will be zero on a ship, and first-order velocity terms will have no effect other than the Eotvos effect. The power series will then be

$$e_s = a_1\langle x''^2\rangle + a_2\langle y''^2\rangle + a_3\langle z''^2\rangle + a_4\langle x'^2\rangle + a_5\langle y'^2\rangle + a_6\langle z'^2\rangle$$
$$+ a_7\langle x''z'\rangle + a_8\langle y''z'\rangle + a_9\langle x''z''\rangle + a_{10}\langle y''z''\rangle + \cdots \qquad (64)$$

where the a's are constants.

In (64) the first two terms have been discussed in connection with stabilized platform performance and the third term has been discussed in connection with gravity meter nonlinearity. The fourth and fifth terms produce no effect on the gravity meter and can be disregarded. The sixth term can occur because of nonlinearity in some types of gravity meters, as has been explained previously. The seventh and eighth terms give the inherent type of cross coupling in overdamped meters, and the ninth and tenth terms give imperfection types of cross coupling. Higher-order terms can probably be disregarded but can be included if necessary.

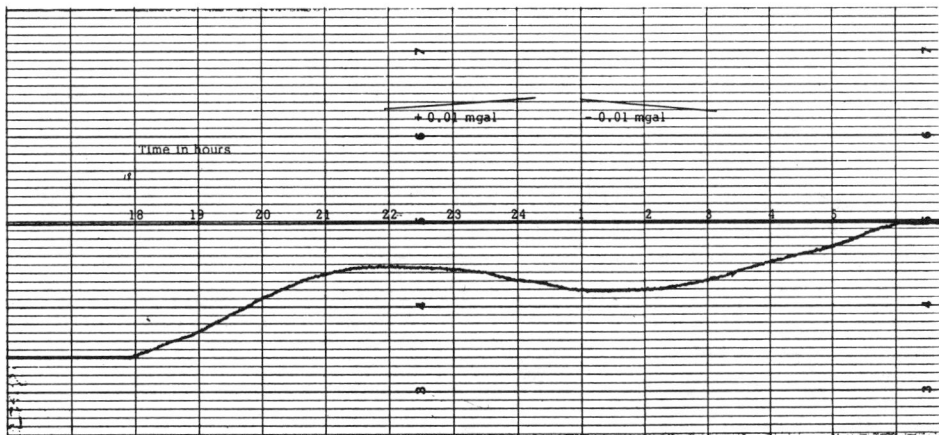

Fig. 20. Earth-tide record obtained with L&R air-sea gravity meter S22. Record shows integral of gravity; therefore, gravity is represented by slope of curve.

LaCoste and Romberg has recently provided means for recording the terms in (64) that might be expected to be significant. A study of the correlation between each term and the error will determine a correction that can either be applied to the data or can be made to the gravity meter. A simple method of making the correction directly to the gravity meter is being used. The final error calibration can then be made at sea rather than in the laboratory, although laboratory calibration is still done as carefully as possible.

Accuracy in the Air

Airborne gravity measurements were first made in 1958 [*Thompson and LaCoste* 1960]. Flights were made in a U. S. Air Force KC-135 jet tanker over Edwards Air Force Base, California. LaCoste and Romberg gimbal supported gravity meter S5 was used. Accurate navigational data were provided by phototheodolites on a tracking range on the ground. It was believed that the phototheodolites gave the airplane altitude to an accuracy of about ±2 feet. The airplane was operated on an automatic pilot, which was adjusted as carefully as possible to give smooth flight. Doppler radar was available for measuring velocity but was not used because data from the tracking range were considered to be more accurate. Altitude variations were measured by a very sensitive pressure transducer or hypsometer. A hypsometer measures pressure by measuring the temperature of the boiling point of a liquid. Flights were made in the morning so as to have smooth air.

The gravity meter performed well in most cases. On a northbound flight the automatic pilot caused the plane to hunt with a 42-sec period, and this caused a 78-mgal error because the periods of the long-period pendulum vertical references were only 1 min. The presence of the 42-sec period was evident from the horizontal acceleration records; therefore, there was no question that the data for that run should be discarded. Vertical accelerations were only of the order of 10 gals. The navigational accuracy was adequate to permit gravity readings to be averaged over times as short as five minutes without a great loss in accuracy. (Velocities and accelerations can be measured more accurately over long times or distances because a given uncertainty in position has a smaller percentage effect if the distance involved is large.) Corrections were made for vertical accelerations determined by the pressure sensitive altimeter. The need for these corrections was shown by the fact that they greatly smoothed the final gravity data.

The accuracy of the results of the airborne measurements was estimated from a comparison with the available ground gravity data after they were corrected for the elevation difference. It appeared that an accuracy of 10 mgal or better was obtained. This accuracy was also borne out by gravity readings at flight crossings. It was not surprising that the gravity meter performed well because the accelerations present in the flights were much smaller than those normally encountered on ships. With these accelerations the gravity meter errors should be considerably smaller than the 10-mgal errors observed. The navigational accuracy is therefore the limiting factor in airborne gravity measurement.

Shortly after the first airplane test a second test was made in a B 17 airplane operated by Fairchild Aerial Surveys [*Nettleton et al.,* 1960]. The gravity meter used was L&R S3, which was loaned by the Institute of Geophysics and Planetary Sciences at the University of California, Los Angeles. The airplane was equipped with a mapping camera to determine its position and an APR precision radar altimeter and a hypsometer for determining the altitude. Again flights were made in the morning in order to have calm air, and the airplane was flown on an automatic pilot which was carefully adjusted to minimize hunting, particularly long period-hunting. Flights were made over the Imperial Valley in California.

Again it was found that good gravity data were obtained on all flights except on one going north when the automatic pilot again caused the airplane to hunt at a period of 1 to 2 min. The problem in the automatic pilot has since been solved; it occurred because of the dip of the magnetic field. The airborne gravity results were evaluated in several ways: (1) by a comparison with a ground gravity contour map, (2) by repeat observations over almost the same course, (3) by comparisons at flight intersections, and (4) by comparisons with values calculated from known ground gravity stations. In correcting ground values for altitude differences, it was necessary to correct for attenuation or smoothing of details at flight altitudes. Again the estimated errors were found to be within 10 mgal. Also, it was found that gravity values could be averaged over times as short as three minutes without losing much accuracy, this corresponds to a distance of about ten miles for the aircraft used.

Thompson and LaCoste [1960] noted that airplane velocity errors have less effect on gravity meter accuracy when the airplane is flying west, because the airplane speed is subtracted from the surface speed of the earth due to its rotation. The resulting reduction in speed not only reduces the centripetal acceleration (which varies with the square of the speed) but also reduces its derivative with respect to speed (which is its sensitivity to errors in speed). *Glicken* [1962] has extended this analysis of the effects of various errors on the accuracy of gravity observations. He gives useful curves.

The first well controlled airborne gravity meter test was made by Fairchild LaCoste Gravity Surveys, Inc., for the U. S. Army Map Service [*Nettleton et al.,* 1962]. The test was made over the triangle formed by Houston, Texas, Shreveport and Baton Rouge, Louisiana. The triangle was flown twice in opposite directions. The reasons for picking this triangle were (1) gravity values on the ground were well known over that region, (2) the topography was relatively flat so that there would be little uncertainty in computing gravity at flight altitudes from the ground data, and (3) good aerial photographs of the region were available so that flight positions could be accurately determined.

The flights were made in a B 17 aircraft with a Bendix autopilot which had been adjusted for small amplitude and short-period hunting. As in the previous test, photography was used for navigation and an APR radar altimeter and hypsometer were used for determining altitude. Doppler radar was carried on the flights in orders to evaluate it. The gravity meter used was L&R S8. It was a gimbal supported gravity meter, but the long-period vertical references had been adjusted to 2-min periods, and the damping had been made 0.7 times critical.

Flights were made at 12000 feet. The airplane altitude was determined with the radar altimeter at the corners of the flight triangle where the elevations of the ground were known. Along the flights the altitude of the airplane was determined by means of the hypsometer by using *Henry*'s [1948] formula for the isobaric surface. Henry's formula is $dh/ds = 0.035A \sin \phi \sin \delta$, where dh/ds is the slope of the isobaric surface in feet per mile, A is the true air speed in statute miles per hour, ϕ is the latitude, and δ is the angle of drift of the aircraft. The airplane was flown in a straight line except for occasional course changes of about 1°.

Most of the flights were made early in the morning to obtain smooth air, but some turbulent air was encountered before flying each complete triangle. The turbulence caused the automatic pilot to hunt at a long enough period to cause small errors even though the periods of the long-period pendulum vertical references had been increased to 2 min. Similar errors were produced when the airplane made 1° course changes. A first-order correction for these errors was made, as explained in the discussion of equation 38. Essentially the correction consisted of applying the recorded horizontal accelerations to the vertical references as tilts, in the laboratory. The difference between the inputs and outputs in the laboratory permitted an extrapolation to the original input that occurred during flight. The application of this first-order correction made the gravity records smooth at the places where course changes were made and reduced turbulent air errors to about the same value as smooth air errors.

The airplane gravity results are compared with the ground gravity data corrected for altitude in Figure 21 for the Houston–Baton Rouge flights. For these two flights the mean error was 2.2 mgal and the rms error was 6.5 mgal. The values for all the flights was +1.55 mgal for the mean error and 6.6 mgal for the rms error. All readings were calculated over 3-min intervals. The small values of the mean error indicate an accurate gravity meter calibration because

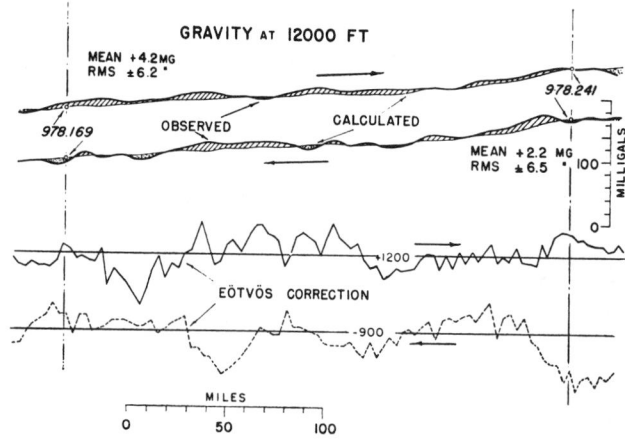

Fig. 21. Results of airplane flight tests of L&R gravity meter between Houston and Baton Rouge on Army Map Service test. (Reprinted from *Journal of Geophysical Research, 67,* p. 4405, 1962.)

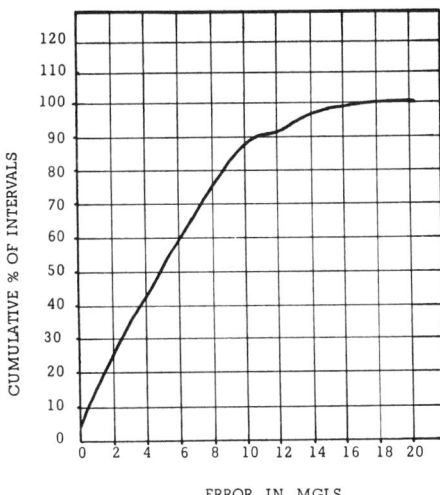

Fig. 22. Cumulative error curve between measured and calculated gravity at 12,000 feet for L&R gravity meter on Army Map Service test. (Reprinted from *Journal of Geophysical Research, 67,* p. 4409, 1962.)

there are large altitude and Eotvos corrections in all flights. A cumulative error curve is shown in Figure 22.

The preceding test and the entire data reduction procedure were independently evaluated by the University of Wisconsin [*Coons et al.,* 1962]. To the extent that it was feasible, the various inputs were digitized and the data reduction was carried out on an electronic computer: Although the digital filtering was somewhat different, the errors were almost the same. The digital analysis gave a mean error of 1.75 mgal instead of 1.55 mgal and an rms error of 6.0 mgal instead of 6.6 mgal.

In 1962 *Thompson* [1965] made over 100 test flights over a calibration range to evaluate (1) a Graf-Askania Gss 2 gravity meter on a stabilized platform, (2) LaCoste and Romberg gimbal supported gravity meter S6, and (3) L&R gravity meter S6 on a stabilized platform. The stabilized platform was an Aeroflex ART-25 camera mount whose accuracy was given as ±5 min of arc by the manufacturer. Gravity meter S6 was not modified for operation on a stabilized platform. The tests were made over the Edwards Air Force Base test range. Navigational data were provided by phototheodolites on the ground as in the first airplane gravity test. Tracking was thought to be accurate to ±2 feet in altitude, ±½ knot in ground speed and, ±½ min of arc in true course.

Unfortunately, phototheodolite tracking was available for only 30 to 40 miles of each flight, which corresponded to only 5 to 6 min of flying time for the C 130 aircraft used. The short length of the flights where navigational control was available could result in errors caused by initial conditions (acceleration) preceding the usable 6-min parts of the flights. Also, since the various 6-min parts of flights are independent of each other, their average is not as meaningful as it would be if they were all consecutive parts of a single flight. In the latter case a periodic acceleration error giving a high gravity value over one part would probably give a low gravity value over the next.

Thompson considered about half of the flights usable and found an rms error of 7.2 mgal for the Graf, an rms error of 8.5 mgal for S6 suspended from gimbals, and an rms error of 7.3 mgal for S6 on the stabilized platform. He found it necessary to make a 5-mgal correction for the long-period pendulum reference for the gimbal supported S6, but after making this correction he found that the mean for each of the three systems tested agreed within 1 mgal with the gravity values computed from ground data. The accuracy with which the mean values checked the ground data might have been somewhat fortuitous because some of the gravity values from individual flights differed over 20 mgal from the ground data.

In view of the results of *Harrison and LaCoste* [1968] at sea it is somewhat surprising that L&R gravity meter S6 performed as well as it did on a stabilized platform in the airplane, because it had not been modified for such use. The answer probably lies in the fact that the accelerations in the airplane were small, generally below 10 gals. Thompson did report errors when the air became turbulent.

It is obvious that gravity values obtained in a fast moving airplane cannot show detail because of the speed of the airplane and the need for an averaging time of a few minutes. Also, detail is lost if the airplane is flying high. Gravity measurements can, however, be obtained rapidly in an airplane and are sufficiently accurate to be useful in geodetic work. Navigational accuracy will have to be improved before over-all accuracy can be improved.

Other Uses of Air-Sea Gravity Meters

The difficulty of obtaining gravity details in airplane measurement can be overcome in some cases by operating in a helicopter rather than in an airplane. A helicopter can be hovered very accurately at 200 feet or less from the ground. This will greatly reduce errors in the Eotvos correction and acceleration corrections for altitude variations. Also, surveying techniques for accurately locating helicopters have been worked out by the U. S. Geological Survey using Telurometers and transits. Tests of air-sea gravity meters have already been made in helicopters by the U. S. Naval Oceanographic Office and the U. S. Army Map Service, and the results have been very good.

A somewhat similar method is to operate air-sea gravity meters in hovercraft, which ride on a cushion of air a few feet above the ground or water. Hovercraft can not operate over rough terrain or over a rough sea, but they have an advantage over helicopters in that they do not sink in case they fall into water. Hovercraft can travel at speeds of 60 miles per hour.

REFERENCES

Allan, T. D., P. Dehlinger, C. Gantar, C. Morelli, M. Pisano, and J. C. Harrison, Comparison of Graf-Askania and LaCoste-Romberg surface ship gravity meters, *J. Geophys. Res., 67,* 5157–5162, 1962.

Beyer, L. A., R. E. von Huene, T. H. McCulloh, and J. R. Lovett, Measuring gravity on the sea floor in deep water, *J. Geophys. Res., 71,* 2091–2100, 1966.

Bower, Donald R., The determination of cross-coupling errors in the measurement of gravity at sea, *J. Geophys. Res., 71,* 487– 493, 1966.

Bower, D. R., and B. Loncarevic, Sea-gravimeter trials on the Halifax test range aboard C.S.S. Baffin, 1963, *Publ. Dominion Astrophys. Obs., 36*(1), Ottawa, Canada, 1967.

Browne, B. C., The measurement of gravity at sea, *Monthly Notices Roy. Astron. Soc., Geophys. Suppl., 4,* 271–279, 1937.

Browne, B. C., and R. I. B. Cooper, The British submarine gravity surveys of 1938 and 1946, *Trans. Roy. Soc. London, A, 242,* 243–310, 1950.

Browne, B. C., and R. I. B. Cooper, Gravity measurements in the English Channel, *Proc. Roy. Soc. London, B, 139,* 426–447, 1952.

Caputo, M., J. C. Harrison, R. von Huene, and M. D. Helfer, Accuracy of gravity measurements off the coast of southern California, *J. Geophys. Res., 68,* 3273–3282, 1963.

Clarkson, H. N., and L. J. B. LaCoste, An improved instrument for measurement of tidal variations in gravity, *Trans. Am. Geophys. Union, 38,* 8–16, 1957.

Coons, R., W. Strange, and G. P. Woollard, Evaluation and study of airborne gravimeter operational test, Reference 62-2, Geophysics and Polar Research Center, Department of Geology, University of Wisconsin, Madison, Wisconsin, 1962.

Dehlinger, P., Reliability at sea of gimbal-suspended gravity meters with 0.7 critically damped accelerometers, *J. Geophys. Res., 69,* 5383–5394, 1964.

Dehlinger, P., R. W. Couch, and M. Gemperle, Surface-ship gravity meter measurements corrected for vertical accelerations, *J. Geophys. Res., 71,* 6017–6023, 1966.

Dehlinger, P., and S. H. Yungul, Experimental determinations of the reliability of the LaCoste and Romberg surface ship gravity meter S-9, *J. Geophys. Res., 67,* 4389–4394, 1962.

Fleischer, U., Surface ship gravity measurements in the North Sea, *Geophys. Prospecting, 11,* 535–549, 1963.

Gilbert, R. L. G., A dynamic gravimeter of novel design, *Proc. Phys. Soc. London, B, 62,* 445–454, 1949.

Glicken, Milton, Eotvos corrections for a moving gravity meter, *Geophysics, 27,* 531–533, 1962.

Goodacre, A. K., A shipborne gravimeter testing range near Halifax, Nova Scotia, *J. Geophys. Res., 69,* 5373–5381, 1964.

Graf, A., and R. Schulze, Improvements on the sea gravimeter Gss 2, *J. Geophys. Res., 66,* 1813–1821, 1961.

Graham, K. W. T., and A. L. Hales, Surface-ship gravity measurements in the Agulhas Banks Area, south of South Africa, *J Geophys. Res., 70,* 4005–4011, 1965.

Harrison, J. C., Tests of the LaCoste-Romberg surface ship gravity meter, 1, *J. Geophys. Res., 64,* 1875–1881, 1959.

Harrison, J. C., The measurement of gravity at sea, in *Methods and Techniques of Modern Geophysics,* edited by S. C. Runcorn, Interscience Publishers, New York, 1960.

Harrison, J. C., and Lucien LaCoste, The performance of LaCoste-Romberg shipboard gravity meters in 1963 and 1964, *J. Geophys. Res., 73,* 1968.

Harrison, J. C., and F. N. Spiess, Tests of the LaCoste and Romberg surface-ship gravity meter, 2, *J. Geophys. Res., 68,* 1431–1438, 1963.

Hayes, D. E., J. L. Worzel, and H. Karnick, Tests on the 1962 model of the Anschütz gyro-table, *J. Geophys. Res., 69,* 749–757, 1964.

Henry, T. J. G., Determination of topographic profiles by use of radar and pressure altimeters (unpublished), Meteorological Division, Department of Transport, Ottawa, Ontario, Canada, 1948.

Jacoby, H.-D., and R. Schulze, New method to eliminate the cross-coupling effect in gravity measurements at sea, *J. Geophys. Res., 72,* 2199–2207, 1967.

LaCoste, L. J. B., *U. S. Patent 2 589 709,* 1952a.

LaCoste, L. J. B., *U. S. Patent 2 589 710,* 1952b.

LaCoste, Lucien, Surface ship gravity measurements on the Texas A and M college ship, the *Hidalgo, Geophysics, 24,* 309–322, 1959.

LaCoste, L. J. B., *U. S. Patent 2 964 948,* 1960.

LaCoste, L. J. B., *U. S. Patent 2 977 799,* 1961.

LaCoste, L., N. Clarkson, and G. Hamilton, LaCoste and Romberg stabilized platform shipboard gravity meter, *Geophysics, 32,* 99–109, 1967.

LaCoste, L. J. B., and J. C. Harrison. Some theoretical considerations in the measurement of gravity at sea, *Geophys. J. Roy. Astron. Soc., 5,* 89–103, 1961.

LaCoste, L. J. B., and A. Romberg, *U. S. Patent 2 377 889*, 1945.

LaFehr, T. R., and L. L. Nettleton, Quantitative evaluation of a stabilized platform shipboard gravity meter, *Geophysics, 32*, 110–118, 1967.

Loncarevic, B. D., Sea gravimeter reliability tests (abstract), *Trans. Am. Geophys. Union, 45*, 34, 1964.

Loncarevic, B. D., *Bulletin d'Information No. 12 of the Bureau Gravimétrique International*, Paris, France, 1966.

Nettleton, L. L., L. J. B. LaCoste, and Milton Glicken, Quantitative evaluation of precision of airborne gravity meter, *J. Geophys. Res., 67*, 4395–4410, 1962.

Nettleton, L. L., L. J. B., LaCoste, and J. C. Harrison, Tests of an airborne gravity meter, *Geophysics, 25*, 181–202, 1960.

Schulze, Reinhard, Automation of the sea gravimeter Gss 2, *J. Geophys. Res., 67*, 3397–3401, 1962.

Schulze, R., E. Brede, and E. Thebis, *German Patent 1 165 290*, 1964.

Talwani, Manik, Some recent developments in gravity measurements aboard surface ships, *Gravity Anomalies: Unsurveyed Areas, Geophys. Monograph 9*, edited by H. Orlin, pp. 31–47, American Geophysical Union, Washington, D. C., 1966.

Talwani, M., W. P. Early, and D. E. Hayes, Continuous analog computation and recording of cross-coupling and off-leveling errors, *J. Geophys. Res., 71*, 2079–2090, 1966.

Thompson, L. G. D., and L. J. B. LaCoste, Aerial gravity measurements, *J. Geophys. Res., 65*, 305–322, 1960.

Thompson, L. G. D., Comparison of LaCoste-Romberg and Askania-Graf gravity meters in gimbal and stabilized mounts, *J. Geophys. Res., 70*, 5599–5613, 1965.

Tsuboi, C., Y. Tomoda, and H. Kanamori, Continuous measurements of gravity on board a moving surface ship, *Geophys. Notes, Tokyo, 14*(2), 1961.

Vening Meinesz, F. A., *Theory and Practice of Pendulum Observations at Sea*, Waltman, Delft, The Netherlands, 1929.

Vening Meinesz, F. A., Gravity measurements at sea, 1, 2, and 3, *Publ. Neth. Geodesy Comm., Delft*, 1932, 1934, and 1948.

Vening Meinesz, F. A., *Theory and Practice of Pendulum Observation at Sea*, vol. 2, Waltman Delft, The Netherlands, 1941.

Wall, R. E., M. Talwani, and J. L. Worzel, Measurement of the cross-coupling effect for gravity measurements at sea using a stable platform (abstract), *Trans. Am. Geophys. Union, 45*, 33, 1964.

Wall, R. E., M. Talwani, and J. L. Worzel, Cross-coupling and off-leveling errors in gravity measurements at sea, *J. Geophys. Res., 71*, 465–485, 1966.

Worzel, J. L., Continuous gravity measurements on a surface ship with the Graf sea gravimeter, *J. Geophys. Res., 64*, 1299–1315, 1959.

Worzel, J. L., and M. Ewing, Gravity measurements at sea, 1947, *Trans. Am. Geophys. Union, 31*, 917–923, 1950.

Worzel, J. L., and M. Ewing, Gravity measurements at sea, 1948 and 1949, *Trans. Am. Geophys. Union, 33*, 453–460, 1952.

Worzel, J. L., G. L. Shubert, and M. Ewing, Gravity measurements at sea, 1950, 1951, 1952, and 1953, *Trans. Am. Geophys. Union, 36*, 326–338, 1955.

(Manuscript received March 28, 1967.)

GEOPHYSICS, VOL. XXVIII, NO. 5, PART I (OCTOBER, 1963), PP. 724-735, 12 FIGS.

AN APPROXIMATE SOLUTION OF THE PROBLEM OF MAXIMUM DEPTH IN GRAVITY INTERPRETATION*

D . C . S K E E L S†

It is assumed that the maximum depth for the mass responsible for a given gravity anomaly is closely approximated by the depth to the top of the vertical-sided mass (prism or cylinder, as the case may be) whose calculated anomaly gives the closest fit to the observed anomaly, and whose density contrast is the maximum permitted from geological considerations. A set of charts is presented by means of which the depth and dimensions of the prism (or cylinder) of "best fit" can be determined quickly from the amplitude, half-maximum, and three-quarter maximum widths of the anomaly, together with the assumed density contrast. Four examples are given of the use of the method with actual data.

INTRODUCTION

It is now generally understood that gravity data cannot be interpreted uniquely in terms of depth; even when the density contrast is specified, a whole family of configurations can be found for the surface of density contrast, at various depths, any one of which will satisfy the observed gravity data. The shallowest possible configuration is at zero depth, that is, at its highest point it coincides with the surface of the ground.

At the other extreme, it can be shown that for any given anomaly and for any assumed density contrast there is some maximum depth, such that no distribution of mass with this density contrast which is entirely below this depth can satisfy the anomaly, while at least one configuration can be found whose top is at this depth which does satisfy the anomaly.

This is a very useful concept in practical gravity interpretation, as it often enables one to distinguish anomalies which must be associated with structure within the sedimentary section (using "structure" in its broadest sense) from those which may be due entirely to density variations within the crystalline basement. If the basement depth is known approximately from drilling or from aeromagnetic data, and if the maximum depth to the source of the anomaly can be shown

to be less than this depth, then at least a part of the anomaly must be due to something more shallow than the basement. If, on the other hand, the maximum depth is greater than the depth to the basement, we can only say that the source *could* be entirely within the basement. It could also be partly in the basement and partly within the sediments, or it could be entirely within the sediments. In some cases this last possibility can be ruled out, because the sedimentary thickness is not sufficient to accommodate a mass large enough to produce the observed anomaly without postulating densities that are geologically unreasonable.

It would obviously be useful to have a quick method of determining or even of approximating the maximum depth for a given anomaly and for a given density contrast. In this paper such a method is given, which was developed by the author several years ago while he was on the staff of the Standard Oil Company (N. J.) in New York. The method has been tested in a great number of applications since then, and has been proven to be practical and to give results consistent with geological data.

The method is based on the observation that when we attempt to calculate configurations at various depths to satisfy a given gravity anom-

* Presented at the 32nd Annual SEG Meeting, Calgary, Alberta, Canada, September 20, 1962. Manuscript received by the Editor April 9, 1963.

† Imperial Oil Enterprises, Ltd., Calgary, Alberta, Canada.

724

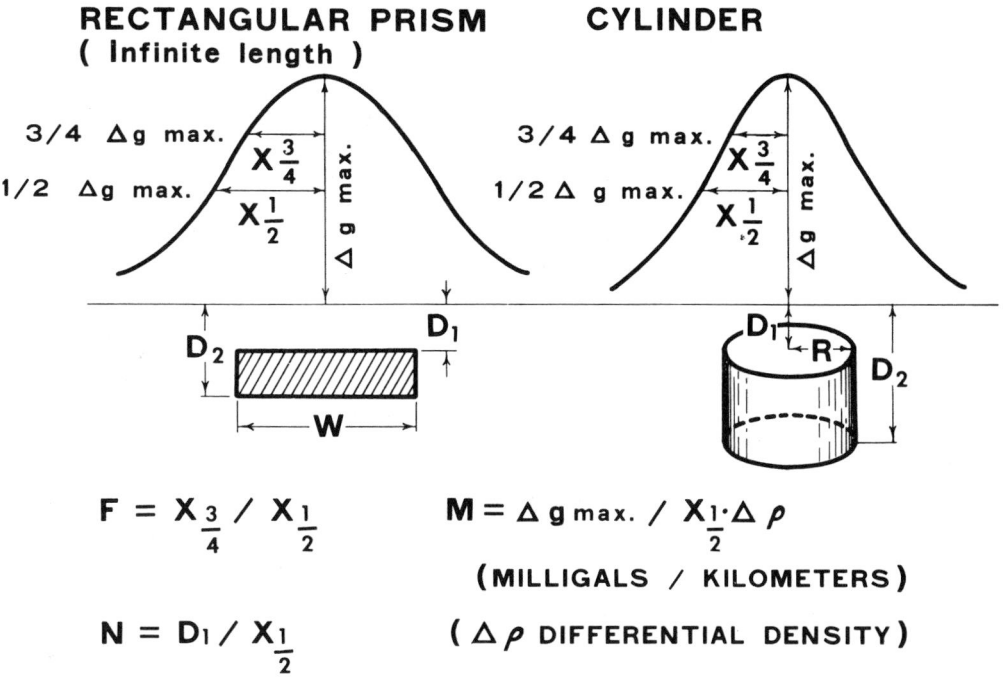

FIG. 1. Parameters for prisms and cylinders.

aly, using either trial-and-error methods, such as graticules, or downward-continuation techniques, we invariably find that as we go deeper we must make the slopes of our configurations larger. This suggests that the deepest possible configuration which can satisfy a given anomaly will be a mass with vertical sides. For an elongated anomaly, this means a two-dimensional prism of rectangular cross-section; for a circular anomaly it means a circular cylinder with vertical axis.

This suggestion leads to our basic assumption, which is: *The maximum depth for a given anomaly is closely approximated by the depth to the top of the vertical-sided mass (prism or cylinder, as the case may be) whose anomaly gives the closest fit to the observed anomaly, when the density contrast used is the maximum permitted by geological considerations.*

This method is referred to as an approximate solution, in recognition of the fact that there are undoubtedly some unusual shapes, in which the mass is concentrated toward the edges of the body, with a deficiency in the center, which could satisfy a given anomaly at a depth slightly greater than that calculated for the prism or cylinder.

DESCRIPTION OF THE METHOD

For purposes of this method, we shall define the prism (or cylinder) giving the "best fit" to a particular anomaly as the prism (or cylinder) whose calculated anomaly, for the assumed maximum permissible density contrast, matches the observed anomaly at the maximum, half-maximum, and three-quarter maximum points. If the observed anomaly is symmetrical, this means that the upper halves of the observed and calculated anomalies are virtually identical. Since there are three parameters to be determined for the prism (depth to top, D_1; depth to base, D_2; and width, W), the three observed values from the anomaly serve to identify the prism uniquely, once the density contrast is assigned. (For circular anomalies the same is true, substituting the radius, R, for the width.)

Rather than using these quantities themselves, it is more practical to combine them as ratios. For the observed anomaly we define two ratios (see Figure 1):

$$M = \Delta g \max / X_{1/2} \Delta \rho$$

(mgal/km/unit density contrast)

and

$$F = X_{3/4}/X_{1/2},$$

where

Δg_{max} is the maximum value of gravity for the anomaly, after correction for regional effects,

$X_{1/2}$ is the horizontal distance from the point of maximum gravity to the point where gravity has half its maximum value,

$X_{3/4}$ is the horizontal distance from the point of maximum gravity to the point where gravity has three-fourths of its maximum value, and

$\Delta \rho$ is the maximum density contrast permitted by geological considerations.

The quantity M has the dimensions of a gradient; we measure it in mgal/km/unit density contrast. The quantity F is a pure number, being the ratio of two distances; it is a measure of the "kurtosis" of the anomaly (to borrow a term from statistics)—or of the flatness or sharpness of the top. It might be called the "flatness factor." We note that both M and F are independent of scale; that is, all prisms for which the ratios D_1/D_2 and W/D_2 are the same, and which have the same density contrast, have the same values of M and F.

Figure 2 is a chart showing M and F, as functions of D_1/D_2 and W/D_2, for two-dimensional rectangular prisms. Note that the horizontal scale changes at $W/D_2=1.0$. Note, also, that the intersection angles of the two families of curves, F and M, are quite good in all parts of the chart, so that the values of D_1/D_2 and of W/D_2 for the prism which fits the anomaly can be determined quite accurately.

Furthermore, it can be shown that for any pair of values M and F there is one and only one pair

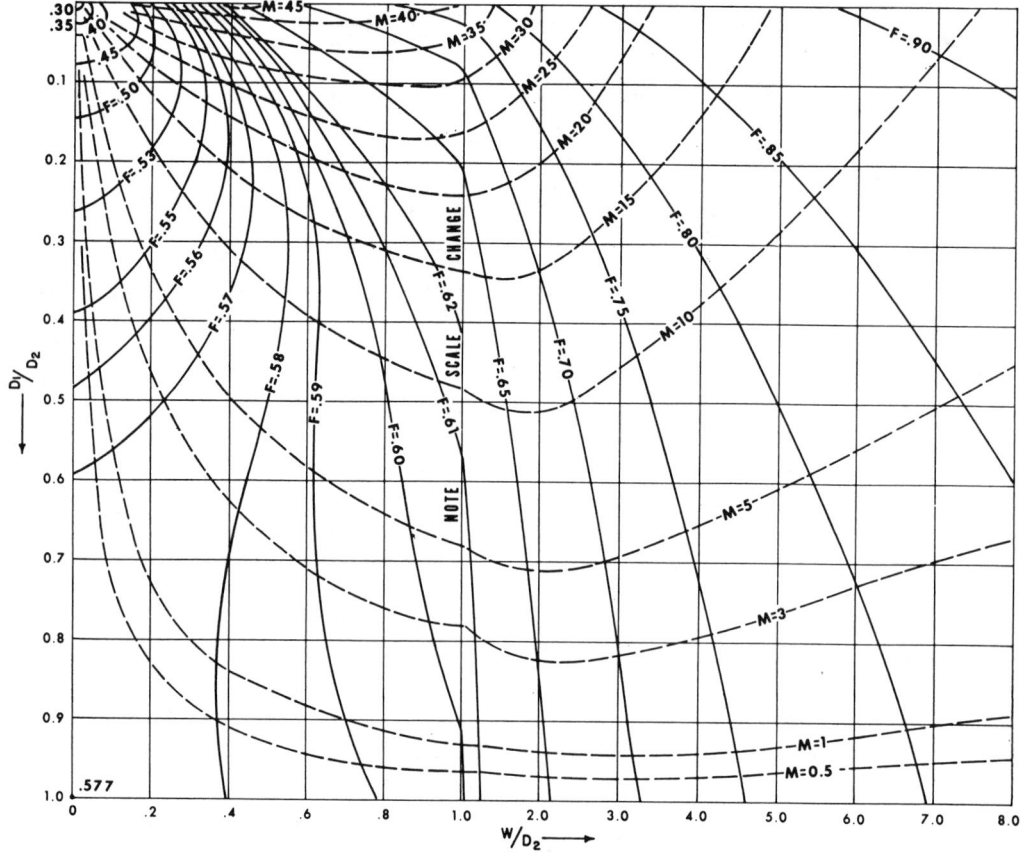

FIG. 2. Two-dimensional prisms—F and M as functions of W/D_2 and D_1/D_2.

of values D_1/D_2 and W/D_2 so that these ratios are uniquely determined by M and F.

To determine the actual depth, D_1, which by our assumption is the maximum depth, we define another ratio, which we call the "depth factor":

$$N = D_1/X_{1/2};$$

that is, N is simply the factor by which we must multiply the half-maximum distance to obtain D_1, the depth to the top of the prism.

Figure 3 shows N as a function of D_1/D_2 and of W/D_2. Therefore, having determined D_1/D_2 and W/D_2 from the values of F and M, using Figure 2, we can find the value of N from Figure 3. Multiplying the half-maximum distance by N we get D_1, which we assume to be the maximum depth. If we divide D_1 by D_1/D_2 we obtain D_2; and if we multiply D_2 by W/D_2 we obtain W. We may, therefore, solve for all of the parameters of the prism whose anomaly satisfies the upper half of the observed gravity anomaly.

The procedure for circular anomalies is com-pletely analogous. Figure 4 gives the values of F and M as functions of D_1/D_2 and R/D_2. Figure 5 gives the values of N as a function of the same ratios. These charts are plotted on a logarithmic scale, which was desirable because of the behavior of the F and M curves for small values of R/D_2 and D_1/D_2 (in the upper left-hand part of the chart). It will be noted that in this area the intersections are not good, in that the F and M curves are virtually parallel. This means that D_1/D_2 and R_1/D_2 are not uniquely determined in this part of the chart. However, this is not as troublesome as it might seem, since the curves for N (Figure 5) have virtually the same trend. This means that although there may be some ambiguity regarding the depth to the base of the cylinder, the depth to the top is fairly accurately determined by the gravity anomaly.

It will also be noted that the upper right-hand corner of these charts for circular anomalies is left blank. This is because the method used to calculate the effects of cylinders is an approximation, in which the cylinder is replaced by one

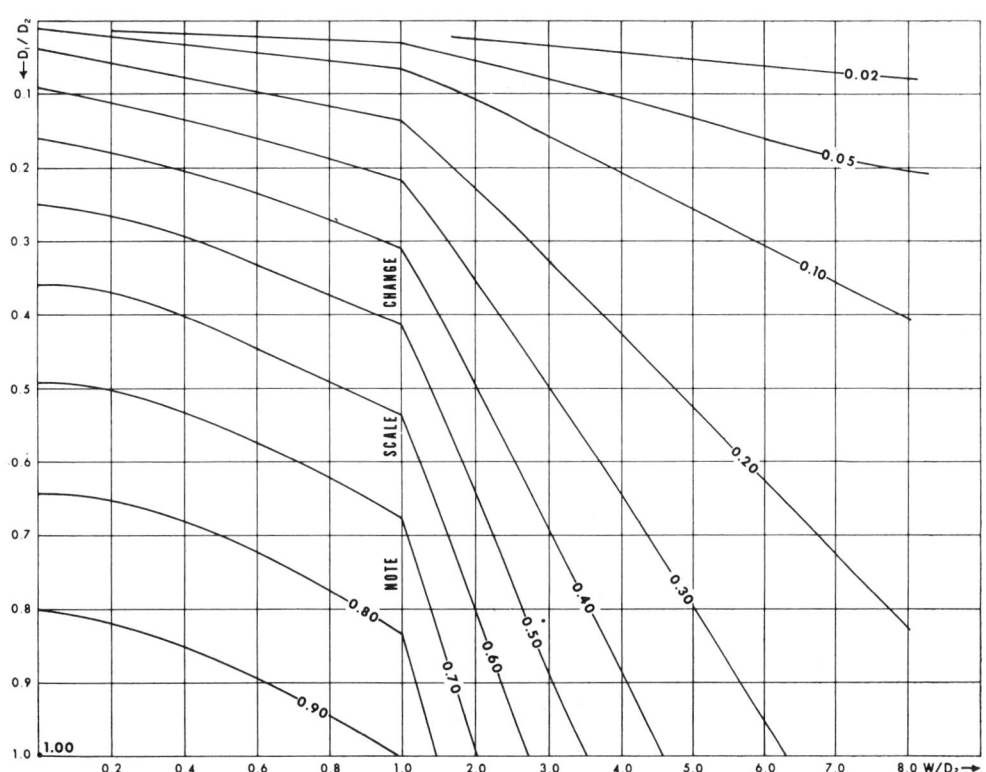

FIG. 3. Two-dimensional prisms—N as a function of D_1/D_2 and W/D_2.

hundred equally-spaced vertical lamellae of infinitesimal thickness. This approximation is very good as long as the distance between lamellae, $R/50$, is small compared to D_1 and D_2, that is, as long as R/D_1 is less than about 30. If it were important to have these missing values they could be computed by another formula, or we could use the same approximation and increase the number of lamellae.

If we are interested only in D_1, the depth to the top of the mass, and not in its dimensions, N may be obtained directly from Figure 6 or Figure 7, which show N as a function of F and M. These charts also give us some insight into the possibilities and limitations of the method. For example, we can see quite readily what effect a change in the assumed value of the density contrast (which affects the magnitude of M) would have on the calculated value of N (and, therefore, on D_1). We can also see how an error in F, due to lack of sufficient control points on the anomaly, would affect the value of N. For example, in Figure 6, if F is less than 0.5 or greater than

0.75, N cannot be greater than 0.4 no matter what the value of M. On the other hand, if M is greater than 20 mgal/km, N cannot be greater than 0.34 no matter what the value of F.

The need to know or to assume a value for the density contrast is a limitation of the method, and it may occur to the reader to ask whether this is really necessary. Theoretically, we should be able to identify the prism which fits the anomaly best without reference to the actual value of maximum gravity, and therefore without reference to density, by measuring another distance on the anomaly, such as the quarter-maximum distance, and using the ratio of this to the half-maximum distance, instead of the index M (which depends upon density). Originally, an attempt was made to do this, but it was found that in many cases prisms which had different values for N and the same values for the three-quarter and half-maximum distances, had values for the quarter-maximum distance that were so similar that this parameter was of little use in distinguishing one from the other. Also, the quarter-maxi-

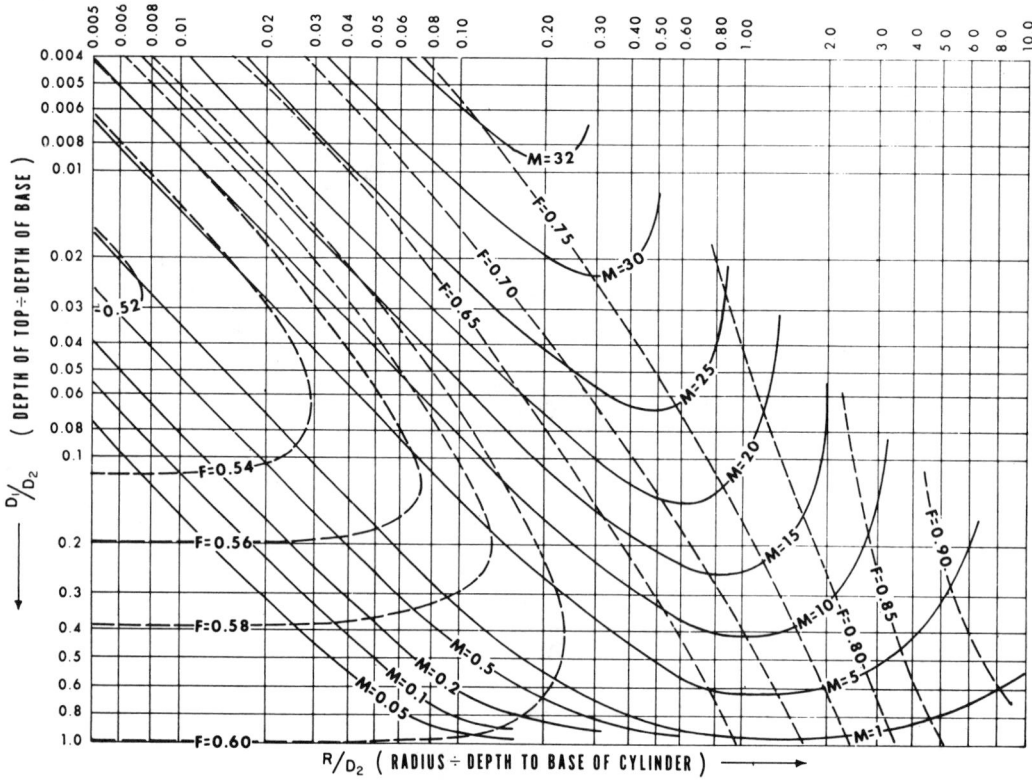

Fig. 4. Values of F and M for vertical circular cylinders.

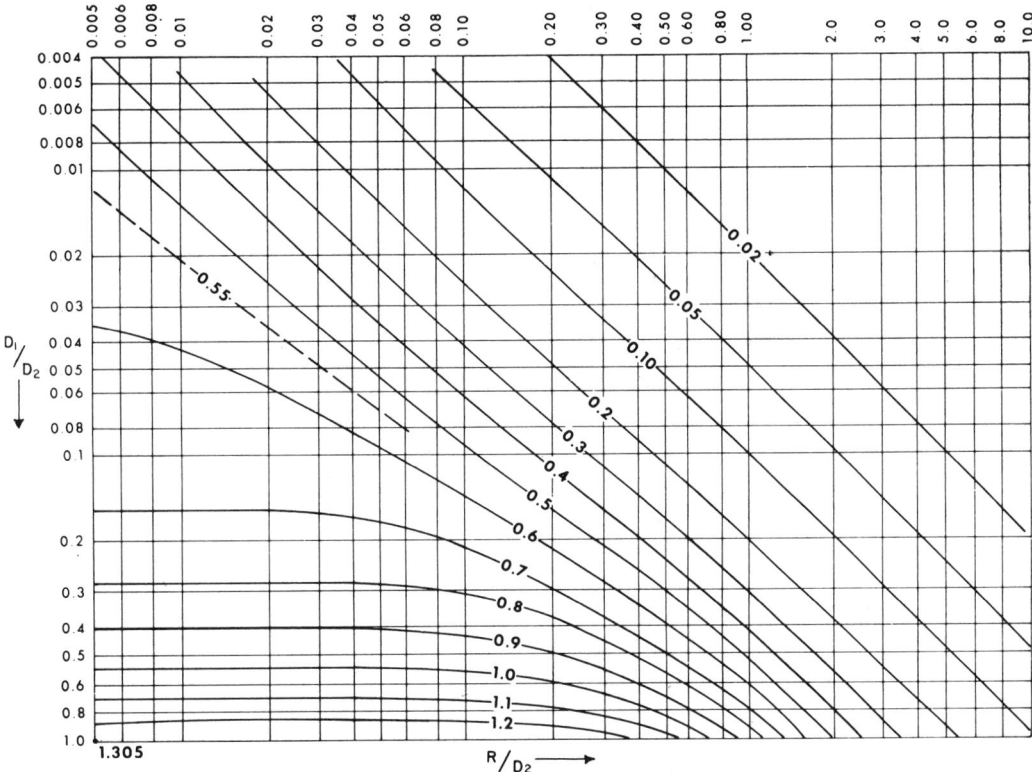

FIG. 5. Values of $N(=D_1/X_{1/2})$ for vertical circular cylinders.

mum distance is much more influenced by errors in estimating the zero base and by effects of adjacent anomalies than are the half-maximum or three-quarter maximum distances. The truth seems to be that even when we assume that the anomaly is due to an elongated prism of rectangular shape of unspecified density contrast, there is still considerable ambiguity from the practical standpoint, if not from the theoretical. Much of this ambiguity can be resolved if the density contrast is known.

Some general instructions and precautions regarding the application of the method may be in order:

1. The interpreter should remember that he is not solving for *the* structure that causes the anomaly, but for a particular structure which differs from the other possible solutions in that it has vertical sides, has the greatest density contrast that is geologically permissible, and is, therefore, by our assumption, deeper than any of the other possible solutions.

2. The anomaly must be isolated from regional effects and, as much as possible, from the effects of other anomalies.

3. Both M and F are influenced by the position of the zero base line; hence this should be estimated with as much care as possible.

4. For anomalies which are neither circular nor greatly elongated (length/width greater than 1.5 but less than 2.5), the best practice is to calculate D_1 using both charts and to average the two values.

5. If the anomaly profile is strongly asymmetric, the steepest side (which will usually give the smallest value for D_1) should be used for the calculations. If the anomaly is only slightly asymmetric, it is probably better to use the values for $X_{1/2}$ and $X_{3/4}$ obtained by averaging the two sides.

6. The accuracy with which D_1 can be calculated obviously depends upon the accuracy and amount of detail of the original gravity data and, also, upon the degree to which the regional effects can be removed. In most

applications an approximate value is sufficient, and that is all that the method is really intended to give.

It should be pointed out that the charts presented here can also be used in reverse; that is, we can use Figure 2 and Figure 3 to calculate the maximum gravity value, and the three-quarter and half-maximum distances for any assumed prism. They can, therefore, be used to determine quickly how large an anomaly could be expected for a given geological structure and, therefore, whether or not the gravity method is applicable to a given problem. The same remarks, of course, apply to Figures 4 and 5 for vertical cylinders; they could be used to calculate the anomalies to be expected from salt domes or igneous plugs.

EXAMPLES OF THE ACTUAL APPLICATIONS

Four examples will be given showing the use of the method for calculating the maximum depth to the sources of actual anomalies. Most of the examples are from company files, and exact location cannot always be given. In all cases in which a pronounced regional anomaly was present this has been subtracted.

Figure 8 is a profile across an elongated anomaly in the northwest of Cuba. Enough was known of the geology to indicate that the crystalline basement was quite deep, and that the main density contrast would be between Tertiary shales and marls and the Cretaceous limestone. The peculiar shape of the anomaly suggested a very shallow origin. The flatness-factor, F, has a very high value, 0.87, and M is quite high, 12.5 mgal/km for $\Delta\rho = 0.3$. D_1 was calculated to be 93 m, or 305 ft; as the Cretaceous limestone was the objective horizon, it was decided that there was not enough cover for a good prospect. To check this, a core hole was put down on the feature; it encountered cavernous Cretaceous limestone at 136 ft. The cavernous nature of the limestone might be responsible for the irregularities on the top of the anomaly; they were smoothed out in the interpretation.

Figure 9 is from Volume I of Geophysical Case Histories; it is a composite profile by Romberg

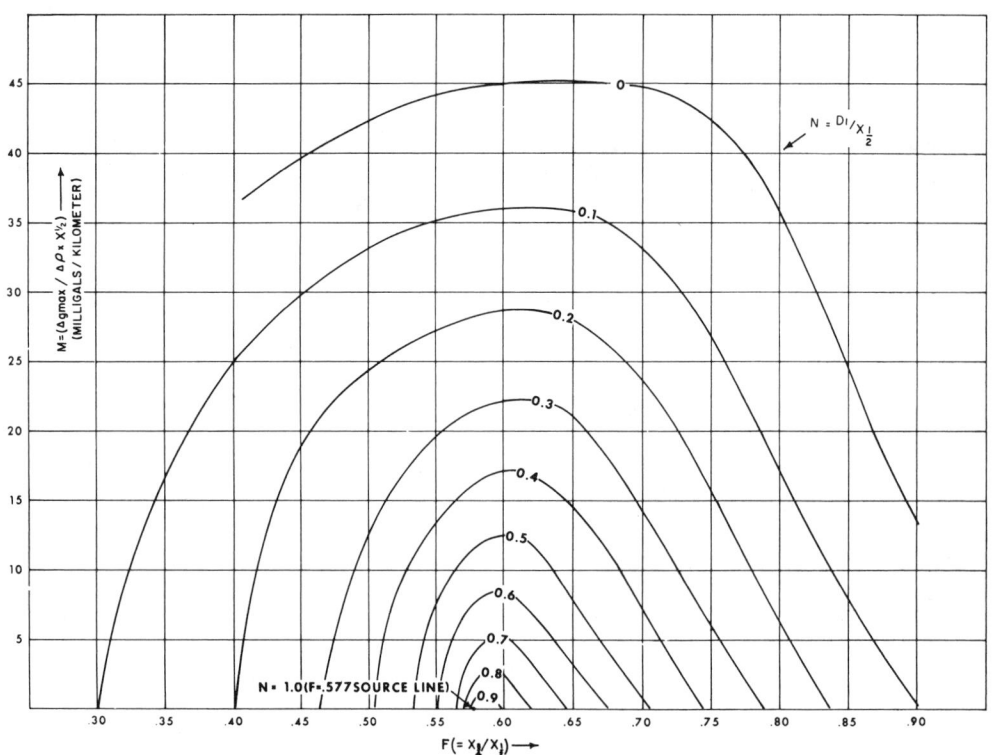

FIG. 6. Two-dimensional prisms—values of N as a function of F and M.

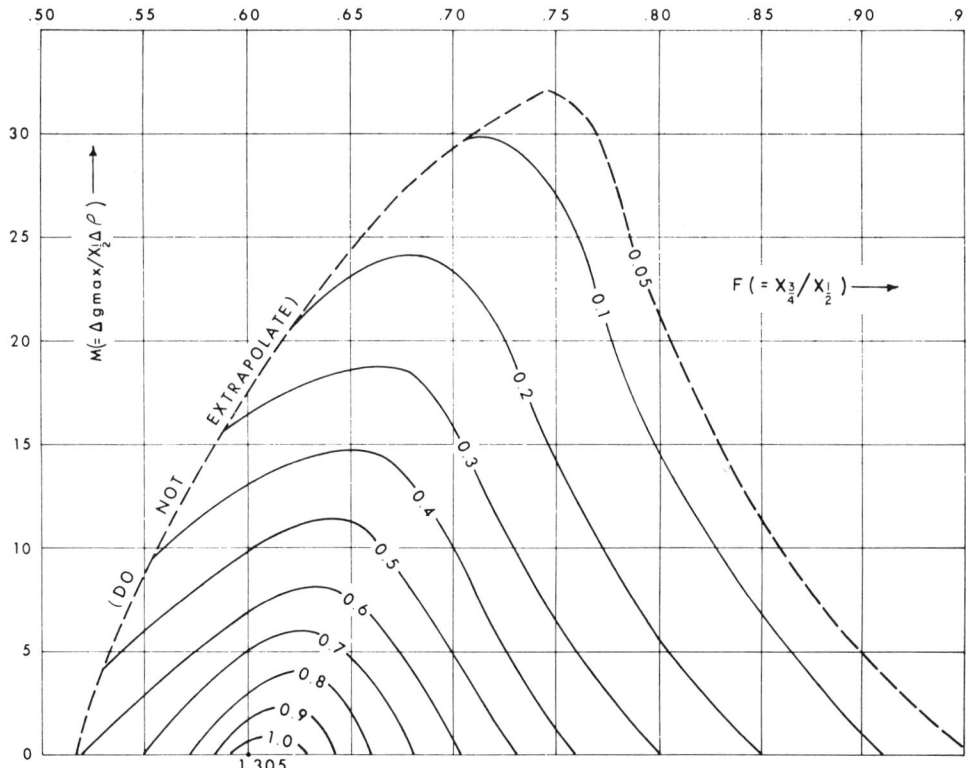

FIG. 7. Circular anomalies—values of $N = (D_1/X_{1/2})$ as a function of F and M.

and Barnes for the Smoothing-Iron Granite mass. Our calculated maximum depth, using the chart for circular anomalies, is 0.375 mile or 1,980 ft. In this example, the known depth to the granite is zero, since it crops out; however, it is easy to imagine a case in which a similar mass might be covered by a thin veneer of alluvium, and in which the knowledge that the top of the mass was less than 2,000 ft would be of considerable importance.

Figure 10 is a profile across an anomaly in central Alberta. This anomaly is neither circular nor greatly elongated; it is roughly elliptical, with the long axis about twice the short axis. For that reason, calculations were made using both the two-dimensional and the circular anomaly charts, and the two values of D_1, D_2, and W were averaged. The mean for the calculated maximum depth D_1 is 1.27 miles, which is almost identical with the basement depth interpolated from nearby wells (6,500 ft or 1.24 miles). Hence, it is concluded that this anomaly could be, and prob-

ably is, caused by a mass entirely within the basement, and that this mass probably has its top at the surface of the basement.

Figure 11 shows a profile across a very unusual anomaly, for which the location cannot be given at this time. The value of F is very low, 0.475, which indicates a vertical rather than a horizontal distribution of mass. The calculated maximum depth is 1.6 km or one mile. If the source of the anomaly is entirely within the basement, the depth to the latter cannot be more than one mile. If the basement surface is deeper than one mile, then a part of the anomaly would have to originate in the sedimentary section. The assumed maximum density contrast for this anomaly was 0.3; this may be too low. If the density contrast were increased to 0.5 (which might be appropriate for periodotite intruded into a granitic crust) the maximum depth, D, would be increased by only about 10 percent (to 1.75); that is, for the values of M and F in question, N is not very sensitive to changes in M.

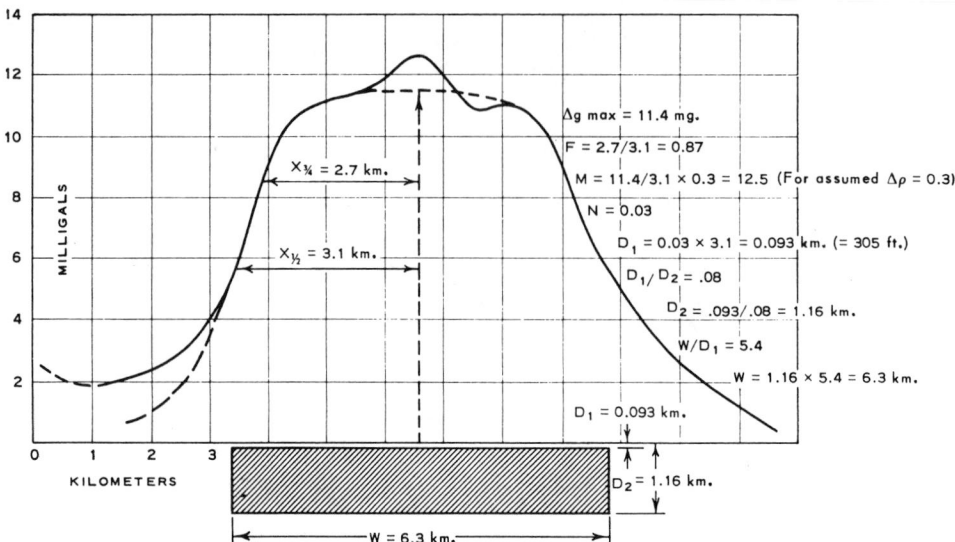

FIG. 8. Fragosa anomaly, north coast of Cuba.

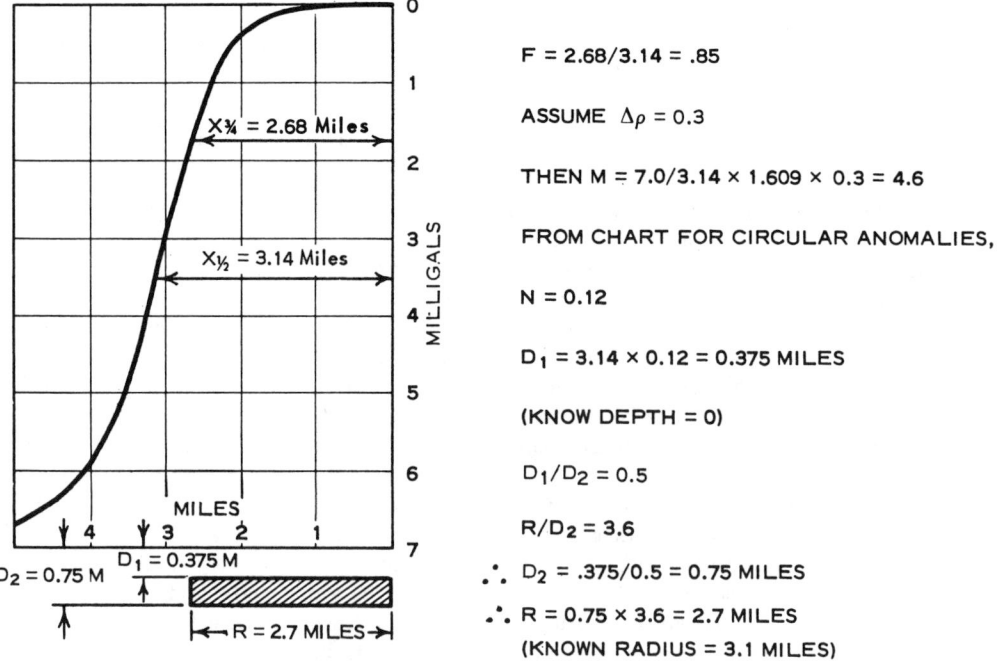

FIG. 9. Composite gravity profile over the Smoothing-Iron Granite mass.

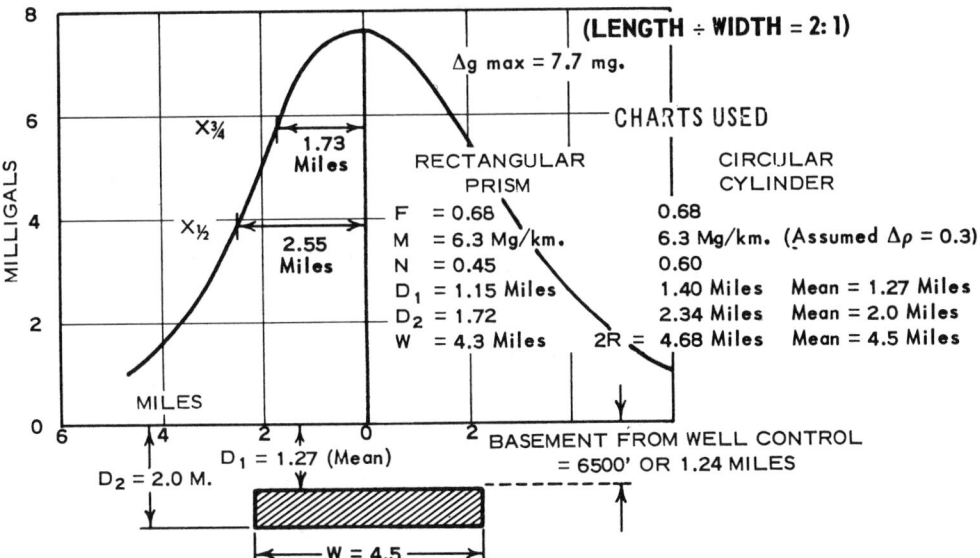

FIG. 10. Profile across a slightly elongated gravity anomaly—central Alberta.

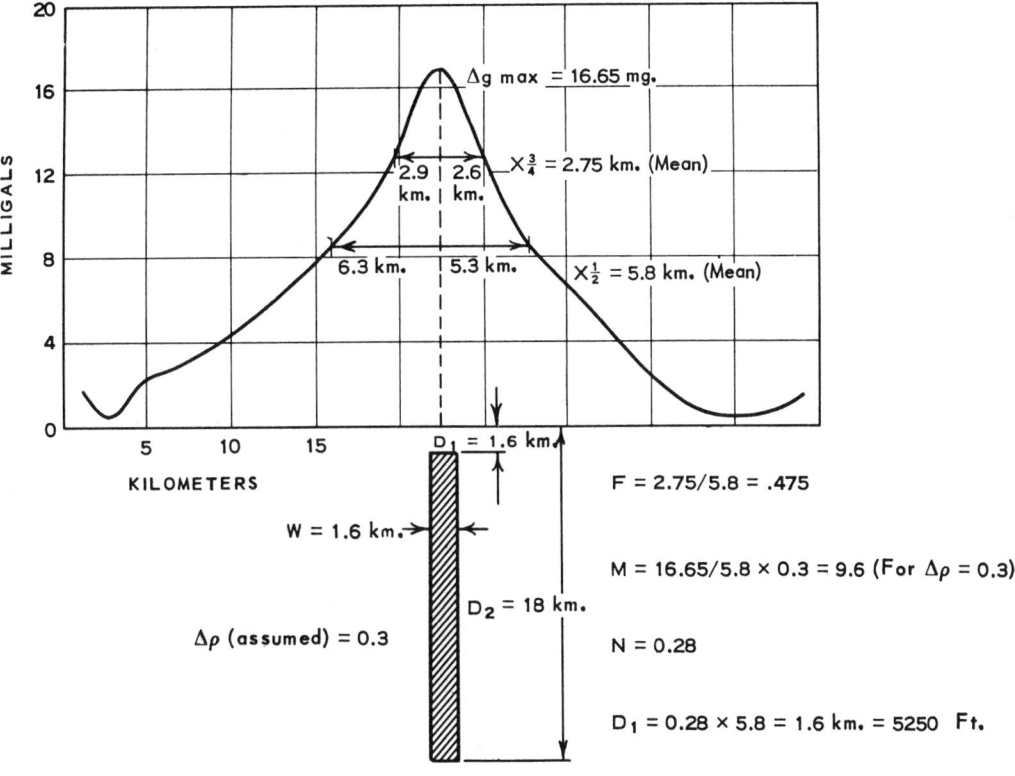

FIG. 11. Profile across an elongated anomaly of unspecified location.

DERIVATION OF THE CHARTS

The curves for F, M, and N were constructed after computing gravity profiles for a large number of prisms and circular cylinders; the values of F, M, and N were computed and plotted against the ratios D_1/D_2 and W/D_2, and the points contoured. For prisms, the formula of Dr. Vening Meinesz (1934) was used. For cylinders, a numerical integration method was used in which the cylinder was approximated by 100 vertical strips of infinitesimal thickness, as mentioned previously.

The first version of these charts was constructed several years ago, when the author was engaged in research on gravity interpretation for the Standard Oil Co. (N.J.). Several man-months of the author's and of an assistant's time were required to produce a set of charts which were known to contain some inaccuracies due to insufficient control points. Recently, the entire set was recomputed using Imperial's IBM 704. The program was written so that the machine not only computed the gravity profiles, but picked the half-maximum and three-quarter maximum distances and calculated the ratios F, M, and N. The total machine time was about $2\frac{1}{2}$ hours and over two hundred control points were computed for each chart.

ACKNOWLEDGMENTS

The author wishes to thank the Standard Oil Co. (N.J.) and Imperial Oil Limited for permission to publish this paper, and Imperial's Electronic Computing and Data Processing Department for their assistance in producing the present set of charts.

REFERENCE

Vening Meinesz, F. A., 1934, Gravity expeditions at sea 1923–1932, v. 2, Delft, Netherlands Geodetic Commission, pp. 23–24.

APPENDIX

COMPARISON WITH DEPTHS CALCULATED FOR HORIZONTAL CYLINDERS AND SPHERES

The author has been asked whether the depth calculated by his method differs significantly from those that would be derived by assuming a circular cylinder with its axis horizontal and of infinite extent (for elongated anomalies) or a sphere (for circular anomalies). The answer depends upon the value of F for the observed anomaly.

For a horizontal cylinder of infinite extent, the value of F is 0.577, no matter what the radius or depth of the cylinder. The depth to the center of the cylinder is equal to the half-maximum distance. A simple calculation shows that the depth to the top of the cylinder is

$$D = X_{1/2}[1 - (M/2\pi K)^{1/2}].$$

Therefore,

$$N = 1 - (M/2\pi K)^{1/2},$$

where N and M are defined as before, and K is the gravitational constant. That is, for the horizontal cylinder, N is a function only of M.

For the sphere, we can show that

$$N = 1.305[1 - (0.575M/\pi K)^{1/3}].$$

In Figure 12, these relations are shown graphically. From this chart, we can read off the value of N for any given value of M, for either of the two shapes.

By comparing values of N in Figure 12 with those in Figures 6 and 7, for given values of M, we can arrive at a quantitative answer to the question that prompted this Appendix. For example, assume that for a given elongated anomaly we have $M = 5$ mg/km. If we assume a horizontal cylinder, we have $N = 0.65$, and multiplying this by the half-maximum distance would give the depth to the shallowest part of the cylinder. On Figure 6, for the same value of M, we have a range of values of N, depending on F; for $F = 0.40$, $N = 0.19$; for $F = 0.60$, $N = 0.70$; for $F = 0.80$, $N = 0.21$. The central value, $N = 0.70$ (for $F = 0.60$), is fairly close to 0.65 for the horizontal cylinder, but the extreme values are each less than one-third the value for the cylinder. We note that $F = 0.60$ is quite close to 0.577, the F-value for the horizontal cylinder, and tentatively conclude that when the observed anomaly has an F-value somewhere near 0.58, the depth calculated for a rectangular prism will not be very different from that calculated for a horizontal cylinder. Comparison of Figures 6 and 12 for other values of M substantiates this conclusion. We may also observe that when the observed value of F is much different from 0.58 (either larger or smaller) the anomaly for the horizontal

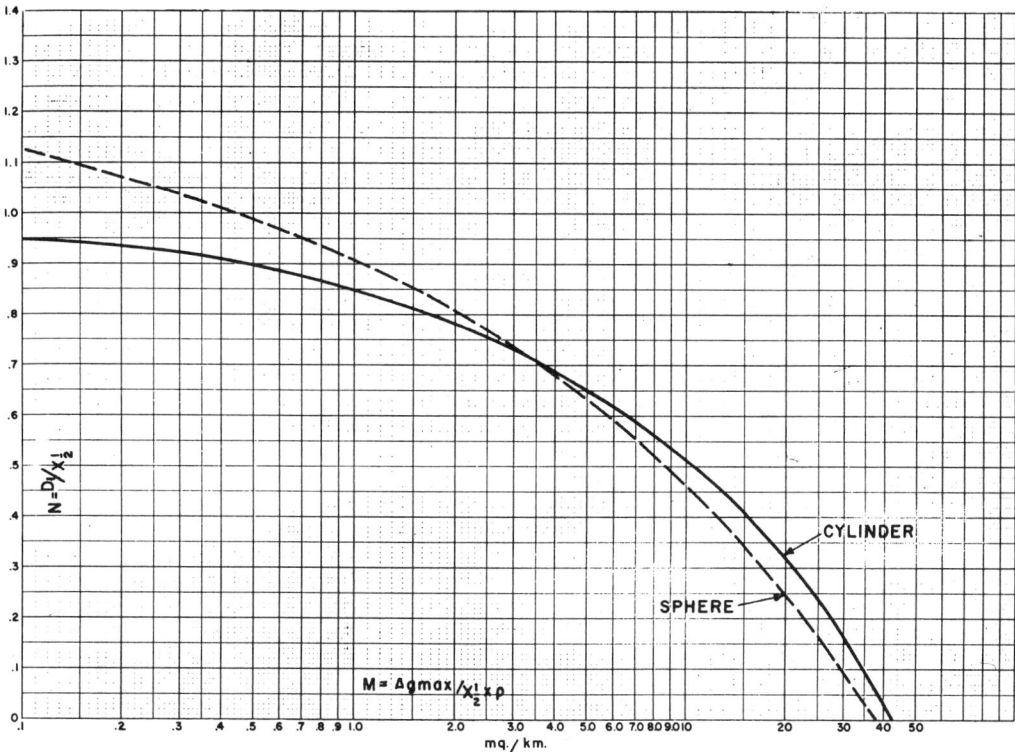

FIG. 12. $N-M$ values for horizontal infinite cylinder and sphere.

cylinder does not really satisfy the observed anomaly very well; the mass which does satisfy it must be shallower and have a more rectangular shape.

The same observation can be made regarding circular anomalies, by comparing Figures 7 and 12; if the value of F is fairly close to 0.60 (the value for the sphere), the depth calculated from Figure 7 will not be very different from that from Figure 12. We may note, in fact, that the tolerance is much greater than in the two-dimensional case. For $M=20$ and $F=0.70$, the value of N for the vertical cylinder is 0.25, while for the sphere it is 0.24, or practically the same. For $M=20$ and $F=0.80$, however, the vertical cylinder gives $N=0.05$, which is only one-fifth the value derived by assuming a sphere.

Thus, we might say that for anomalies that can

be satisfied by a sphere or horizontal cylinder, the present method offers only slight improvement over the calculations based on those simpler shapes, but that for anomalies that are noticeably flat-topped or sharp-topped the method described in this paper will nearly always give a more restricted, and, hence, a more useful value for the maximum depth.

An additional argument in favor of this method is that the shapes assumed, rectangular prisms and vertical cylinders, are shapes that are more often met in geology than are horizontal cylinders and spheres. Finally, it should be pointed out that it takes very little more time to derive a depth by this method than it does for the simpler figures, and the interpreter has the satisfaction of knowing that three parameters of the observed anomaly are satisfied instead of only two.

American Association of Petroleum Geologists Memoir 16.
*Stratigraphic Oil and Gas Fields—Classification, Exploration
Methods, and Case Histories*, edited by Robert E. King,
copyright 1972, p. 244-251.

Use of Gravity, Magnetic, and Electrical Methods in Stratigraphic-Trap Exploration[1]

L. L. NETTLETON

Independent, Houston, Texas 77027

Abstract Gravity, magnetic, and electrical methods of
prospecting can be useful in exploration for pinchout
and reef traps. A pinchout zone might produce a recognizable gravity anomaly as a result of the density contrast (1) between the porous sandstone and the adjacent
beds or (2) between the water-saturated and oil-saturated
parts of the sandstone section. Reefs also can produce
gravity anomalies as a result of differences in density
between the reef material and the laterally adjacent
beds—whether these beds are salt, limestone, or shale.
Where the reef is in shale, compaction of the shale in
the area adjacent to and above the reef is an important
factor. Magnetic methods are useful where basement
structure is a factor in formation of stratigraphic traps.
The use of electrical methods in exploring for stratigraphic traps is much less promising. Overall, however,
these potential methods have possibilities for application
in the search for certain types of stratigraphic traps.

INTRODUCTION

Gravity and magnetic surveys make measurements, at or above the surface, of an "action at a distance." If "stratigraphic" oil accumulations are defined in a broad sense, these
methods are useful in locating certain types of
such traps. The measured variations in the intensity of the fields are caused by the distribution of mass or of magnetized rock below the
surface. For all such potential[2] methods, the
measured natural fields decrease with distance.
In gravity measurements, the intensity of the
field of a unit disturbance decreases as the
square of the distance. The measured fields become weaker and also smoother and less detailed as the depth below the surface of measurement increases.

Electrical geophysical methods also may be
considered as potential methods in that the theory of their application involves potential fields.
They differ, however, from the gravity and
magnetic methods in that the effects measured
are induced by the application of electric currents at the surface (except for the "self-poten-
tial" method, which measures natural fields).

For the purposes of this discussion, "stratigraphic" traps will be classified as (1) pinchout and (2) reef traps. Though there is not
general agreement that reefs are classifiable as

[1] Manuscript received, October 23, 1970.

[2] "Potential" is used here in the mathematical sense.

stratigraphic traps, they are so classified in this
volume and, thus, in this paper.

UPDIP-PINCHOUT STRATIGRAPHIC TRAPS

In this type of oil accumulation, updip migration of oil is interrupted by a structural or
lithologic barrier. A structural barrier occurs
where a dipping porous layer is truncated at its
updip edge. The lithologic barrier may be a facies change such as a porous sandy phase grad-
ing updip into a "tight" shale. By this definition
the stratigraphic trap differs from the structural
trap in being homoclinal and without dip reversal. From the standpoint of geophysical detection, the monoclinal dip and the absence of
structure make its recognition very difficult.

Perhaps the best-known example of an oil
field of this kind is the East Texas field in
northeast Texas (Minor and Hanna, 1933;
1941). The trap (Fig. 1) is formed by the up-
dip pinchout of the Woodbine Sandstone
against the overlying unconformity. The total
thickness of the Woodbine Sandstone west of
the oil field and the pinchout is less than 200 ft
(60 m; Minor and Hanna, 1933, Fig. 8). The
maximum thickness of the zone of saturation is
approximately 100 ft (30 m), but only 30–40
percent of this total is saturated with oil.

There are two ways in which the pinchout
zone might produce a recognizable gravity
anomaly. First, the porous sandstone section itself, without regard to the oil saturation, could
have a density contrast relative to adjacent
beds. The porous sandstone would be expected
to be of lower density than the underlying shale
and limestone and the overlying Austin Chalk.
The average porosity of the sandstones is about
20 percent (Minor and Hanna, 1933, p. 783).
For quartz sandstone with grain density of
2.65, saturated with water, the density would
be about 2.3 (Fig. 2). The density of the rocks
above and below may be about 2.5, which
would give a density contrast of 0.20. In the
Longview area (Minor and Hanna, 1933, Fig.
8), the pinchout from about 200 ft (60 m) of
sandstone to zero thickness occurs in a horizontal distance of approximately 5 mi (8 km), and

244

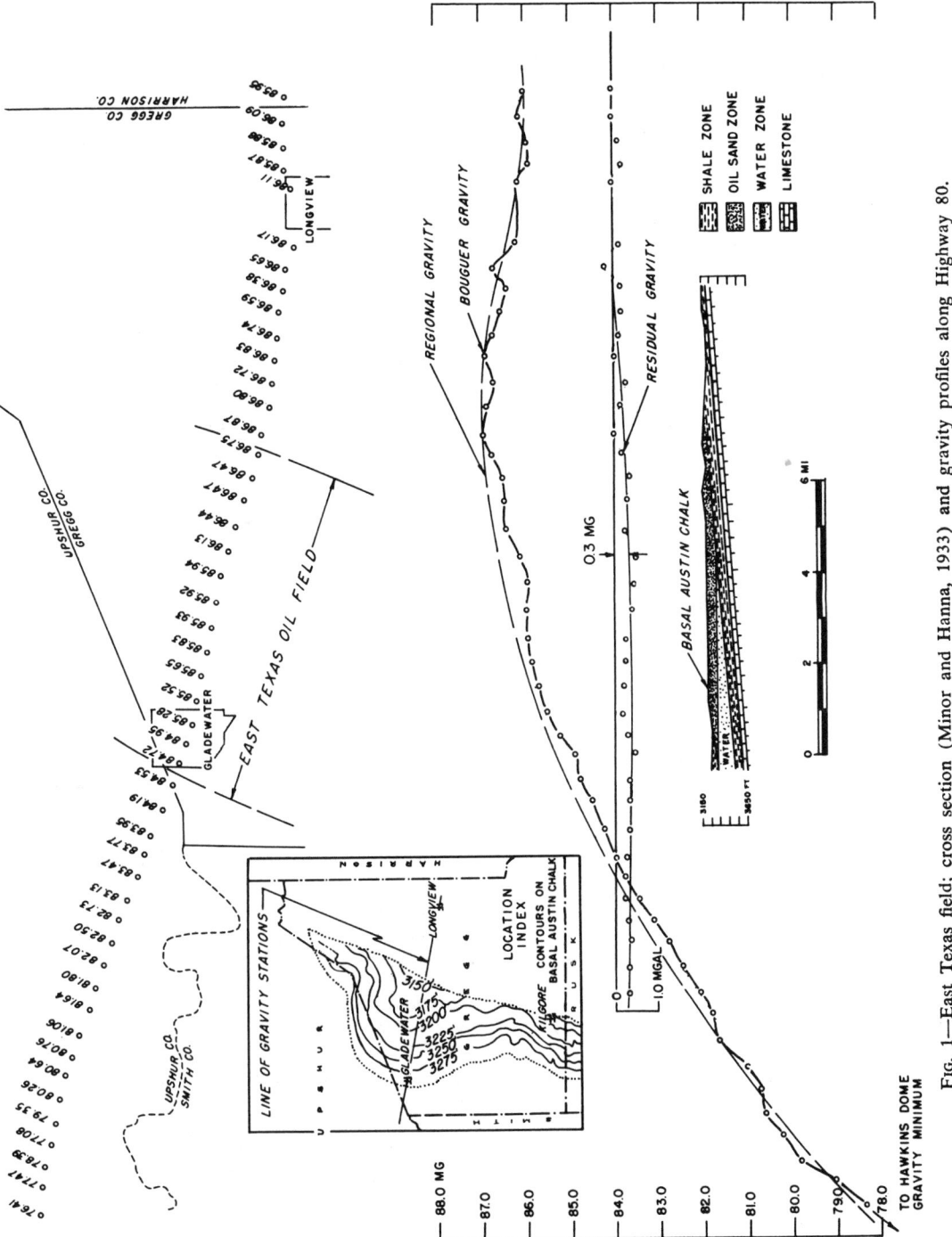

Fig. 1—East Texas field; cross section (Minor and Hanna, 1933) and gravity profiles along Highway 80.

the average depth of the sandstone layer is about 3,500 ft (1,070 m).

For the estimated density contrast of 0.20 and for a thickness change from 200 ft (60 m) to zero across the pinchout, the expected negative gravity effect is approximately 0.4 mgal, spread over the 5-mi (8 km) zone of thinning.

The second, but much smaller, possible gravity effect could result from the difference in density between the water-saturated and oil-saturated parts of the sandstone section. The density contrast would be that between salt water, with a density of about 1.1, and oil, with a density of about 0.8, or a density contrast of about 0.3. This contrast would be applicable only to the pore space of the sandstone. Thus, an average thickness of saturated sandstone of less than 30 ft (9 m) with porosity of 20 percent would produce a gravity effect of only about 0.02 mgal, which would be undetectable.

Figure 1 includes a profile from a high-quality gravity traverse along old Highway 80, which crosses the northern part of the East Texas field at its maximum width. A smooth "regional" curve can be drawn, leaving a broad but rather irregular negative anomaly as indicated. The definition of the regional curve is not clear because of local disturbances on the east side and the marked curvature into the large negative anomaly due to the Hawkins field on the west. The ideal anomaly would be a broad, one-sided negative feature beginning at the updip edge of the wedge of porous sandstone and increasing in magnitude westward to

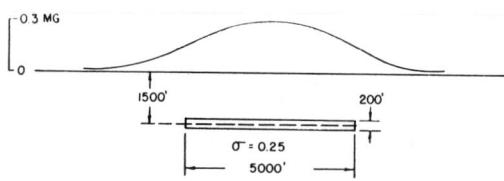

FIG. 3—Calculated effect of circular mass simulating a "pinnacle" reef.

the point where the top of the sandstone is present beneath the unconformity; the magnitude would be maintained westward where the sandstone thickness is uniform. The regional curve as drawn shows the residual anomaly—pictured above the geologic section—to have the general character expected and an irregularly defined amplitude of 0.3 to 0.4 mgal, which is consistent with the calculations given.

Admittedly, this "hindsight" analysis, particularly in choosing the regional, has been greatly influenced by the form and magnitude of the expected anomaly. The eastern limit, which corresponds generally to the updip limit of the sandstone, is reasonably objective, partly because the regional is flat in that area. The continuation of the anomaly to the west, beneath the markedly curving regional, is a somewhat artificial interpretation, but a reasonable one. In general, this analysis suggests the possible presence of a gravity anomaly with form, magnitude, and horizontal position which are consistent with expectations based on the geologic section and with reasonable density assumptions. It is doubtful that the same result would be obtained by an objective interpretation if there were no reason to suspect the presence of a gravity anomaly.

The East Texas example has a relatively simple background gravity pattern and apparently favorable density conditions. Nevertheless, the presence of a very low-relief anomaly spread over several miles is questionable. A series of parallel profiles with consistent anomalies on each would be necessary to make a convincing indication.

Geologic conditions more favorable for gravity detection would be greater sandstone thickness, greater density contrast, and a narrower zone of thinning. If the background gravity pattern were not simple and regular, a low-relief anomaly would be hard to recognize. In general, it cannot be said that the gravity method offers favorable possibilities for finding pinchout stratigraphic traps. The writer knows of no ex-

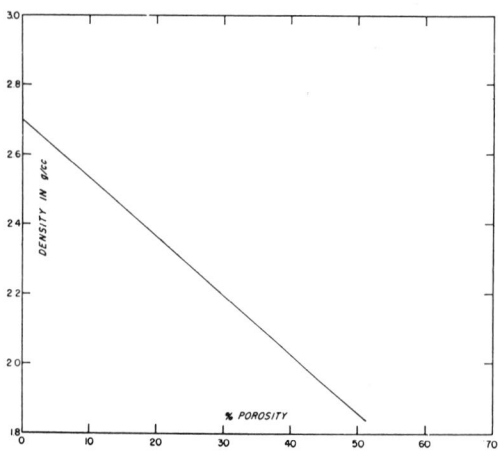

FIG. 2—Relation of density to porosity for water-saturated sediments, calculated for grain density of 2.70.

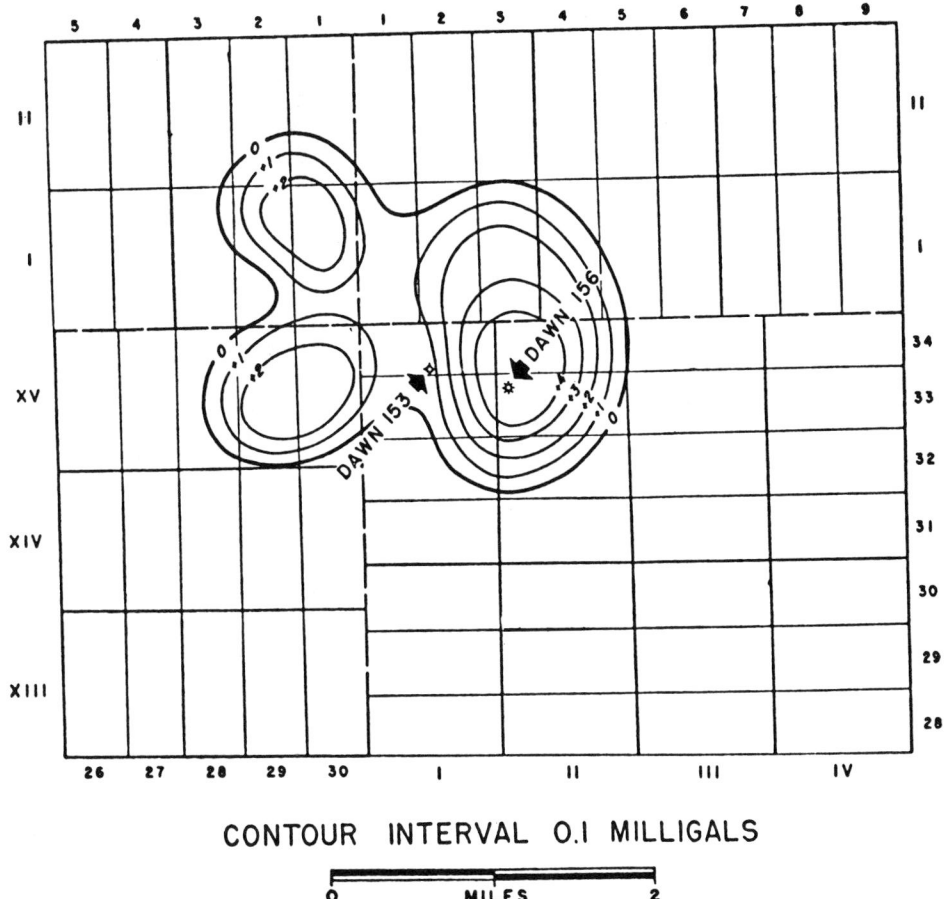

CONTOUR INTERVAL 0.I MILLIGALS

Fig. 4—Residual gravity map, Dawn 156 pool (Pohly, 1956).

ample of an oil field in such a trap being found by gravity surveys.

APPLICATION OF GRAVITY METHOD TO FINDING REEF RESERVOIRS

Many oil accumulations are found in porous reefs. Reefs are formed in a variety of ways but may be included in the "stratigraphic" category because a reef may be considered as a local horizontal facies change—that is, a change in stratigraphy.

Some types of reef produce gravity anomalies directly from a density contrast between the reef material and the laterally adjacent beds. An example is the more or less circular "pinnacle" reefs in the Niagaran in southwestern Ontario and southeastern Michigan. The reef material is of higher density than the adjacent salt. In other places the reef limestone is pres-

ent locally in shales.

The gravity method is the only one of the potential methods which, on the basis of detection of effects from reefs, has led to the discovery of oil fields in reefs. Careful surveys have been made over oil fields in which the accumulations are in reefs, and published accounts are available of many which have definite gravity anomalies associated with the reefs. Reasonable explanations of the gravity anomalies can be derived from the sedimentary section and the geometry of the reef and adjacent strata as known from drilling. Detailed examples are given in the two papers by Ferris in this volume (p. 252–267 and p. 460–471).

There are several ways in which the materials of a reef can cause a gravity anomaly. First, the reef material may be heavier than the laterally adjacent beds. The most conspicuous and interesting examples are the Niagaran reefs in salt in southwestern Ontario and southeastern Michigan. Here, the reef growth has occurred in such a way that the roughly circular reef masses are within the normal evaporite section and the reef replaces the salt. Furthermore, the apparently sharp and steep boundaries between the reef and the salt result in a marked density contrast between reef material with a density of the order of 2.5 and adjacent salt with a density of 2.2 to 2.3. The reefs are present at depths of about 1,000 ft (305 m) and deeper, and those that cause the most conspicuous gravity anomalies, which are clearly associated with the reef-salt density contrast, are relatively shallow, *i.e.,* less than 2,500 ft (760 m). As an example of the gravity magnitude to be expected, Figure 3 shows the calculated gravity from a reef represented as a circular disc which is 200 ft (60 m) thick and 5,000 ft (1,525 m) in diameter and has a positive density contrast of 0.25. The maximum anomaly calculated over the center of this feature is 0.30 mgal. Gravity anomalies of this magnitude and with a total area of 1 sq mi (2.6 sq km) have been found as a result of careful gravity surveys over known reefs of this kind, and also have led to the discovery of several reef oil fields.

The expected anomaly relief of a few tenths of a milligal requires that the gravity surveys be carried out very carefully to yield accurate results and that they use very close station density, commonly a concentration of 25 stations per square mile. The close station spacing is required to distinguish the reef anomaly from other disturbances, particularly if the survey is made in a glaciated area, because gravity measurements commonly are disturbed by inhomogeneities in the glacial drift itself or, more seriously, by buried irregular topography at the base of the drift. For all of these reasons—the small area, the small relief of the expected gravity anomaly, and the environment of disturbing, shallow density irregularities—the reef anomalies rarely are cleanly expressed. Their recognition and mapping may depend to a considerable extent on the skill and experience of the interpreter.

Porous reefs which occur in a limestone also may produce gravity anomalies. In this case, the porous reef material would be expected to be of lower density than the surrounding limestone and, therefore, to produce a negative gravity anomaly. With the porosity difference of 15 percent between the reef material and the adjacent limestone, the density contrast would be 0.27 (calculated using a grain density of 2.7 and assuming that pore space is saturated with water; Fig. 2), which is about the same as that used for the example of reef in salt, but the density contrast is negative instead of positive. Although such a situation seems to be a theoretical possibility, the writer is not aware of an example of a negative anomaly over a reef oil field.

A third type of reef is a buildup in shale. In this case, the expected density contrast between the porous reef and the shale is probably smaller than in the preceding example, but gravity surveys over several reefs of this type show definite positive anomalies with magnitudes of a few tenths of a milligal. Careful mapping of the stratigraphy over some of these features, as shown in following examples, demonstrates distinct compaction of the shales in the area adjacent to and above the reef. The resultant structural deformation over the reef may be evident, but with gradually reduced amplitude, to a considerable height above the reef itself. The compacted sediments are of slightly higher density than the noncompacted, laterally adjacent, similar material; therefore, there is a small density contrast which carries upward throughout the zone of compaction above the reef. It is the accumulation of the gravity effects arising from these zones which produces the total effect, rather than the direct effect of the density contrast of the reef itself with respect to the adjacent material.

The magnitude and general form of anomalies produced by reefs are very similar whether the reef buildup is surrounded by shale or by salt. Reef buildups in shale are much more common, however. Several discoveries of "reef" oil fields are attributed directly to small positive anomalies caused apparently by compaction over the reef (Ferris, this volume, p. 253, Table 1).

A gravity anomaly produced by compaction over a reef is no different in principle from a gravity anomaly produced by normal structure. Quite commonly, a positive gravity anomaly is produced by a structural uplift but there is no definite horizon at which a density contrast occurs. The anomaly is simply the sum of the results of small elements of density contrast produced by the local uplift of a sedimentary column in which density gradually increases with depth. Thus, in the uplifted area, heavier materials are present where the laterally adjacent rocks have slightly lower densities.

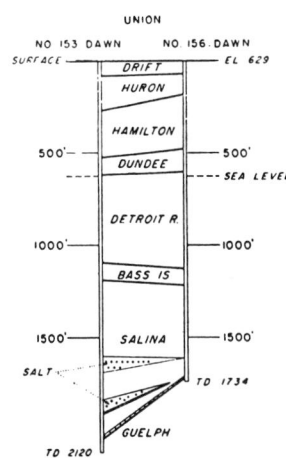

Fig. 5—Cross section through Dawn 153 and Dawn 156 (Pohly, 1956).

Application of Magnetic Methods to Finding Reef Reservoirs

The materials which form reefs are not magnetic, nor are the sedimentary beds in which they occur; therefore, it is doubtful that any magnetic survey could find a reef as a result of direct magnetic effects of the reef itself. There are, however, reefs in the lower part of the geologic column and near the basement which are controlled by the configuration of the basement rocks. Some of these appear to be controlled, in part at least, by details of the basement surface such as a change of slope. Some of the productive reefs of northwestern Alberta appear to be of this type. There are examples where careful interpretation of airborne-magnetometer surveys has led to subtle indications of basement features possibly related to oil accumulations in reefs. Possibly, some especially favorable conditions exist in which the association of oil accumulations in reefs with features of the basement, determined from analysis of magnetic surveys, may lead to discoveries of such accumulations in new areas. Any such use of magnetics depends very much upon skillful and imaginative correlations of sedimentary and basement features. Instances of such interpretations have been reported, but no examples are available.

Electrical Methods

Electrical methods have possible application to the search for oil accumulations in stratigraphic traps. Such application probably would not be in the detection of geologic conditions or boundaries associated with the stratigraphic trap, whether reef or pinchout. Throughout the history of the geophysical search for oil, there has been hope that electrical methods might lead to the detection of oil because of the higher resistivity of saturated oil sandstones compared to that of water-saturated strata. If successful, this method might be used in the search for extensive stratigraphic accumulations of the East Texas type.

Much literature is available on electrical prospecting for oil. A recent extensive review (Keller, 1968), including a large bibliography, covers a rather highly mathematical body of theory on this subject. There are several claims of measurable electrical effects related to structure or to oil occurrence, but these are not generally accepted as reliable. Apparently no active attempts are presently being made in this country or the free world to apply electrical methods on the principle that they might find oil directly. Considerably more work of this kind has been done in Russia, but the writer knows of no readily available example of its success.

Keller (1968) summarized the problem as follows:

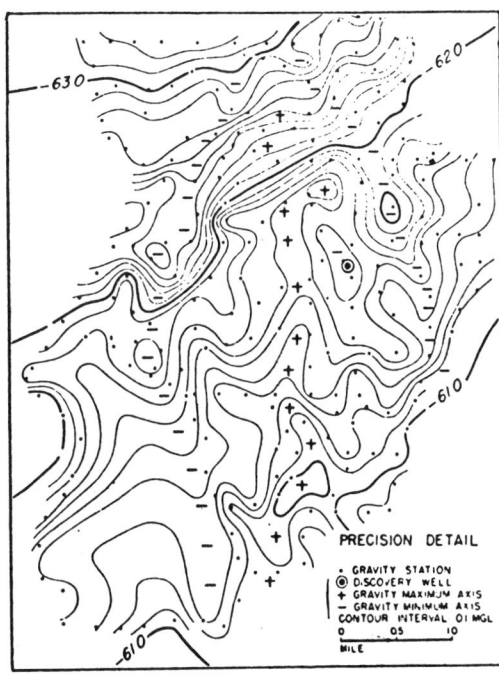

Fig. 6—Precision detail gravity map of Jameson area, resolving north-south trends. Contour interval is 0.10 mgal (Brown, 1949).

In the application of electrical methods for the direct location of oil fields, the question is not so much whether the accuracy of the measurements may be improved to the point where the direct anomaly can be detected, but rather, whether an anomaly of one-half percent in DC resistivity or 500 parts per million in AC coupling can be recognized as representing oil saturation. It is readily apparent that other changes in a section will cause measured field quantities to vary by more than these minimum amounts even when oil is not present. Direct location of oil with electrical methods would seem to be feasible only in cases where the approximate location of an oil field was known beforehand, both in plan and in depth. For example, it may be known that oil pools occur in a given reservoir horizon, along a trend line, and that otherwise, the character of the section varies in a logical manner. One would then prospect for small differences in resistivity associated with that specific depth. In such cases, an anomaly of a half percent in DC resistivity or 500 parts per million in AC coupling might be recognized as representing oil saturation, even though changes in other parts of the section would cause considerably larger changes in observed resistivity.

EXAMPLES OF REEF GRAVITY ANOMALIES

There are many published examples of gravity surveys over reef-type anomalies. Details from several such surveys are given by Ferris (this volume, p. 252–267 and p. 460–471), and others are listed in the references at the end of this paper. Two papers of particular interest are reviewed briefly in the following paragraphs.

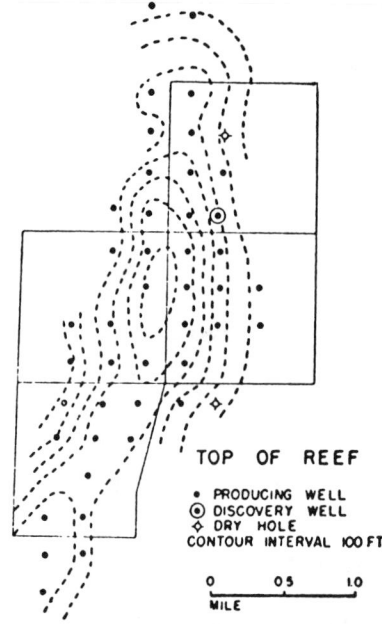

FIG. 8—Contours of Jameson area showing top of reef, about 6,000 ft (1,830 m) deep, as defined by all field wells drilled as of February 1949. This map is directly comparable with Figures 6 and 7 (Brown, 1949).

Dawn 156 Field

Pohly (1956) presented a good example from southern Ontario of a local positive anomaly with relief of about 0.4 mgal caused by reef limestone in salt. The anomaly (Fig. 4) was developed by additional control with more closely spaced stations (not shown) to give much more detail within a much larger positive anomaly. The discovery well of the Dawn 156 oil field was drilled on this positive anomaly. A comparison with the Dawn 153 well, located about 0.5 mi (0.8 km) west (Fig. 5), shows that the producing well drilled into the reef whereas the dry hole encountered approximately 150 ft (45 m) of salt not present in the producing well. Thus, it seems quite probable that the replacement of the salt by the reef limestone was largely responsible for the gravity anomaly, as its magnitude is generally similar to that of the calculated example (Fig. 3) for an assumed reef in salt with similar dimensions.

Jameson Field

Brown (1949) was one of the first to look for the possible gravity expression of reefs by investigation of the Jameson area in Coke

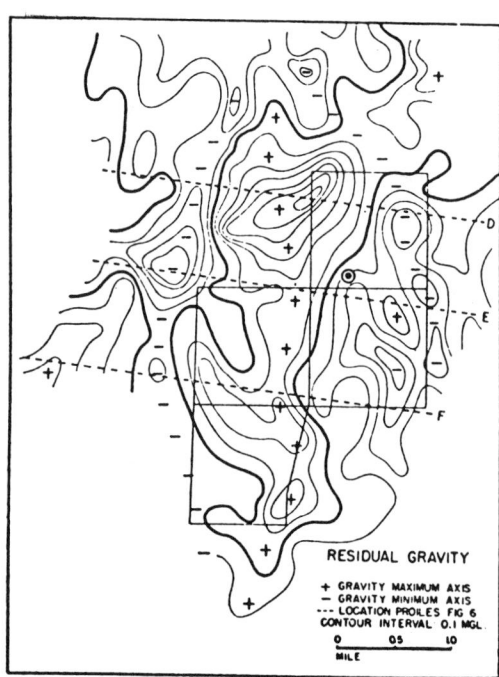

FIG. 7—Residual gravity map of Jameson area as computed from Figure 6 (Brown, 1949).

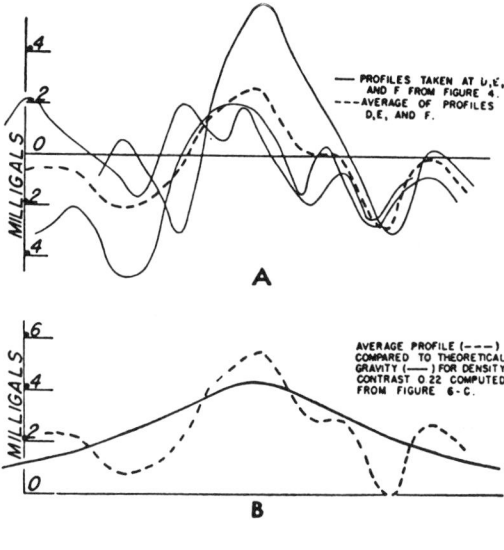

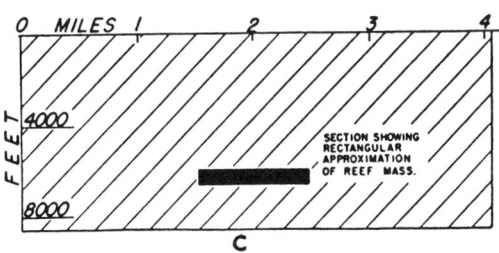

FIG. 9—Theoretical gravity computed from an approximate reef form, compared with gravity profiles as measured at Jameson (Brown, 1949).

County (West Texas), where one of the first "reef" fields in the area was found in 1948. A reconnaissance survey with fairly wide station spacing and, later, a conventional survey with about 0.5 mi (0.8 km) station spacing along the roads had failed to show any definite feature that could be correlated with the reef oil field then being developed. A very carefully conducted survey with an average of about 15 stations per square mile, mostly on 0.25-mi (0.4 km) spacing, followed. This work developed an irregular gravity pattern (Fig. 6) from

which a regional was subtracted to give a north-south-trending maximum with flanking minima (Fig. 7). The discovery well of the oil field is on the east edge of this maximum. The structure at the top of the reef (Fig. 8) coincides well with the ragged maximum developed from the residual gravity map. The general relief of the gravity anomaly is about 0.5 mgal; the relief of the reef structure is about 600 ft (180 m) at a depth of a little over 6,000 ft (1,830 m), and its width is slightly more than 1 mi (1.6 km). If some allowance is made for smoothing of effects from shallow sources, this anomaly is in reasonable accord with the calculated effect of a geometric approximation of the form of the reef body with a density contrast of 0.22 (Fig. 9). Thus, the density contrast that apparently explains the gravity anomaly is that of the reef limestone itself with respect to the laterally adjacent shales. No compaction structure above the reef has been invoked to explain the gravity anomaly. The logs of the wells drilled in the field at the time of the gravity survey gave no indication, down to within a few hundred feet of the top of the reef, of the presence of a compaction structure. The Jameson example, therefore, is one of an observable gravity anomaly which is a direct effect of the density contrast between the reef itself and the surrounding sedimentary beds.

REFERENCES CITED

Brown, Hart, 1949, A precision detail survey of the Jameson area, Coke County, Texas: Geophysics, v. 14, p. 535–542.
Ferris, Craig, 1972a, Boyd-Peters reef: this volume.
——— 1972b, Use of gravity meter in search for stratigraphic traps: this volume.
Keller, G. V., 1968, Electrical prospecting for oil: Colorado School Mines Quart., v. 63, no. 2, 268 p.
Minor, H. E., and Hanna, M. A., 1933, East Texas oil field: Am. Assoc. Petroleum Geologists Bull., v. 17, no. 7, p. 757–792.
——— and ——— 1941, East Texas oil field, Rusk, Cherokee, Smith, Gregg, and Upshur Counties, Texas, p. 600–640 in A. I. Levorsen, ed., Stratigraphic type oil fields: Am. Assoc. Petroleum Geologists, 902 p.
Pohly, R. A., 1956, Gravity case history—Dawn No. 156 pool, Ontario, in P. L. Lyons, ed., Geophysical case histories, v. 2: Soc. Exploration Geophysicists, p. 179–187.

THE DIRECT APPROACH TO MAGNETIC INTERPRETATION
AND ITS PRACTICAL APPLICATION*

LEO J. PETERS†

ABSTRACT

This paper discusses the solution of the inverse potential problem and its practical application
in the interpretation of field data which have a scalar potential distribution. The discussion will be in
terms of the interpretation of magnetic data. Among the topics discussed are: the direct calculation
of basement relief, the derivation of the potential and the horizontal components of the field from
the vertical intensity, the continuation of the field upward, the continuation of the field downward
towards its source, the calculation of derivatives of the vertical intensity with special attention to the
second and fourth, and the estimation of depths to igneous basement rocks. The uses of these tools
and the information of practical value which can be obtained by their use are discussed and illustrated. Methods of rapidly making calculations using magnetic field data are given.

Several volumes could be written on the subject of the interpretation of
gravity and magnetic data. It is therefore a difficult task to write a single paper
covering the subject, and severe limitations must be placed on the choice of topics
to be covered and the detail in which they are discussed. This paper will present
and discuss some of the special methods developed at Gulf Research Laboratories
starting some nineteen years ago to aid in an interpretation of magnetic and
gravity data. The discussion will be in terms of magnetic interpretation and functions, but its extension to the gravitational case can be made in most cases as
both are potential fields.

In 1929 when the writer joined the Gulf organization, it was common practice
to make gravity and magnetic calculations by assuming a distribution of mass or
of magnetic material, calculating the field due to this assumed distribution, and
modifying the distribution until the calculated field fit as closely as desired the observed data. We at the Gulf Laboratories undertook to solve the inverse problem.
That is, we started with the observed field and undertook to calculate directly a
distribution of matter that would account for this field. Unfortunately, this problem has no unique solution. It can be made unique only by making assumptions as
to the nature of the distribution causing the field. It is our inability to know just
what these assumptions should be that leads to the ambiguity in the interpretation of gravity and magnetic data. This lack of uniqueness applied also to the indirect method of interpretation and the direct method of attack—as we shall
show—led to some powerful tools which are an aid in the solution of the uniqueness problem.

As happens at all large research institutions, the solution of a problem is
carried forward by team work among a group of individuals. In addition to the

* Presented at the St. Louis Meeting of the Society, March 16, 1949. Manuscript received by the
Editor March 28, 1949.

† Gulf Research & Development Company, Pittsburgh, Pennsylvania.

290

writer, those who contributed to the development of the material covered in this paper are: John Bardeen, H. M. Evjen, T. A. Elkins, James Affleck, and H. H. Evinger. Of those mentioned, the contribution of John Bardeen was outstanding. The writer wishes to express his appreciation to Messrs. Affleck and Elkins for their aid in preparing the present paper.

The problem of magnetic interpretation is to deduce geological subsurface information from magnetic measurements made at the surface or above the surface of the earth. In most cases, but not in all, the sediments are made up of non-

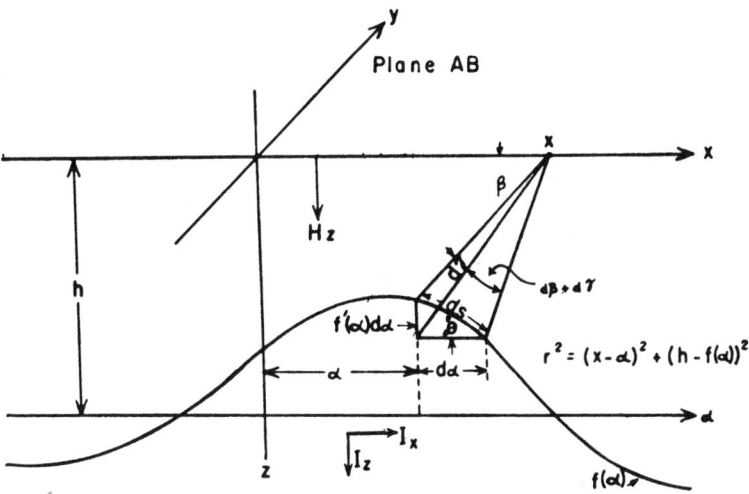

FIG. 1. Plane *AB* on which measurements of the vertical intensity of the magnetic field have been made.

magnetic rocks, so in most areas the problem of magnetic interpretation is to deduce as much information as possible concerning the nature and the attitude or convolutions of and the depth to the igneous basement rocks.

As stated above, the problem of the direct calculation of the distribution of magnetic material from the observed field was attacked in 1929. By the first part of 1930, direct two-dimensional structure calculations were being made as a routine matter of interpreting field data. This early solution of the direct calculation of structure is given in sketch form below:

In Figure 1, *AB* is a plane on which measurements of the vertical intensity of the magnetic field have been made. The curve centering on the line $z = h$ represents the convolutions of the magnetically active rocks, which are assumed to have a constant polarization. The contribution of the element ds to the vertical intensity, $H_{z_1}(x)$, at x is:

$$dH_{z_1}(x) = -2[-I_z \cos \theta + I_x \sin \theta] \frac{\sin \beta}{r} \cdot ds. \qquad (1)$$

113

Now Equation (1) can be written in the form

$$dH_{z_1}(x) = 2I_z(d\beta + d\gamma) + 2I_x f'(\alpha)\frac{h - f(\alpha)}{r^2}\,d\alpha. \tag{2}$$

Since

$$\frac{\cos\theta\,\sin\beta ds}{r} = (d\beta + d\gamma), \qquad \sin\theta ds = -f'(\alpha)d\alpha,$$

and

$$\sin\beta = \frac{h - f(\alpha)}{r}\,.$$

The sign of $d\gamma$ is the same as the sign of $(x-\alpha)f'(\alpha)$ and is negative in Figure 1.

Now the term $2I_z d\beta$ integrates to $2\pi I_z$ since I_z is constant and represents the vertical field if no structure is present. If this constant term is disregarded the anomaly field is given by the relation

$$H_z(x) = 2\int_{-\infty}^{\infty}\frac{I_z(x - \alpha) + I_x(h - f(\alpha))}{(x - \alpha)^2 + (h - f(\alpha))^2}\,f'(\alpha)d\alpha. \tag{3}$$

Equation (3) relates the vertical magnetic intensity on the plane AB to the structure of the magnetic rocks producing the field. It may be considered as an integral equation for determining $f(\alpha)$ when $H_z(x)$ is known. The integral equation is nonlinear. It can be made linear under the condition that the structure is small compared to the depth of burial, that is if $f(\alpha) << h$.

Since our magnetic data were known only at discrete points along AB, the equation was solved under the condition $f(\alpha) << h$ by replacing the integral by the sum

$$H_z(x_j) = 2\sum_{i=-n}^{i=+n}\frac{I_z(x_j - \alpha_i) + I_x h}{(x_j - \alpha_i)^2 + h^2}\,a_i \tag{4}$$

where

$$a_i = \{f'(\alpha)d\alpha\}_i.$$

For the case in which the rocks are vertically polarized only, the solution takes the form

$$a_0 = \frac{b}{2I_z}\sum_i B_i(H_{+i} - H_{-i}). \tag{5}$$

Table I gives a set of coefficients B_j to use with this equation; all measurements are in terms of the depth of burial where b is the separation between points

Table I
Two Dimensional Structure Coefficients

$$a_0 = \frac{b}{2I_z} \sum_j B_j (H_{+j} - H_{-j})$$

$(H_{+j} - H_{-j})$	B_j	$(H_{+j} - H_{-j})$	B_j
$H_{+\frac{1}{2}} - H_{-\frac{1}{2}}$	1.86864	$H_{+8} - H_{-8}$	0.04307
$H_{+1} - H_{-1}$	−0.62953	$H_{+10} - H_{-10}$	0.21834
$H_{+2} - H_{-2}$	0.09034	$H_{+12} - H_{-12}$	−0.18946
$H_{+3} - H_{-3}$	0.00850	$H_{+15} - H_{-15}$	−0.01879
$H_{+4} - H_{-4}$	0.01505	$H_{+18} - H_{-18}$	−0.01504
$H_{+5} - H_{-5}$	0.00088	$H_{+24} - H_{-24}$	−0.00526
$H_{+6} - H_{-6}$	0.04647		

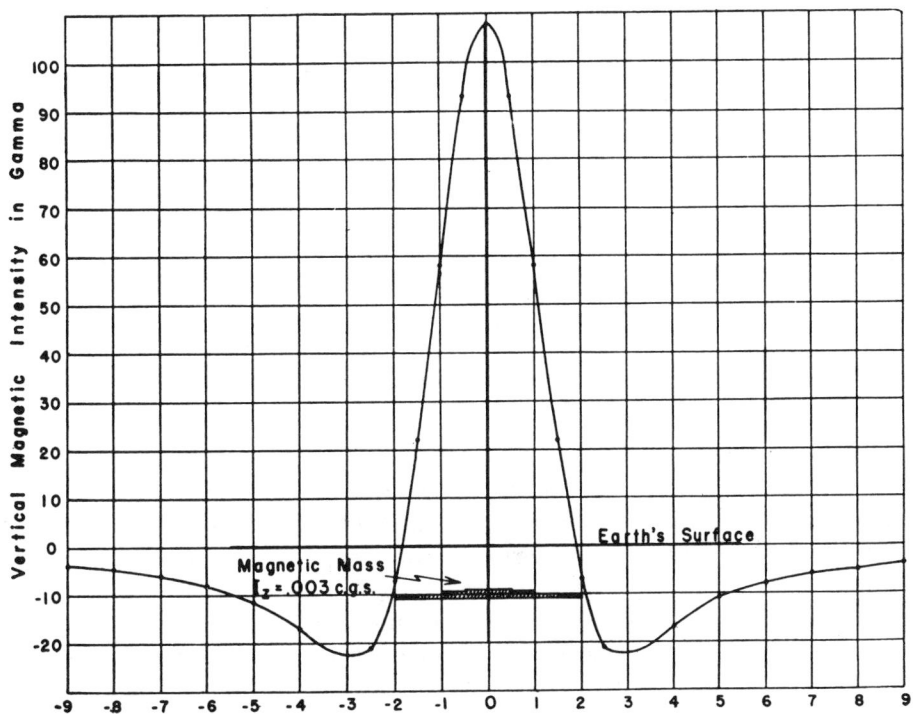

FIG. 2. A structure and the calculated magnetic anomaly produced by it.

at which calculations are made. The quantity a_0 gives the rise of the structure between two consecutive points as this quantity represents $f'(x)dx$. Figure 2 shows a structure and the calculated magnetic anomaly which it produces. Figure 3 shows the structure as calculated by using the coefficients of Table I and this magnetic curve.

We next proceed to show how, given the magnetic intensity on the surface of the earth or on a plane above the earth, the intensity can be projected upward to

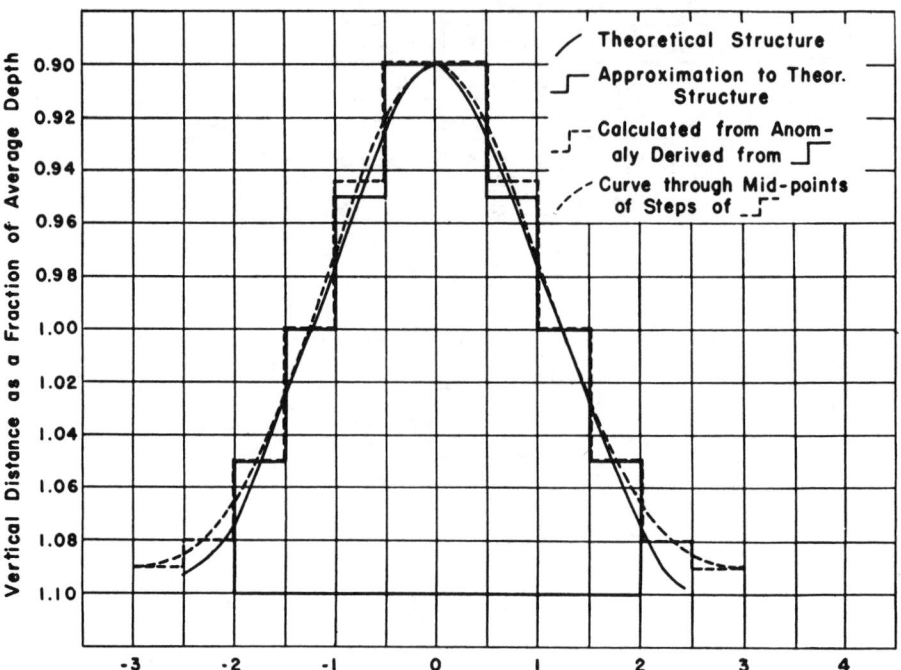

FIG. 3. Structure as calculated by using the coefficients of Table I and the magnetic curve.

higher levels or projected downward to lower levels approaching or coincident with the rocks producing the field. The projection of the field upward is old. The problem of the analytical continuation of the field towards its source was solved at the Gulf Laboratories in late 1930 and early 1931. It is a valuable tool in the interpretation of all geophysical data which have a potential field distribution. We discuss first the problem of continuation upward.

Let the vertical magnetic intensity be known on any plane $z = 0$. This plane may be the surface of the earth or a plane above the surface of the earth. Let z be positive downward in a direction towards the source of the field. Then the continuation upward away from the source of the field is given by the well-known equation:

$$H_z(x, y, z) = \frac{-z}{2\pi} \int_{-\infty}^{\infty} \int_{-\infty}^{\infty} \frac{H(\alpha, \beta, 0)d\alpha d\beta}{[(x-\alpha)^2 + (y-\beta)^2 + z^2]^{3/2}} ; \quad z \gtrless 0. \tag{6}$$

Since the measured field is not known as a mathematical function, we wish to devise a practical scheme for the numerical evaluation of the integral. We proceed as follows:

Let

$$\alpha - x = r\cos\theta; \quad \beta - y = r\sin\theta$$

$$\overline{H}_z(r) = \frac{1}{2\pi} \int_0^{2\pi} H_z(r, \theta)d\theta$$

$$H_z(x, y, z) = -\int_0^{\infty} \frac{\overline{H}_z(r)zrdr}{(z^2 + r^2)^{3/2}} \tag{7}$$

where $\overline{H}_z(r)$ is the average value of $H_z(\alpha, \beta, 0)$ around a circle of radius r centered at $(x, y, 0)$.

Let the height of the continuation upward be $z = -h$. Then Equation (7) becomes:

$$H_z(x, y, h) = \int_0^{\infty} \frac{\overline{H}_z(r)hrdr}{(h^2 + r^2)^{3/2}} . \tag{7a}$$

The integral is replaced by the sum

$$H_z(x, y, h) = \frac{\overline{H}(0) + \overline{H}(b_1)}{2} \int_0^{b_1} \frac{hrdr}{(h^2 + r^2)^{3/2}}$$

$$+ \frac{\overline{H}(b_1) + \overline{H}(b_2)}{2} \int_{b_1}^{b_2} \frac{hrdr}{(h^2 + r^2)^{3/2}} + \cdots \text{ etc.} \tag{7b}$$

Upon carrying out the integration and collecting terms, this becomes:

$$\overline{H}_z(x, y, h) = \frac{\overline{H}(0)}{2}\left[1 - \frac{h}{(h^2 + b_1^2)^{1/2}}\right] + \frac{\overline{H}(b_1)}{2}\left[1 - \frac{h}{(h^2 + b_2^2)^{1/2}}\right]$$

$$+ \frac{\overline{H}(b_2)h}{2}\left[\frac{1}{(h^2 + b_1^2)^{1/2}} - \frac{1}{(h^2 + b_3^2)^{1/2}}\right]$$

$$+ \frac{\overline{H}(b_3)h}{2}\left[\frac{1}{(h^2 + b_2^2)^{1/2}} - \frac{1}{(h^2 + b_4^2)^{1/2}}\right] + \cdots \text{ etc.} \tag{8}$$

Magnetic data are generally presented in the form of a map with readings on some sort of a grid. Let the grid of Figure 4 represent such a map with the vertical intensity values known at every corner. The continuation upward is to be calculated at the point "o." It is convenient to use the grid spacing as the unit of measure. In this measure draw a circle of radius unity around the point o. Four magnetic values fall on this circle, and the average value around the circle of $\overline{H}_z(b_1)$ is the sum of these four values divided by four. If the intensity is to be calculated one unit above the plane $z=0$, then $h=1$, $b_1=1$ and the term multiplying $\overline{H}_z(0)$ is $\frac{1}{2}[1-(1/\sqrt{2})]=0.1464$. Next draw a circle with radius $\sqrt{2}$ about o. Four values lie on this circle and $\overline{H}(b_2)$ is one-fourth the sum of these four values and $b_2=\sqrt{2}$. The term multiplying $\overline{H}(b_1)$ is $\frac{1}{2}[1-(1/\sqrt{3})]=0.2113$. The value of b_3 is taken as $\sqrt{5}$. Referring again to Figure 4, the circle of radius b_3 passes

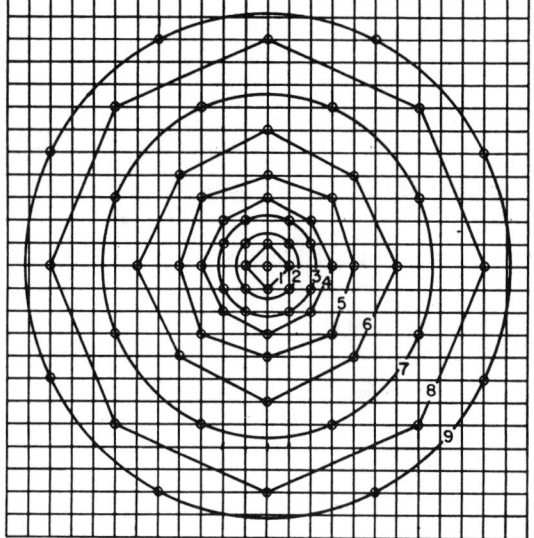

Circle No.	Average Radius
0	0
1	1
2	$\sqrt{2}$
3	$\sqrt{5}$
4	$\sqrt{8.5}$
5	$\sqrt{17}$
6	$\sqrt{34}$
7	$\sqrt{58}$
8	$\sqrt{99}$
9	$\sqrt{125}$

FIG. 4. Chart for evaluating $\overline{H}(r)$.

through the eight points on the corners of the octagon and the values at these eight points are used to obtain the average value of $\overline{H}(b_3)$ on the circle for which $b_3=\sqrt{5}$. The grid of Figure 4 shows the octagons connecting points whose values are used in obtaining the average values of the intensity on successive circles. Some of the points used do not fall exactly on circles. In those cases, the average distance to the points on the octagon is used as the radius.

Table II gives two sets of coefficients to be used with this chart. One set is for continuation upward a distance of one grid spacing (two of these coefficients were calculated above) and the other for continuation up a distance of two grid spacings. In practice, an opaque card is placed on the map from which calculations are being made. The card has openings exposing the numbers which fall at the points used in the calculations. The computer runs off the sums for each circle on an adding machine. The coefficients are divided by a number equal to the number of points on the associated octagon so the sums can be used directly to multiply these divided coefficients. The coefficients for continuation upward should add to unity. The coefficient for the largest circle is adjusted in the sets given to meet this requirement.

The grid and coefficient system of evaluating the surface integral for continuation upward has been discussed here because the same method can be used for making many of the calculations presented in this paper.

In the above discussion, $\overline{H}(r)$ was considered a constant over each interval of integration. More accurate coefficients may be obtained by interpolation formulae for $\overline{H}(r)$. For instance, $\overline{H}(r)$ might be considered linear between b_i and

Table II

Coefficients for Continuation Upwards

$$H_0(-h) = \sum \overline{H}(n) b_n$$

Circle No.	Average Radius	b_n for height h = 1 spacing	b_n for height h = 2 spacings
0	0	.1464	.0528
1	1	.2113	.0918
2	$\sqrt{2}$	.1494	.1139
3	$\sqrt{5}$	.1264	.1153
4	$\sqrt{85}$	.0862	.1151
5	$\sqrt{17}$	.0778	.1207
6	$\sqrt{34}$	.0528	.0902
7	$\sqrt{58}$	.0346	.0637
8	$\sqrt{99}$	.0206	.0400
9	$\sqrt{125}$	.0949	.1965

Coefficient for 9th circle is obtained by subtracting the sum of the first nine coefficients from unity.

b_{i+1} or may be represented by a second order equation over each of these intervals.

CONTINUATION DOWNWARD

The anomalies in the magnetic field of the earth may be considered as arising from two causes, lateral changes in the magnetic polarization of the basement rocks and the topography or structure of the basement surface. It is topography of the basement surface and its depth that is of importance in oil exploration. Before structure calculations can be made, the field due to changes in polarization must be removed. When the basement rocks are buried one mile or more, the anomalies due to oil structure are of the order of 0 to 150 gamma, with most of them lying below 50 gamma. At the same depth of burial, polarization anomalies may have a relief of several thousand gamma with the majority being of the order of several hundred gamma or lower. There is no unique criterion for removing the field due to polarization changes. However, if we could calculate the field at a level near or on the basement surface, we would increase the relief of the magnetic field over sharp structures greatly while increasing the relief over broad polarization changes but little. Anomalies would also be separated more clearly and their individual study made easier. We therefore proceed to give the solution of the problem of analytic continuation downward. In addition to enabling us to calculate the field near the basement surface from measurements made at the surface, it is also the key to the complete solution of the inverse potential problem. That is, it enables us to calculate directly from the field observed at the surface a distribution of magnetically active rocks which will account for this field.

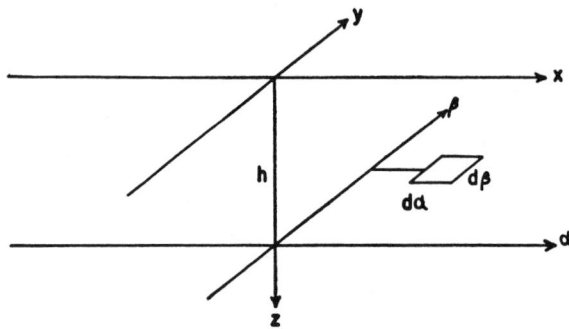

FIG. 5. Continuation downward.

The problem of analytical continuation downward towards the source of the potential field has been solved in a number of ways. In Figure 5 let the vertical component of the magnetic intensity on the plane $z = h$ be represented by $F(\alpha, \beta)$. To calculate the field at planes above $z = h$, this field can be considered as replaced by a vertically polarized plane in which the polarization is $H(\alpha, \beta)$ divided by 2π. We therefore write

$$H(x, y) = \frac{1}{2\pi} \int_{-\infty}^{\infty} \int_{-\infty}^{\infty} \frac{hF(\alpha, \beta)d\alpha d\beta}{[(x - \alpha)^2 + (y - \beta)^2 + h^2]^{3/2}}. \tag{9}$$

This equation is identical with Equation (6) for continuation upward. It may be considered as an integral equation for determining $F(x, y, h)$ given $H(x, y, 0)$. This integral equation was solved at the Gulf Laboratories in the Fall of 1930 by an application of Fourier's Integrals and by a method of successive approximations. It can also be solved by replacing the integral equation by an associated set of linear algebraic equations.

In January, 1931 the problem of analytical continuation was solved at the Gulf Laboratories by the use of Taylor's Expansion. Since this is by far the simplest method of solving the problem and the solution is readily amenable to numerical calculations from the type of data obtained in gravity or magnetic surveys, it is the method which will be discussed here.

Since the vertical magnetic intensity is analytic in the space above magnetic rocks, which we will consider to be the igneous basement rocks, this intensity may be expanded in terms of the intensity on the surface of the earth in the form

$$H(x, y, h) = H(x, y, 0) + hH^I(x, y, 0) + \frac{h^2}{2!} H^{II}(x, y, 0)$$

$$+ \frac{h^3}{3!} H^{III}(x, y, 0) + \cdots \text{etc.} \qquad (10)$$

where $H^I(x, y, 0)$ represents

$$\frac{\partial H}{\partial z}(x, y, z)$$

taken at $z = 0$.

There are several ways of calculating the derivatives in Equation (10). Since the calculation of even derivatives is more direct than the calculation of odd derivatives, we proceed as follows to obtain an expansion in terms of even derivatives. The continuation upward to a height $z = -h$ can be written

$$H(x, y, -h) = H(x, y, 0) - hH^I(x, y, 0)$$

$$+ \frac{h^2}{2!} H^{II}(x, y, 0) - \frac{h^3}{3!} H^{III}(x, y, 0)$$

$$+ \frac{h^4}{4!} H^{IV}(x, y, 0) + \cdots \text{etc.} \qquad (11)$$

Upon adding Equations (10) and (11) and solving for $H(x, y, h)$, we obtain:

$$H(x, y, h) = -H(x, y, -h) + 2\left[H(x, y, 0) + \frac{h^2}{2!} H^{II}(x, y, 0) \right.$$

$$+ \frac{h^4}{4!} H^{IV}(x, y, 0) + \frac{h^6}{6!} H^{VI}(x, y, 0) + \cdots \left. \text{etc.} \right]. \qquad (12)$$

The term $H(x, y, -h)$ is the continuation upward to a height h and its calculation has been discussed above. We therefore proceed to develop methods of calculating the series in brackets, which we will designate as $F(x, y, h)$ so that

$$H(x, y, h) = 2F(x, y, h) - H(x, y, -h). \qquad (13)$$

From Laplace's equation

$$\frac{\partial^2 H}{\partial z^2} = -\left[\frac{\partial^2 H}{\partial x^2} + \frac{\partial^2 H}{\partial y^2} \right]. \qquad (14)$$

Since $\partial^2 H/\partial z^2$ is also a potential function it also satisfies Laplace's equation and we have

$$\frac{\partial^4 H}{\partial z^4} = \frac{\partial^4 H}{\partial x^4} + 2\frac{\partial^4 H}{\partial x^2 \partial y^2} + \frac{\partial^4 H}{\partial y^4} = \left(\frac{\partial^2}{\partial x^2} + \frac{\partial^2}{\partial y^2}\right)^2 H \tag{15}$$

and $F(x, y, h)$ can be written in the symbolic form:

$$F(x, y, h) = H(x, y, 0) - \frac{h^2}{2!}\left(\frac{\partial^2}{\partial x^2} + \frac{\partial^2}{\partial y^2}\right) H(x, y, 0)$$

$$+ \frac{h^4}{4!}\left(\frac{\partial^2}{\partial x^2} + \frac{\partial^2}{\partial y^2}\right)^2 H(x, y, 0)$$

$$- \frac{h^6}{6!}\left(\frac{\partial^2}{\partial x^2} + \frac{\partial^2}{\partial y^2}\right)^3 H(x, y, 0) + \cdots \text{etc.} \tag{16}$$

In order to make Equation (16) easier to handle without any sacrifice of generality, we transform to polar coordinates and calculate $F(0, 0, h)$. As before we define

$$\overline{H}(r) = \frac{1}{2\pi}\int_0^{2\pi} H(r, \theta)d\theta. \tag{17}$$

It can be shown that continuation downward depends only on the average value of the potential function being continued around circles of radius r. We may therefore write Laplace's equation in the form.

$$\frac{\partial^2 \overline{H}}{\partial z^2} = -\left(\frac{\partial^2 \overline{H}}{\partial r^2} + \frac{1}{r}\frac{\partial \overline{H}}{\partial r}\right) \tag{18}$$

and Equation (16) becomes

$$F(0, 0, h) = \left[\overline{H} - \frac{h^2}{2!}\left(\frac{\partial^2}{\partial r^2} + \frac{1}{r}\frac{\partial}{\partial r}\right)\overline{H} + \frac{h^4}{4!}\left(\frac{\partial^2}{\partial r^2} + \frac{1}{r}\frac{\partial}{\partial r}\right)^2 \overline{H}\right.$$

$$\left. - \frac{h^6}{6!}\left(\frac{\partial^2}{\partial r^2} + \frac{1}{r}\frac{\partial}{\partial r}\right)^3 \overline{H} + \cdots \text{etc.}\right]_{r=0|z=0} \tag{19}$$

where

$$\left(\frac{\partial}{\partial r^2} + \frac{1}{r}\frac{\partial}{\partial r}\right)^i$$

means that the operation is to be applied i times in succession. It can be shown, by an application of Green's Theorem, that $\overline{H}(r)$ is an even function, and it may be written in the form

$$\overline{H}(r) = b_0 + b_2 r^2 + b_4 r^4 + b_6 r^6 + \cdots \text{etc.} \tag{20}$$

If we carry out the operations indicated in Equation (19) we find

$$\left(\frac{\partial^2}{\partial r^2} + \frac{1}{r}\frac{\partial}{\partial r}\right)\overline{H} = 4b_2$$

$$\left(\frac{\partial^2}{\partial r^2} + \frac{1}{r}\frac{\partial}{\partial r}\right)^2\overline{H} = 64b_4$$

$$\left(\frac{\partial^2}{\partial r^2} + \frac{1}{r}\frac{\partial}{\partial r}\right)^3\overline{H} = 2304b_6.$$

Equation (13) for the continuation downward becomes

$$H(o, o, h) = 2\left[b_0 - 2b_2h^2 + \frac{8}{3}b_4h^4 - \frac{16}{5}b_6h^6 + \cdots \text{etc.}\right]$$
$$- H(o, o, -h). \tag{21}$$

By experience, we find that, in general, the inaccuracies in the observed intensity limit us to the use of the first three terms of the series in Equation (21), but the convergence is rapid enough so that these are fairly adequate for practical purposes.
Therefore

$$H(o, o, h) \cong 2\left[b_0 - 2b_2h^2 + \frac{8}{3}b_4h^4\right] - H(o, o, -h). \tag{22}$$

The values for b_0, b_2, and b_4 are obtained by a least squares solution of the abbreviated form of Equation (20).

$$\overline{H}(r) \cong b_0 + b_2r^2 + b_4r^4. \tag{23}$$

Using average values of H around circles of radius o, 1, $\sqrt{2}$, $\sqrt{5}$, $\sqrt{8.5}$, $\sqrt{17}$, $\sqrt{34}$, $\sqrt{58}$, $\sqrt{99}$ we find by the least squares method,

$$b_0 = 0.2471\overline{H}(o) - 0.2351\overline{H}(1) + 0.2234\overline{H}(\sqrt{2}) + 0.1874\overline{H}(\sqrt{5})$$
$$+ 0.1521\overline{H}(\sqrt{8.5}) + 0.0717\overline{H}(\sqrt{17}) - 0.0449\overline{H}(\sqrt{34})$$
$$- 0.1095\overline{H}(\sqrt{58}) + 0.0500\overline{H}(\sqrt{99}) \tag{24}$$

$$b_2 = -0.0119\overline{H}(o) - 0.0105\overline{H}(1) - 0.0091\overline{H}(\sqrt{2}) - 0.0053\overline{H}(\sqrt{5})$$
$$- 0.0011\overline{H}(\sqrt{8.5}) + 0.0077\overline{H}(\sqrt{17}) + 0.0192\overline{H}(\sqrt{34})$$
$$+ 0.0218\overline{H}(\sqrt{58}) - 0.0108\overline{H}(\sqrt{99}) \tag{25}$$

$$b_4 = 0.00010\overline{H}(o) + 0.00009\overline{H}(1) + 0.00007\overline{H}(\sqrt{2}) + 0.00004\overline{H}(\sqrt{5})$$
$$- 0\overline{H}(\sqrt{8.5}) - 0.00009\overline{H}(\sqrt{17}) - 0.00020\overline{H}(\sqrt{34})$$
$$- 0.00020\overline{H}(\sqrt{58}) + 0.00020\overline{H}(\sqrt{99}) \tag{26}$$

and these, substituted in Equation (22), give coefficients for the circles. By combination with the coefficients for continuation upward $H(o, o, -h)$ given in

Table II, coefficients for continuation downward for $h=1$ and $h=2$ are obtained. These are shown in Table III and are for use with the chart of Figure 4. The coefficients for the last circle are adjusted so that the summation of the coefficients is unity. They are used in the manner described earlier for continuation upward. Other methods, some involving the use of weighting factors for the various circles, can be used for the solution of Equation (22). The derivatives in Equation (12) can also be evaluated graphically as will be shown in the next section.

Table III

Coefficients for Continuation Downward

$$H(o,o,h) = C_0 \, H(o) + C_1 \, H(1) + C_2 \, H(\sqrt{2}) + \quad etc. = \sum_{0}^{9} CH$$

Circle No.	Radius	$h = 1$ Coeff.	$h = 2$ Coeff.
0	0	.3969	.4197
1	1	.3026	.3532
2	$\sqrt{2}$	.3356	.5460
3	$\sqrt{5}$	.2749	.4071
4	$\sqrt{8.5}$	.2234	.2668
5	$\sqrt{17}$	.0356	−.0442
6	$\sqrt{34}$	−.2194	−.3762
7	$\sqrt{58}$	−.3413	−.6236
8	$\sqrt{99}$	.1248	.3130
9	$\sqrt{125}$	−.1331	−.1395
	$\sum$	1.0000	1.0000

$$\text{Coefficient for } H(\sqrt{125}) = 1 - \sum_{n=0}^{n=8} C_n$$

Examples of continuation are shown later in this paper.

The development of digital calculating machines which will solve twelve or more simultaneous linear equations will make it a simple matter to calculate coefficients for continuation downward directly from the integral Equation (9).

DERIVATIVE CALCULATIONS

As pointed out above there are two general types of magnetic anomalies present in any magnetic survey. One of these types consists of broad anomalies of considerable magnetic relief, measured usually by hundreds of gamma, while the second type is that composed of sharper anomalies measured usually in tens or units of gamma. From normal considerations of magnetization of basement rocks, it can be shown that the larger anomalies must be caused in large part by variations of magnetization within the magnetic basement, while the smaller features may be due either to smaller magnetization contrasts within the basement or to relief of the basement surface. Obviously there is a great amount of uncertainty, particularly in intermediate types. Whatever the cause of the differentiation, it is essential that methods be made available for the separation of the smaller anomalies from the more prominent anomalies. One of these methods is continuation downward and was discussed above.

Another method which has been found to be useful for this purpose is that of calculating various derivatives and making contour maps of these derivative values. Smaller anomalies superposed on broad features are identifiable by a change of curvature along any profile, and curvature evaluations depend largely upon the second derivative. For this reason, the second derivative is particularly valuable. In some instances, where used judiciously, the fourth derivative is also valuable.

In the development of analytical continuation, the following expansion was introduced

$$\overline{H}(r) = b_0 + b_2 r^2 + b_4 r^4 \tag{23}$$

and a least squares solution gives the results shown by Equations (25) and (26). When these are substituted in the relationships

$$\frac{\partial^2 H}{\partial z^2} = -4b_2, \quad \text{and} \quad \frac{\partial^4 H}{\partial z^4} = 64b_4 \tag{23a}$$

for the particular circles used in the analytic continuation chart of Figure 4, sets of coefficients for second and fourth derivative calculations are obtained. However, it has been found unnecessary to carry the computations out so far in the r direction. With sufficient precision for nearly all purposes, a method using circles at radii $r_0 = 0$; $r_1 = 1$; $r_2 = \sqrt{2}$; $r_3 = \sqrt{5}$ and $r_4 = \sqrt{9.23}$ has been developed. The value of $\overline{H}(r)$ on the circle of radius r_4 is obtained by taking a weighted average of circles at radii $\sqrt{8}$, $\sqrt{9}$, $\sqrt{10}$, containing 4, 4, and 8 points, respectively. By a least squares solution for Equation (23) and by substitution in the relationships (23a) we obtain

$$\frac{\partial^2 H}{\partial z^2} = 1.156\overline{H}(0) + 0.256\overline{H}(1) - 0.445\overline{H}(\sqrt{2}) - 1.359\overline{H}(\sqrt{5})$$

$$+ 0.392\overline{H}(\sqrt{9.23}) \tag{27}$$

$$\frac{\partial^4 H}{\partial z^4} = 1.753\overline{H}(0) + 0.170\overline{H}(1) - 1.036\overline{H}(\sqrt{2}) - 2.384\overline{H}(\sqrt{5})$$

$$+ 1.497\overline{H}(\sqrt{9.23}). \tag{28}$$

These equations give sets of coefficients for the calculation of second and fourth derivatives.

Another method of calculating derivatives is the graphical one. In this method the operations are under close control of the operator, and it is therefore very advantageous when the accuracy of the data is subnormal. It is also very rapid. The method is based on exactly the same fundamentals as those discussed above. By means of an adding machine, the summations of magnetic values around the circles described above are obtained. Data for the construction of a curve representative of Equation (23) are thus available. However, if we make the substitution $\rho = r^2$ in Equation (23) we have

$$\overline{H}(\rho) = b_0 + b_2\rho + b_4\rho^2 \tag{29}$$

and

$$\frac{\partial \overline{H}}{\partial \rho} = b_2. \tag{30}$$

By plotting the average of the magnetic values around the various circles against the square of the radii as shown in Figure 6, the curve for Equation (29) is available, and the tangent at the origin is the measure of b_2, and $-4b_2 = \partial^2 H/\partial z^2$. By

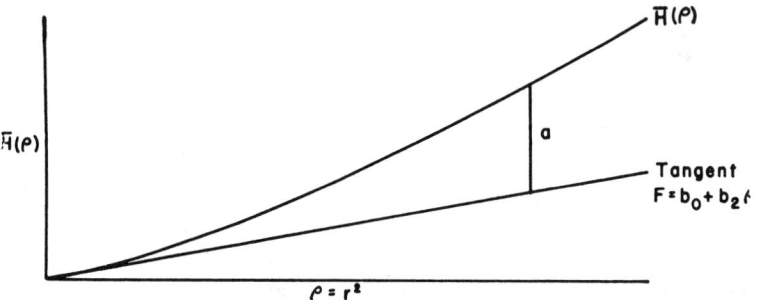

FIG. 6. Graphical computation of derivatives.

proper selection of scales, it is possible to read the derivative directly. The tangent line has the form

$$F = b_0 + b_2\rho. \tag{31}$$

The vertical measurement from any convenient point on the tangent to the curve is therefore equal to $b_4\rho^2$. This vertical distance divided by ρ^2 or r^4 is then proportional to $\partial^4 H/\partial z^4$, since $\partial^4 H/\partial z^4 = 64 b_4$, and again a careful selection of scales can permit a direct reading of the fourth derivative.

These methods, being approximations, give results which depend to a considerable extent on the spacings used. We have found it advantageous to use a standardized spacing so that comparisons from area to area are possible. The results are more qualitative than quantitative in value, but these methods have proven to be some of the better tools available for mass production analysis. It is apparent that these methods are applicable without modifications to gravity data.

DERIVATION OF THE ENTIRE FIELD FROM THE VERTICAL INTENSITY

The magnetic potential at any point in the space above the surface of the earth or on this surface can be obtained by multiplying the integral of Equation

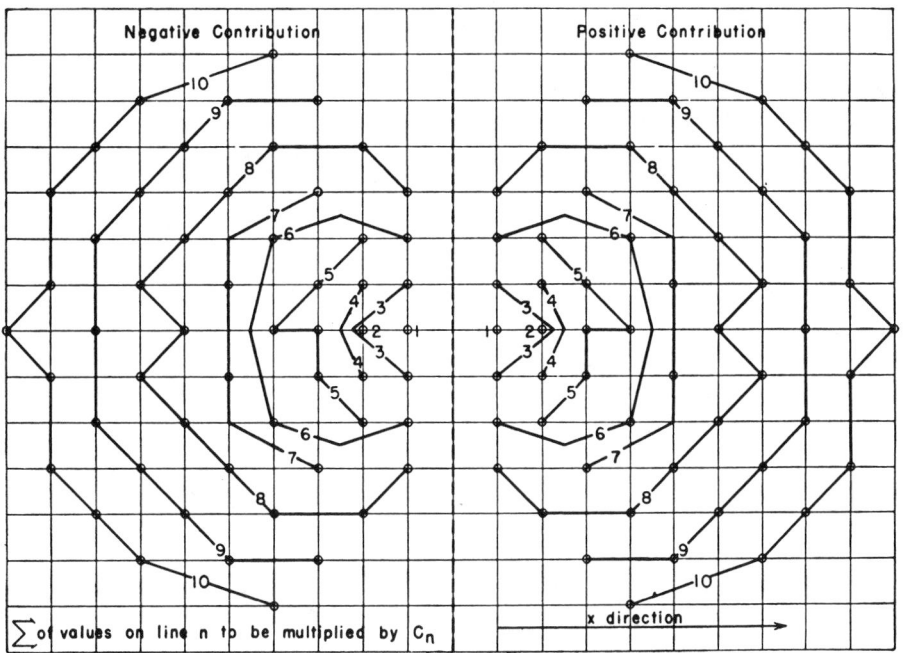

FIG. 7. Grid for calculation of H_x from H_z.

(6) by dz and integrating from $z = -\infty$ to $z = z$. If this operation is carried out we obtain:

$$\phi(x, y, z) = -\frac{1}{2\pi} \int_{-\infty}^{\infty} \int_{-\infty}^{\infty} \frac{H_z(\alpha, \beta, 0)d\alpha d\beta}{[(x-\alpha)^2 + (y-\beta)^2 + z^2]^{1/2}}. \quad (32)$$

The horizontal components of the field are then given by

$$H_x = -\frac{\partial \phi}{\partial x} = \frac{1}{2\pi} \int_{-\infty}^{\infty} \int_{-\infty}^{\infty} \frac{(\alpha - x)H_z(\alpha, \beta, 0)d\alpha d\beta}{R^3} \quad (33)$$

$$H_y = -\frac{\partial \phi}{\partial y} = \frac{1}{2\pi} \int_{-\infty}^{\infty} \int_{-\infty}^{\infty} \frac{(\beta - y)H_z(\alpha, \beta, 0)d\alpha d\beta}{R^3} \quad (34)$$

where

$$R = [(x-\alpha)^2 + (y-\beta)^2 + z^2]^{1/2}.$$

The vertical gradient of the vertical intensity can be obtained by taking the derivative of Equation (6) with respect to z. This operation yields:

$$\frac{\partial H_z}{\partial z} = -\frac{1}{2\pi} \int_{-\infty}^{\infty} \int_{-\infty}^{\infty} \left[\frac{1}{R^3} - \frac{3z^2}{R^5} \right] H_z(\alpha, \beta, 0)d\alpha d\beta. \quad (35)$$

These integrals may all be evaluated numerically by methods similar to those already discussed. Coefficients for the calculation of the horizontal intensity are given in Table IV and the grid with which they are used is given by Figure 7. The calculation of the potential will be discussed in the next section.

All of these quantities have been found useful in the study and analysis of specific anomalies. The potential has a special significance because of its relation to structure calculations which we will proceed to discuss in the next section.

Table IV

Calculation of Horizontal Intensity from Vertical Intensity

$$H_x = \frac{1}{2\pi}\sum c_n H_n$$

n	c_n
1	1.5637
2	0.2663
3	0.4400
4	0.1893
5	0.1219
6	0.0951
7	0.0812
8	0.0473
9	0.0315
10	0.0224

THE DIRECT CALCULATION OF STRUCTURE

The direct calculation of basement structure or relief may be approached in several ways. A two-dimensional solution of the problem was given at the beginning of the paper. We can proceed in a similar manner to solve the integral equation relating basement relief to observed intensity. However, in this section we will proceed in a different way.

In Figure 8 let the wavy line represent the basement surface and the horizontal line at $z = h$ represent the average depth to the basement. Further, let $f(x, y)$ be small compared to h and let the polarization be vertical. Under these conditions, it can be shown that the structure of the basement may be replaced by a double layer of strength $I_z f(x, y)$. The magnetic potential of this double layer

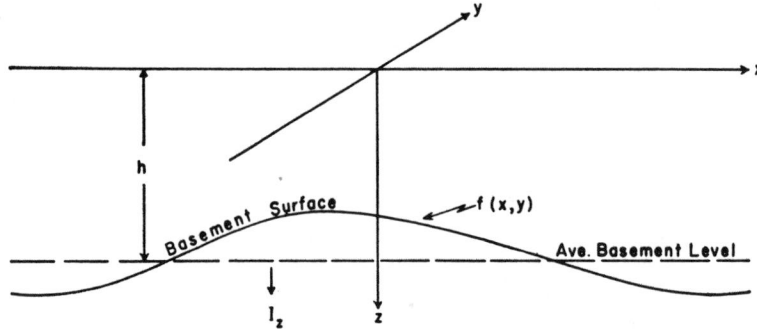

Fig. 8. Method for direct calculation of basement structure or relief.

is given by the equation:

$$\phi(x, y, z) = - I_z \int_{-\infty}^{\infty} \int_{-\infty}^{\infty} \frac{f(\alpha, \beta)(h - z)d\alpha d\beta}{[(x - \alpha)^2 + (y - \beta)^2 + (z - h)^2]^{3/2}} \cdot \quad (36)$$

As z approaches h, the integral approaches $2\pi f(x, y)$ and we have

$$\phi(x, y, h) = - 2\pi I_z f(x, y) \quad (37)$$

or

$$f(x, y) = \frac{-1}{2\pi I_z} \phi(x, y, h). \quad (38)$$

Basement relief may thus be calculated either by continuing to the basement level and calculating the potential by the use of Equation (32) using the values

Table X

Coefficients for Calculation of Structure

$$f(o) = \frac{d}{4\pi^2 I_z \times 10^8} \sum c_n H_n$$

Where f is basement relief above mean plane,

d is grid spacing in same units as f.

n	c_n
0	3.670
1	0.985
2	0.910
3	0.427
4	0.382
5	0.267

of H_z at the basement level or by calculating the potential from H_z on the surface where measurements are made and continuing this potential to the basement level.

Before structure calculations can be made, it is necessary to remove the magnetic variations due to polarization changes in the basement rocks. The portion of the field due to these changes can be estimated more easily from the magnetic intensity at or near the basement. Therefore, in general, structure calculations are made using the magnetic intensity at basement level. If one remembers that $f(x, y)$ is being calculated on the same plane on which the magnetic intensity is known, Equation (32) and (38) combine to give

$$f(x, y) = \frac{1}{4\pi^2 I_z} \int_{-\infty}^{\infty} \int_{-\infty}^{\infty} \frac{H_z(\alpha, \beta, h)}{[(x - \alpha)^2 + (y - \beta)^2]^{1/2}} \, d\alpha d\beta. \tag{39}$$

Table V gives a set of coefficients for the evaluation of this integral and the grid with which they are to be used is shown in Figure 9. These coefficients can also be used with proper modification of multiplying factor to calculate the

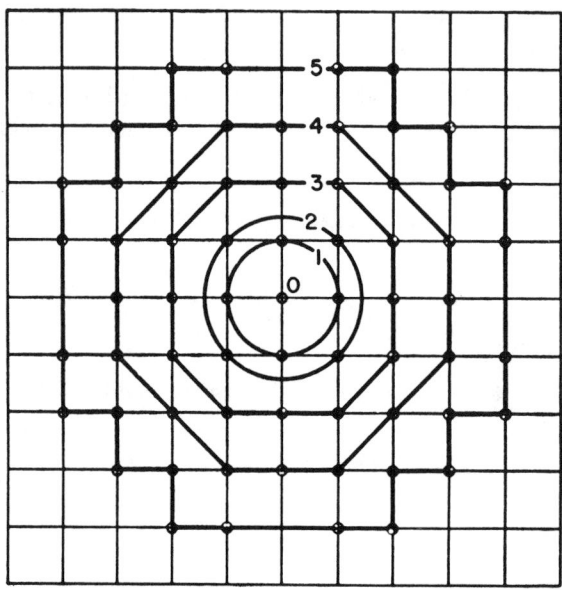

F$_{IG}$. 9. Grid for calculation of structure using coefficients of Table V.

potential from the vertical intensity on any plane where this intensity is known.

Long, uniform regional slopes contribute practically nothing to the magnetic anomaly field. The structure calculated as outlined above will therefore have to have the regional slope added to it. This regional can be obtained from well information or from magnetic depth estimates. The method of making these estimates is the subject of the next section.

The above theory can be modified to handle cases in which the polarization is not vertical.

ESTIMATION OF DEPTHS TO MAGNETIC BASEMENT

This subject is probably one of the most important phases of magnetic interpretation. The results of a series of depth estimates supply information concerning the thickness of the sedimentary section, and it is therefore possible to delineate the configuration of a sedimentary basin at low cost and with reasonably high accuracy.

The magnetic potential due to a magnetized body can be written as

$$\phi = \int_{-\infty}^{\infty}\int_{-\infty}^{\infty}\int_{0}^{z}\left[I_x\,\frac{\partial}{\partial\alpha} + I_y\,\frac{\partial}{\partial\beta} + I_z\,\frac{\partial}{\partial\gamma}\right]$$

$$\cdot\frac{d\alpha d\beta d\gamma}{\{(x-\alpha)^2 + (y-\beta)^2 + (z-\gamma)^2\}^{1/2}} \qquad (40)$$

and

$$H_z = -\frac{\partial\phi}{\partial z}.$$

If it can be assumed that no structure is present (depth constant), then $(z-\gamma)^2 = z^2 = h^2$ for a particular anomaly. If I_x, I_y, and I_z are constants for an anomalous mass, then the integral is a function of the geometric dimensions, except for multiplicative constants, and the solution for depth is unique. However,

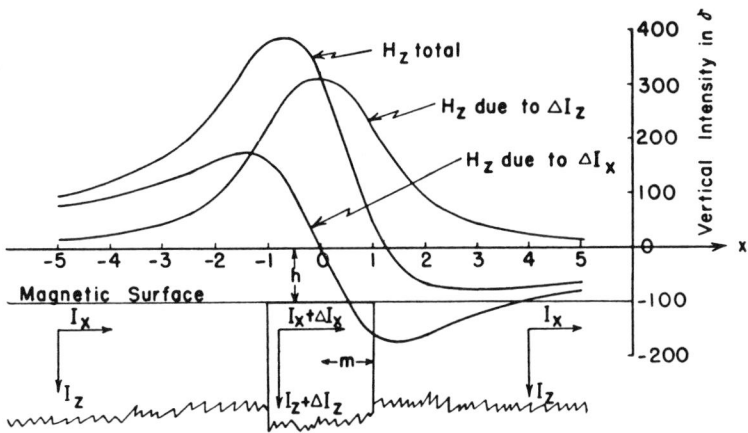

Fig. 10. Method for estimation of depth to magnetic basement.

if I_x, I_y, and I_z are unknown variables in an anomalous mass, it is apparent that there is a complete lack of uniqueness in the solution. A certain degree of uniqueness can be introduced, however, by means of the making of reasonable assumptions. For example, if it be assumed that I_x, I_y, and I_z remain constant over the extent of the geometric mass, and that the magnetization of the surrounding mass is different but uniform, a unique solution can be obtained. In our experience, this form of assumption is applicable in a large number of cases. In Figure 10, the mass is supposed to extend to infinity vertically and in the $\pm y$ directions. The contacts between the masses are vertical. Then

$$H_z = 2\Delta I_z \left[\tan^{-1} \frac{x+m}{h} - \tan^{-1} \frac{x-m}{h} \right]$$

$$+ \Delta I_x \log_e \frac{(x+m)^2 + h^2}{(x-m)^2 + h^2} + \text{const.} \qquad (41)$$

In many cases, particularly in high latitudes, ΔI_x can be taken as zero. This condition can be recognized readily, since when $\Delta I_x = 0$ the anomaly is symmetrical. In either case, however, the solution for h can be obtained in an indirect manner by a series of successive approximations, in each case recalculating the anomaly to determine the accuracy of the approximation. It will be found that it is necessary to match the observed curve to within 1γ or 2γ before a sufficiently accurate depth is obtained. When $0 \geqq (m/h) \geqq 0.5$ the anomaly is very insensitive to variations in h. Better results are obtainable when $0.5 \geqq (m/h) \geqq 1.1$ and excellent results are obtainable when $(m/h) > 1.1$. Usually when suitable anomalies are available, depths accurate to about $\pm 10\%$ are obtained.

We have found that the assumptions described above are applicable to anomalies of elliptical form, where the major axis of the ellipse is at least three times the magnitude of the minor axis. The profile on which the calculation is made is always taken perpendicular to the major axis and through the apex of the anomaly.

Other forms of anomalies, such as those for right circular cylinders, masses with sloping contacts, etc., are used in a similar manner, but many of these lead to less reliable results, due to the presence of additional unknown parameters. Difficulties encountered in this method of solution are that in many instances the anomaly is disturbed by neighboring masses. Therefore, isolated anomalies should be chosen.

Several short-cut methods for the estimation of depths have been published, mostly for use in the estimation of depths to ore bodies. They are usually based on a set of simple assumptions. For the anomaly derived from a single pole, the depth to the pole is 1.3 times the horizontal distance from the apex of the anomaly to the point where the anomaly has half of its maximum value. For a sphere, the depth to the center is twice this horizontal distance. All of these methods involving half maximum values have a weakness in that it is usually difficult to obtain the value of the anomaly remote from the mass due to the presence of other anomalies, and, therefore, it is not possible to determine the half maximum value with sufficient precision.

A better rule of thumb, for use with the anomaly described by Equation (41) when $\Delta I_x = 0$ is as follows: draw in the maximum slope (at the point of inflexion); draw a line whose slope is one-half this maximum slope; draw two lines parallel to this line, tangent to the anomaly curve. The horizontal distance D between the points of tangency is approximately equal to $1.6h$. Actually, when $m/h = 0$, $D = 1.2h$ and when $m/h = \infty$, $D = 2.0h$. For intermediate values of m/h, $D = 1.6h$

gives a fair estimate. This method can be applied to magnetic contour maps. On the flanks of each elongated anomaly, a band of fairly uniformly spaced contours will be found. The width of this uniform band is a measure of D. Great care must be used in the selection of anomalies for these calculations. For quick interpretations, this technique has been adequate.

Direct methods of depth estimation also can be made. These methods are based on analytic continuation. They are limited largely by inaccuracies in the observed data and, of course, by the lack of uniqueness that applies to all potential function data. The continuation of the vertical intensity to any depth gives to within a constant multiplier a distribution of magnetic material which will account for the observed field. As the continuation is carried deeper and deeper, a point is reached at which the simple, sharp distributions are inadequate to reproduce the observed field, and the continued field begins to oscillate violently. This oscillation is indicative that the maximum depth has been passed at which a simple distribution of magnetic material will produce the observed field. Broad anomalies, which probably originate from within the basement complex, can be continued further than the sharper ones. As in the case of the indirect method of making depth estimations, we have found that a proper choice of anomalies leads to surprisingly accurate depth calculations.

One way of systematizing depth calculations is by the use of a technique called the "error curve" method.

If continuation is carried downward to a depth h and then back up again by the methods discussed above, fairly accurate results are obtained on theoretical data such as the arc tangent until the depth of burial is approached. Beyond this depth the error increases rapidly. To apply the method the following procedure is followed.

Let H_a be the recalculated surface intensity from continuation downward at depth h.

Then

$$H_a = \sum C_n \overline{H}(b_n h). \qquad (42)$$

where $\overline{H}(r)$ is the average value of the observed magnetic intensity around a circle of radius r, and b_n is a constant.

By a combination of methods described earlier, the values of Table VI have been obtained.

On a reasonably strong anomaly, make an approximate depth estimate by one of the short-cut methods. The accuracy of this estimate need not be better than about 50%. Select the point of greatest curvature on the anomaly, by inspection. No great accuracy is required here. Using this point as a center, describe circles of radius $b_n h$ where h is the approximate depth estimate. Multiply the average values around these circles by the appropriate value of C_n, and sum the results. Repeat the process, replacing h by $h/2$, $3h/2$, and $2h$ in order. A curve is then plotted, E against h, as in Figure 11. The particular curve in Figure 11 was that

Table VI

Coefficients for the Calculation of Error in Continuation

for Use in Estimation of Depths

$$E = H_0 - H_e = H_0 - \sum C_n \bar{H}(b_n h)$$

b_n	C_n
0.1	0.3001
0.3	0.5102
0.5	0.2622
0.7	0.0311
0.9	-0.0378
1.1	-0.0348
1.3	-0.0190
1.5	-0.0087
1.7	-0.0035
1.9	-0.0011
2.25	0.0000
2.75	0.0007
3.25	0.0003
3.75	0.0003

obtained over a theoretical polarization contrast anomaly, in which case the true value of h is known. The sharp bend point would be selected as the depth of burial in a practical solution.

All of these methods must be used with great care, and isolated anomalies should be chosen, since in every case the effect of neighboring anomalies is detrimental to accuracy.

Some other useful tools have been developed. Where a magnetization contact lies at one depth of burial or farther from the center of an anomaly, the second

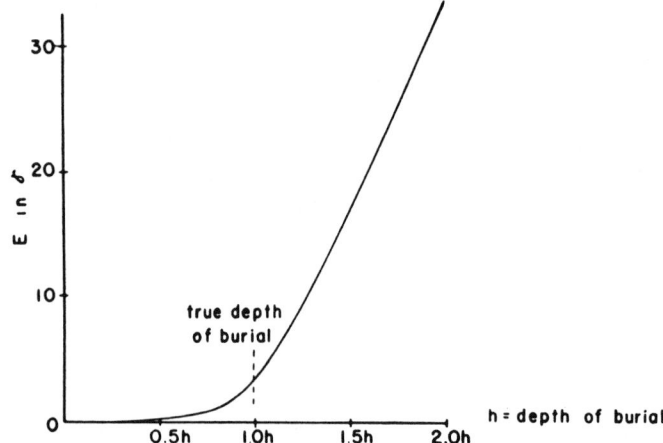

FIG. 11. Error curve for an abrupt polarization change obtained by means of coefficients of Table VI.

derivative in the vertical direction is approximately equal to zero at the contact. This gives a measure of one of the parameters, and is therefore useful in making first approximations to the dimensions of a disturbing mass. First and second derivatives can be taken graphically in the horizontal direction by measurements of slopes, and these are often useful in depth estimate work, since in some instances the horizontal derivatives of functions to be matched are more amenable to solution than the functions themselves.

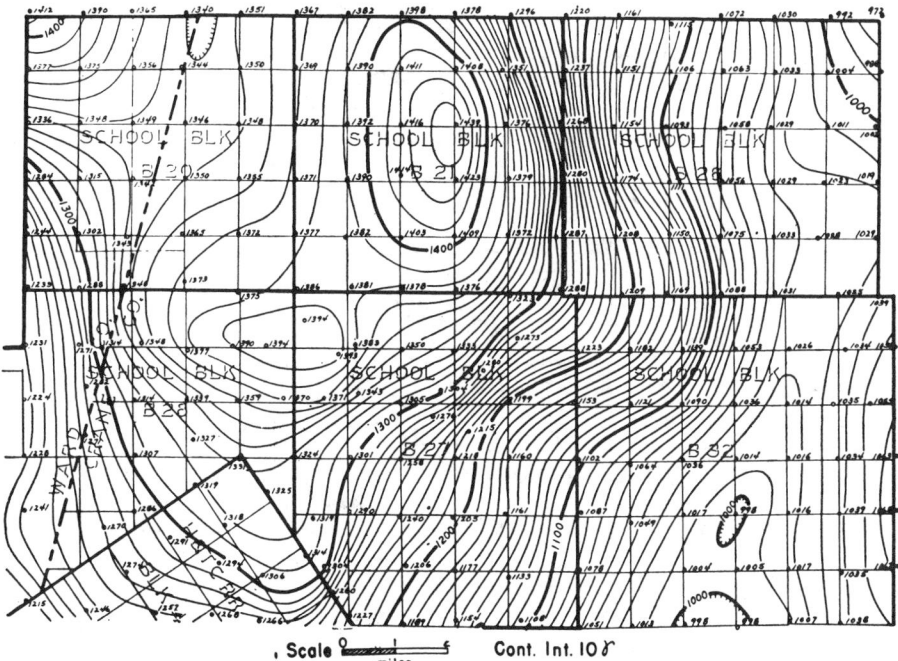

Fig. 12. Observed vertical magnetic intensity portion of Crane and Ward Counties, Texas.

In order to illustrate the practical applications of the theory developed above, a series of six maps is presented. Figure 12 is a map presenting the observed magnetic intensity in a portion of Ward and Crane Counties, Texas, covering the general Sand Hills Area. The observations were made on the corners of a mile grid, and they have been corrected for the normal northward increase of the earth's field.

Figure 13 shows the second vertical derivatives of magnetic intensity. Since these derivatives are a measure of the changes of magnetic gradients, the derivative anomalies should coincide with the nonlinear parts of the observed intensity map. The wells shown on this map and others to follow are those which reached the Ellenburger Limestone, in Lower Ordovician. In almost every instance, it is noticeable that the Ellenburger production is on the flank of one of these anomalies. The Sand Hills Field is the outstanding example. The whole anomaly correlates closely with the outline of Permian production and the Ellenburger production is on the north flank. This may mean that the Ellenburger Lime is thin or absent over the top of this anomaly. The same applies quite well to the other anomalies. This second derivative map alone would have been quite sufficient to localize these structures and to guide the seismograph with a high degree of precision.

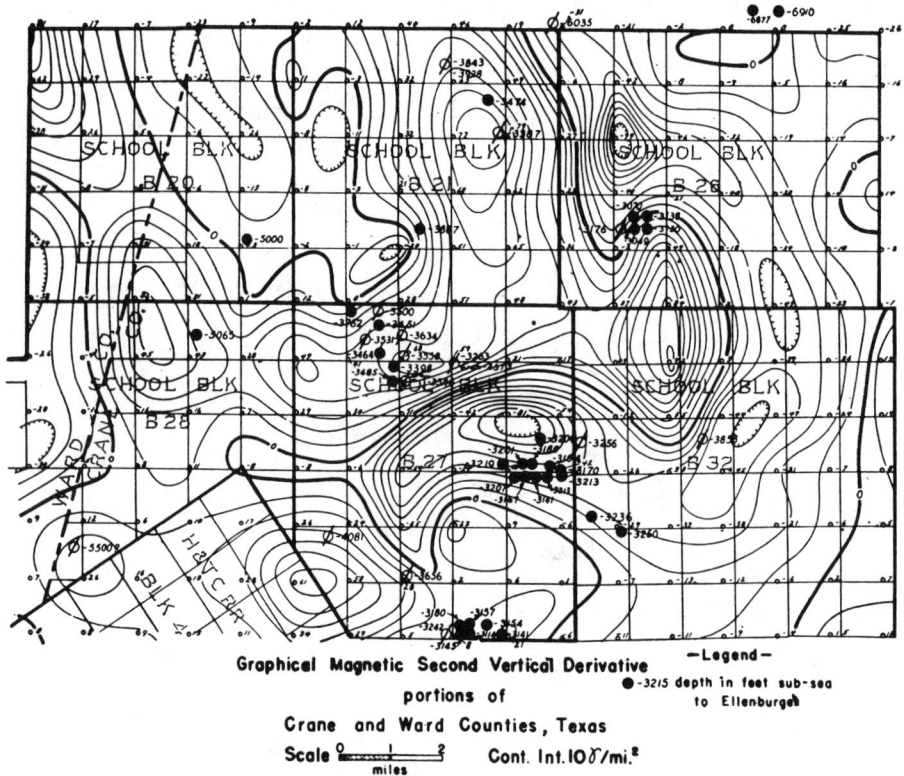

Graphical Magnetic Second Vertical Derivative

portions of

Crane and Ward Counties, Texas

Scale 0 1 2 miles   Cont. Int. 10 δ/mi.²

—Legend—
● -3215 depth in feet sub-sea to Ellenburger

FIG. 13. Graphical magnetic second vertical derivative portions of Crane and Ward Counties, Texas.

Figure 14 is the fourth derivative map. Fourth derivatives tend to oscillate rather widely unless excellent data are available, and they should be used with considerable discretion. In this area, however, the data are good and the fourth derivative calculations are justified. Observe the increased detail on the anomalies and the fault indications, particularly on the east flank of the Sand Hills structure and on the east flank of the McKnight Field. The flanking lows are merely parts of the derivative anomalies caused by the fault. Note that the production on the "Sand Hills Ordovician Field" at the center of the map is definitely on the west flank of the feature. Faulting is suggested on the north and south flanks. The production on the anomaly at the bottom of the map is on the west flank of the feature and faulting is indicated. Derivative maps are normally used in a qualitative manner. If it is deemed advisable or necessary to calculate the structural relief, another technique is necessary.

Figure 15 shows the vertical magnetic intensity on the basement surface, calculated from the observed intensity on the earth's surface. This is the "Analytic Continuation Downward." The appearance of this map is quite similar to that of the derivative maps. However, these values are in units of intensity, and the

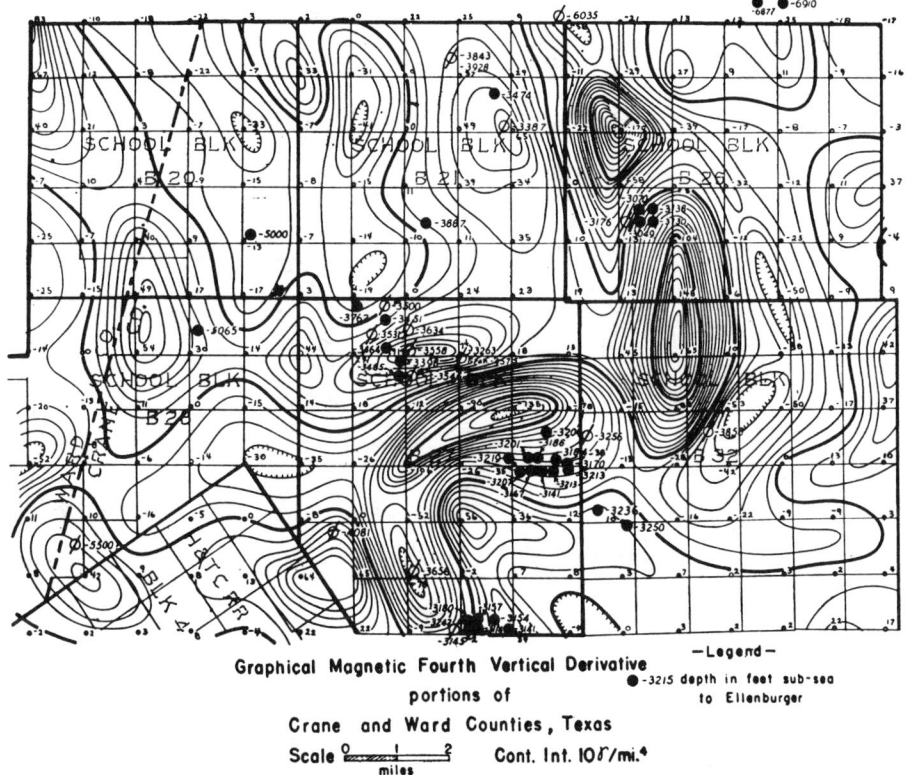

FIG. 14. Graphical magnetic fourth vertical derivative portions of Crane and Ward Counties, Texas.

structural relief can be derived from these values. The details of the anomalies are not quite so apparent as they are on the derivative maps, due to the presence of a broad regional effect. As mentioned earlier, part of the magnetic relief is derived from the more highly magnetized rocks underlying the Basin Platform.

The light contours on the map of Figure 16 show the magnetic relief which, in our judgment, is caused by the more highly magnetized rocks underlying the platform. It is a matter of interpretation and not of calculation, of course, and it is probable that other equally valid assumptions could be made. The sharp anomalies are obtained by subtracting this broad regional from the values shown on the previous map, and the anomalies represent the magnetic intensity at the average basement level caused by structure deformations.

LEO J. PETERS

In Figure 17 the values and contours represent the calculated structure. Since broad regional structures give only a few gamma relief, and since uniform slopes give no magnetic relief, this picture presents only the sharp undulations of the basement. Since the spacing of control points is not much smaller than the depth to the basement rocks, there is considerable flattening over what we conceive the actual structure to be. The true picture would probably show much steeper

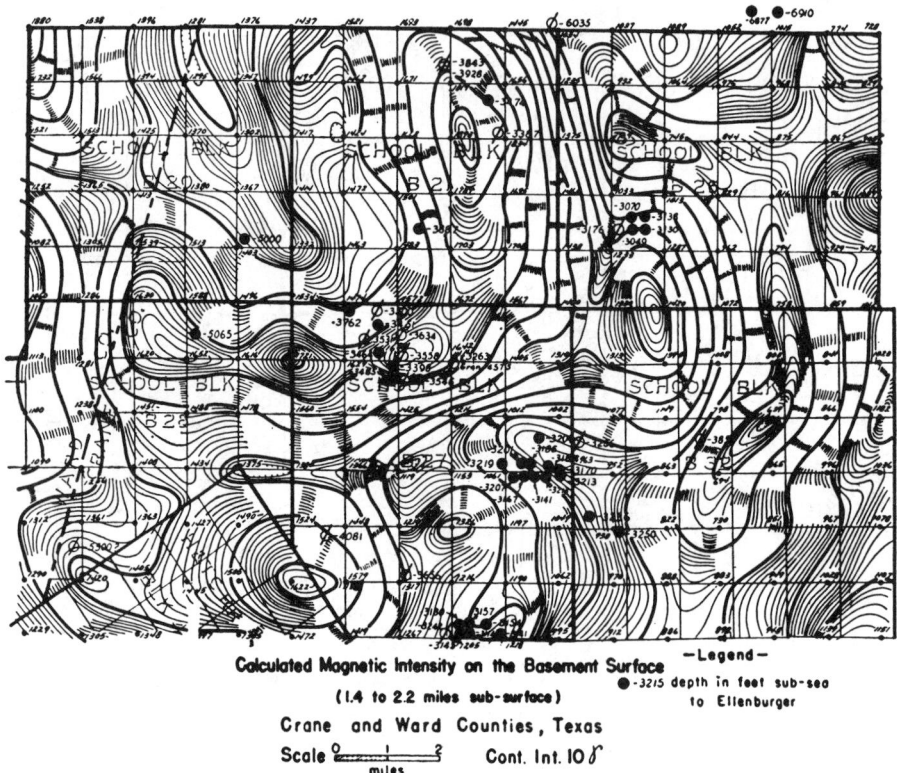

Calculated Magnetic Intensity on the Basement Surface
(1.4 to 2.2 miles sub-surface)
Crane and Ward Counties, Texas
Scale |———|———| Cont. Int. 10 γ
miles

—Legend—
● -3215 depth in feet sub-sea
to Ellenburger

FIG. 15. Calculated magnetic intensity on the basement surface
(1.4 to 2.2 miles sub-surface).

slopes or faults in many places, such as on the east flank of the Sand Hills Field and on the east flank of the structure in Block 21. The polarization used in making the structure calculations was about 0.001 c.g.s. units.

In Figure 18 the regional contours were derived largely from well information and represent that part of the relief which the magnetic method cannot show, since it involves gentler, more uniform slopes. The contour interval is 500 feet. By the addition of this regional to the values of the previous map, we obtain the picture shown by the sharper contouring. This is the final phase of the routine interpretation and represents the calculated relief of the basement. The datum is in feet subsea and the contour interval is 100 feet. The picture could be improved by making the flanks steeper, and by sketching in faults where indicated by the derivative and continuation maps.

Our seismograph work in this area shows excellent correlation with the magnetic results, and it has been found possible to use the magnetic information as an aid in determining which of two possible seismograph interpretations should be used. Here and elsewhere in this vicinity, the seismograph and magnetic fault indications correspond almost exactly.

We have shown no examples of depth estimates in this paper. They are, however, one of the most valuable uses of magnetic data and would in themselves justify the surveying of new areas even if no other information could be derived from the survey. A careful selection of features on which to make depth estimates

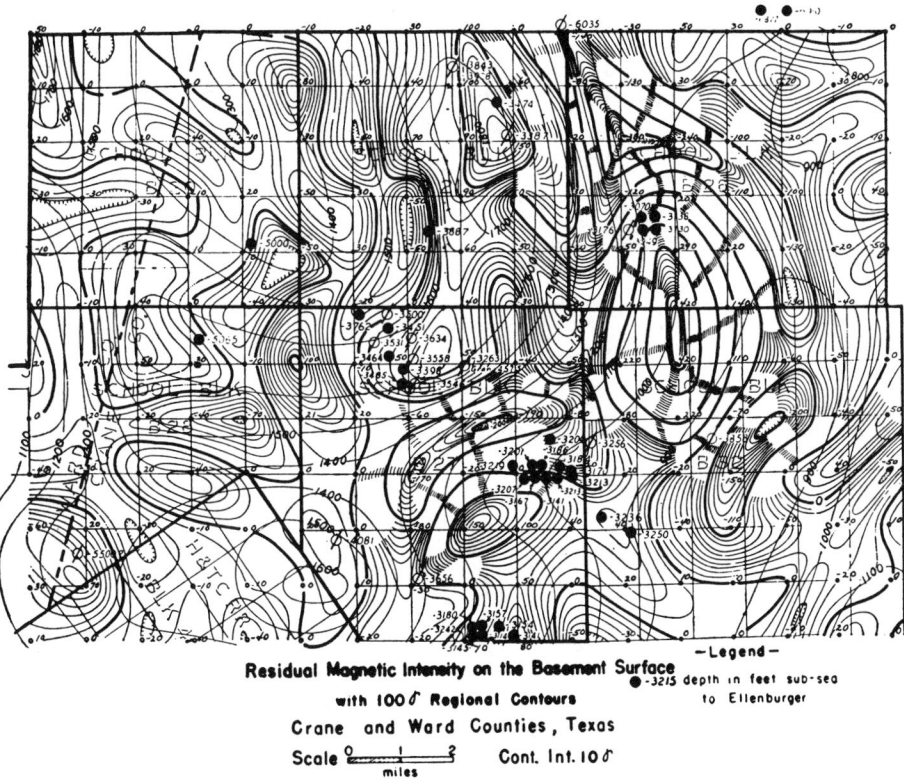

Residual Magnetic Intensity on the Basement Surface

with 100 δ Regional Contours

Crane and Ward Counties, Texas

Scale [====] miles   Cont. Int. 10 δ

—Legend—
●-3215 depth in feet sub-sea to Ellenburger

FIG. 16. Residual magnetic intensity on the basement surface with 100 r regional contours—
Crane and Ward Counties, Texas.

enables one to outline basins, and to map on a large scale basis with surprising accuracy the thickness of the sedimentary column and to outline major uplifts in basins. Over the past 12 or 15 years some 90% of our depth estimates are probably accurate to better than ± 20% and some 75% are probably accurate to within ± 10%. The word "probably" is used above because only a few of our estimates have been checked by wells which reached the igneous basement. The accuracy of the others can only be inferred from other information.

If over an area the anomalies are small compared to the total field of the earth so that the magnetic vector changes its direction but little, airborne data in which

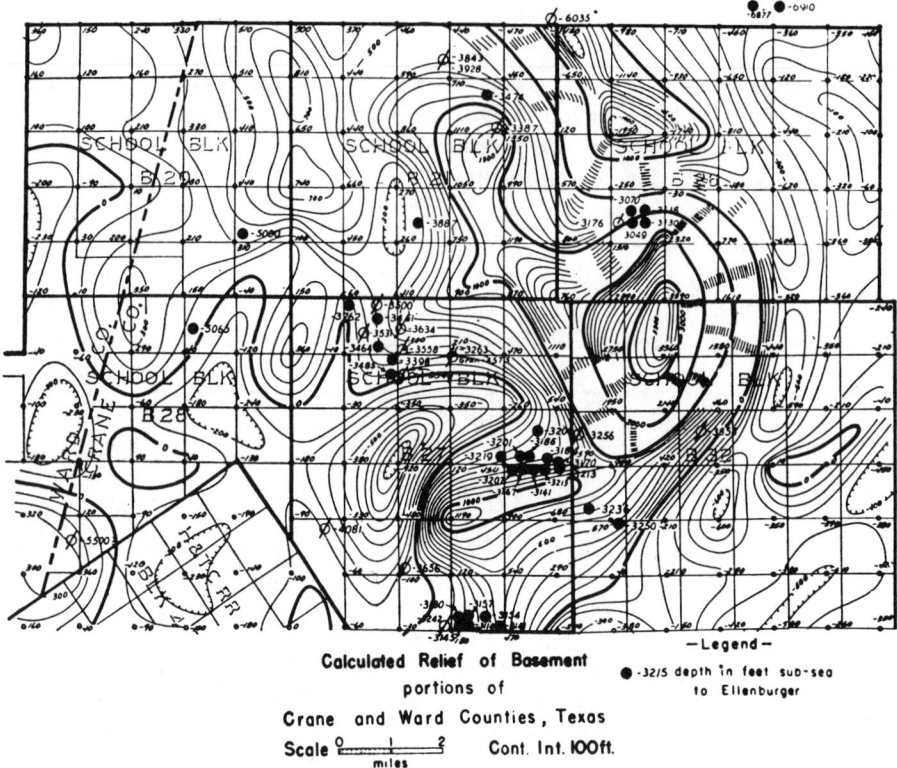

Calculated Relief of Basement

portions of

Crane and Ward Counties, Texas

Scale 0 1 2 miles   Cont. Int. 100 ft.

— Legend —

● -3215 depth in feet sub-sea
to Ellenburger

FIG. 17. Calculated relief of basement portions of Crane and Ward Counties, Texas.

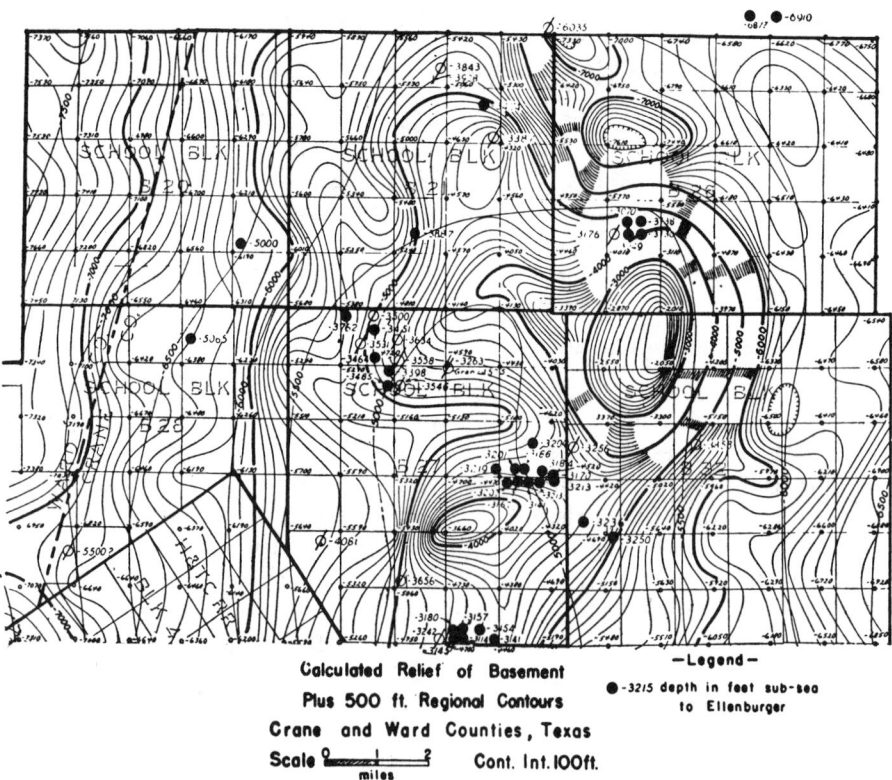

Calculated Relief of Basement

Plus 500 ft. Regional Contours

Crane and Ward Counties, Texas

Scale 0 1 2 miles   Cont. Int. 100 ft.

— Legend —

● -3215 depth in feet sub-sea
to Ellenburger

FIG. 18. Calculated relief of basement plus 500 ft. regional contours—Crane and Ward Counties, Texas.

the total field is measured can be considered to approximate a potential function. This implies that the coefficients for continuation upward, for continuation downward, and the techniques for calculating derivatives can be applied without modification to these types of data. The theory covering potential and structure calculations must be appropriately modified before being applied.

In order that this paper may be of maximum value to geophysicists interested in the interpretation of data depending on potential fields, there is given below a rather complete bibliography on the subject.

BIBLIOGRAPHY
Interpretation of Potential Fields

H. Bateman. Some integral equations of potential theory. *Journal of Applied Physics*, 17, 91–102 (1946).

E. C. Bullard and R. I. B. Cooper. The determination of the masses necessary to produce a given gravitational field. *Proceedings of the Royal Society, Series A*, 194, 332–347 (1948).

H. M. Evjen. The place of the vertical gradient in gravitational interpretations. *Geophysics*, 1, 127–136 (1936).

Sigmund Hammer. Investigations of the vertical gradient of gravity. *Transactions of the American Geophysical Union, Nineteenth Annual Meeting*, 1938, pp. 72–82 (1938).

C. W. Horton. Interpretation of isostatic anomalies south of Java, using integral equations and crustal deformation theories. *Geophysics*, 11, 183–194 (1946).

H. Herbert Howe. Height of magnetic anomalies. *Transactions of 1940 of the American Geophysical Union*, pages 309–311 (1940).

D. S. Hughes. The analytic basis of gravity interpretation. *Geophysics*, 7, 169–178 (1942).

D. S. Hughes and W. L. Pondrom. Computation of vertical magnetic anomalies from total magnetic field measurements. *Transactions, American Geophysical Union*, 28, 193–197 (1947).

N. Malkin. On the application of Neumann's problem to the investigation of the magnetic field at altitudes lying above the regions of anomalies. *Transactions of the Central Geophysical Observatory*, Vol. 1, No. 3, Central Institute of Terrestrial Magnetism and Atmospheric Electricity, No. 1, pp. 28–43 (1934). In Russian, with English summary.

N. Malkin. On the variations of the elements of terrestrial magnetism with the height. *Transactions of the Central Geophysical Observatory*, Vol. 1, No. 3, Central Institute of Terrestrial Magnetism and Atmospheric Electricity, No. 1, pp. 24–27 (1934). In Russian, with English summary.

Takesi Nagata. Magnetic anomalies and the corresponding subterranean mass distribution. *Bulletin of the Earthquake Research Institute*, Tokyo Imperial University, 16, 550–577 (1938).

Takesi Nagata. Magnetic anomalies and corresponding subterranean structure. *Proceedings of the Imperial Academy of Tokyo*, 14, 176–181 (1938).

B. Numerov. Interrelation between local gravity anomalies and the derivatives of the potential. *Zeitschrift für Geophysik*, 5, 58–62 (1929).

Maunu Puranen. The calculation of the strength of the anomaly of a magnetic ore at a given distance above the surface of the earth. *Bulletin de la Commission Géologique de Finlande*, No. 138, pp. 21–26 (1946).

H. Rainbow. The interpretation of torsion balance data. World Petroleum Congress, July 19–25, 1933, *Proceedings*, Vol. 1, pp. 143–146 (1934).

K. Ramsayer. Die Änderung magnetischer Störgebiete mit der Höhe und ihr Einfluss auf die Flugnavigation. *Beiträge zur angewandten Geophysik*, 9, 65–97 (1941).

A. K. Sen'ko. K. interpretatsii gravitatsionnikh nabliudenii po metodu integral'nykh uravnenii. (On the interpretation of gravity surveys by the method of integral equations.) *Biulleten' Neftianoi Geofiziki*, No. 3, pp. 151–152 (1936).

D. C. Skeels. Ambiguity in gravity interpretation. *Geophysics*, 12, 43–56 (1947).

Chûji Tsuboi. Gravity anomalies and the corresponding subterranean mass distribution. *Proceedings of the Imperial Academy of Tokyo*, 14, 170–175 (1938).

Chûji Tsuboi and Takato Fuchida. Relations between gravity values and corresponding subterranean mass distributions. *Bulletin of the Earthquake Research Institute*, Tokyo Imperial University, 15, 636–649 (1937).

Chûji Tsuboi and Takato Fuchida. Relation between gravity anomalies and the corresponding subterranean mass distribution. II. *Bulletin of the Earthquake Research Institute*, Tokyo Imperial University, 16, 273–284 (1938).

Chûji Tsuboi. Relation between the gravity anomalies and the corresponding subterranean mass distribution. III. *Bulletin of the Earthquake Research Institute*, Tokyo Imperial University, 17, 351–384 (1939).

Chûji Tsuboi, Tetuiti Kaneko, Setumi Miyamura, and Tokutaro Yabasi. Relation between the gravity anomalies and the corresponding subterranean mass distribution. IV. Isostasy in the United States of America. *Bulletin of the Earthquake Research Institute*, Tokyo Imperial University, 17, 385–410 (1939).

Chûji Tsuboi. Relation between the gravity anomalies and the corresponding subterranean mass distribution. V. Isostatic anomalies and the undulation of the isostatic geoid in the United States of America. *Bulletin of the Earthquake Research Institute*, Tokyo Imperial University, 18, 384–400 (1940).

Chûji Tsuboi and Seiti Yamaguti. Relation between the gravity anomalies and the corresponding subterranean mass distribution. VI. *Bulletin of the Earthquake Research Institute*, Tokyo Imperial University, 19, 26–38 (1941).

E. H. Vestine and N. Davies. Analysis and interpretation of geomagnetic anomalies. *Terrestrial Magnetism and Atmospheric Electricity*, 50, 1–36 (1945).

E. H. Vestine. On the analysis of surface magnetic fields by integrals. Part I. *Terrestrial Magnetism and Atmospheric Electricity*, 46, 27–41 (1941).

Seiti Yamaguti. The structure of the earth's crusts near the Japan Trench, off Sanriku, and also near the Inland Sea of Japan. *Bulletin of the Earthquake Research Institute*, Tokyo Imperial University, 17, 429–442 (1939).

A. Zamorev. On the definition of derivatives of gravitation potential and relations between the moments of perturbing masses by a derivative given on a plane. *Bulletin de l'Académie des Sciences de l'URSS*, Série géographique et géophysique, No. 3, pp. 275–286 (1939).

MAGNETOTELLURIC METHODS

Reprinted by permission of the Society of Exploration
Geophysicists from *Geophysics*, v. 18, no. 3 (1953), p.
605-635.

BASIC THEORY OF THE MAGNETO–TELLURIC METHOD
OF GEOPHYSICAL PROSPECTING*†‡

LOUIS CAGNIARD§

ABSTRACT

From Ampere's Law (for a homogeneous earth) and from Maxwell's equations using the concept of Hertz vectors (for a multilayered earth), solutions are obtained for the horizontal components of the electric and magnetic fields at the surface due to telluric currents in the earth. The ratio of these horizontal components, together with their relative phases, is diagnostic of the structure and true resistivities of subsurface strata. The ratios of certain other pairs of electromagnetic elements are similarly diagnostic.

Normally, a magneto-telluric sounding is represented by curves of the apparent resistivity and the phase difference at a given station plotted as functions of the period of the various telluric current components. Specific formulae are derived for the resistivities, depths to interfaces, etc. in both the two- and three-layer problems.

For two sections which are geometrically similar and whose corresponding resistivities differ only by a linear factor, the phase relationships are the same and the apparent resistivities differ by the same proportionality constant which relates the corresponding true resistivities. This "principle of similitude" greatly simplifies the representation of a master set of curves, such as is given for use in geologic interpretation.

In addition to the usual advantages offered by the use of telluric currents (no need for current sources or long cables, greater depths of investigation, etc.), the magneto-telluric method of prospecting resolves the effects of individual beds better than do conventional resistivity methods. It seems to be an ideal tool for the initial investigation of large sedimentary basins with potential petroleum reserves.

INTRODUCTION

There is no doubt that the first positive success in geophysical prospecting was obtained by electrical methods. These have always appeared promising both for oil and mineral prospecting because one can usually expect large resistivity contrasts in earth materials. Moreover, in the case of horizontal bedding, electrical prospecting can give information at locations where neither magnetic nor gravity anomalies can exist. The equipotential method, which involves the mapping of the equipotential lines on the earth's surface when current is introduced into the ground through two point electrodes, usually failed because of difficulty in analyzing the diagnostic features. In spite of the simplicity of Ohm's law, the theory of current flow in the earth is very complex. One may resort to experiments on scale models, but these preserve many of the shortcomings of the theoretical approach when applied to a practical situation.

In general, petroleum and mining geologists were not satisfied with the am-

* Manuscript received by the Editor September 1, 1952

† Translated from the French by a professional translator for the Magnolia Petroleum Company.

‡ Translation edited by M. B. Dobrin, R. L. Caldwell, and R. Van Nostrand, Field Research Laboratories, Magnolia Petroleum Company.

§ Professor at the Sorbonne, Paris, Past Director of the Société de Prospection Géophysique and of the Compagnie Générale de Géophysique.

biguous interpretations which geophysicists could offer them on the basis of equipotential data. The use of alternating current is even less desirable in this respect
because Maxwell's equations are considerably less manageable than is Ohm's
law.

The introduction of resistivity methods was a step in the right direction,
chiefly because the "apparent resistivity" of a section whose structure is not too
complicated can actually be calculated, or at least estimated, without too much
risk of error. However, these new methods, especially with respect to depth determination, have not proved to be as spectacular as they first appeared. Even
for the two-layer case, a large amount of labor is involved in developing a master
set of curves and one is seldom able to match his experimental curve with any
of the curves in his catalogue, extensive as it might be. Moreover, the useful
depth of investigation is limited to a few hundred meters in the case of direct
current and even less in the case of alternating current, especially at the higher
frequencies. In order to investigate to a reasonable depth, it is necessary to use
direct current with such great electrode separations that the method no longer
has the advantage of being inexpensive.

It is thus evident that electrical sounding, at least in petroleum exploration,
originally promised much more than it has realized. However, the relatively
recent discovery of the telluric method, although little known and little used
outside of France, offers more favorable prospects. Although the principles involved were recognized about 30 years ago by Conrad Schlumberger,* no practical application was made until a few years before World War II. The telluric
method has several advantages in that it does away with a current source and the
associated long leads, combines flexibility, rapidity, and low cost, and reaches
much greater depths of penetration than do ordinary resistivity methods. In spite
of its fundamental advantages, however, the telluric method seems to represent
only a temporary stage in the development of more advanced methods. The
magneto-telluric method, which is the subject of this paper, answers the ever-
increasing need for quantitative results. Actually, it is not a strictly electrical
method, but rather a combination of telluric and magnetic methods, a combination from which the name of the technique has been derived.

Essentially, the magneto-telluric method involves the comparison, preferably at one and the same place, of the horizontal components of the magnetic
and electric fields associated with the flow of telluric currents. The new method
offers all the advantages of the telluric method and even improves on it with respect to flexibility, speed, and economy. In addition, it offers the inestimable
benefit of making possible, in most cases where the bedding is horizontal, a truly
quantitative interpretation. Also, the method can be applied without particular
difficulty to submarine prospecting.

* E. G. Leonardon, "Some Observations Upon Telluric Currents and Their Applications to
Electrical Prospecting," *Terrestrial Magnetism and Atm. Electr.* 33 (1928), pp. 91–94. A presentation of a report on work dating back to 1921 under the direction of Conrad Schlumberger.

SKIN EFFECT AND ITS CONSEQUENCES. HARMONIC SHEET OF TELLURIC CURRENTS IN AN ELECTRICALLY HOMOGENEOUS EARTH

By way of introduction to the analysis of the magneto-telluric method let us consider a schematic and ideal sheet of telluric current which we shall suppose to be uniform, harmonic, of period T, flowing in a soil electrically homogeneous, of conductivity σ.

During this study, we shall only use electro-magnetic units, both for electric dimensions and magnetic dimensions. Let us choose a rectangular coordinate system o, x, y, z (Fig. 1) such that the origin is on the surface of the ground and oz is the descending vertical. One will notice that on the ground the angle ox, oy is equal to $-(\pi/2)$ for an observer who normally stands with his feet on the ground and his head straight up. It is also useful to remember that, if a current circulates in the ground along ox, oy is at the left of the Amperian man looking up at the sky.

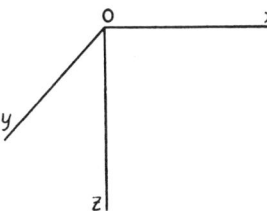

FIG. 1. Coordinate system. xy plane represents earth's surface. z is positive downward.

It is particularly useful when one employs Maxwell's equations and considers a harmonic phenomenon, to bring in the Hertz vector and to make use of imaginary notation. I shall use this approach later, but to handle this first particularly easy case, I prefer to remain as elementary as possible in order to be understood by those who are not familiar with Maxwellian analysis and who are eager to understand the principles of the proposed method.

The term "uniform" when applied to the telluric sheet we want to consider is rather inaccurate. As a matter of fact, there is uniformity only parallel to the surface of the ground, and not along a vertical line. If the density of the current is represented on the surface of the ground, for $z=0$, by

$$I_x = \cos \omega t, \qquad I_y = I_z = 0, \tag{1}$$

the laws of physics show that at depth z one must have

$$I_x = e^{-z\sqrt{2\pi\sigma\omega}} \cos (\omega t - z\sqrt{2\pi\sigma\omega}), \qquad I_y = I_z = 0, \tag{2}$$

e designating the base of natural logarithms. Formula (2) holds for what is called the skin effect. When z increases, one notices an exponential decrease with respect to z at the same time that the phase retardation progressively increases.

Under the conventional name of "depth of penetration" (understood as relating to a layer of conductivity σ and to a telluric sheet of period T) we shall define a term which we are going to use constantly. It designates the depth p when the amplitude is reduced to the fraction $1/e$ of what it is on the surface.

$$p = \frac{1}{\sqrt{2\pi\sigma\omega}} = \frac{1}{2\pi}\sqrt{\frac{T}{\sigma}} . \tag{3}$$

As for the phase, it is retarded one additional radian each time that z is increased by p.

It is obvious that for z infinite, the amplitude of the magnetic field is annulled; otherwise the density of the current could not be zero. At the same time, sym-

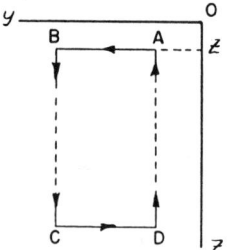

FIG. 2. Path of integration to apply Faraday's Law.

metry requires that the magnetic field be horizontal everywhere, parallel to oy. Let us now apply the theorem of Ampere to a rectangle $ABCD$ with sides AB CD parallel to oy and of unit length, with side AB situated at depth z and with side CD put at infinite depth. It reads

$$H_x = 0,$$

$$H_y(z) = 4\pi \int_z^{+\infty} I_x dz = 2\sqrt{\frac{\pi}{\sigma\omega}} e^{-z\sqrt{2\pi\sigma\omega}} \cos\left(\omega t - z\sqrt{2\pi\sigma\omega} - \frac{\pi}{4}\right). \tag{4}$$

In particular, on the surface of the earth, where $z=0$,

$$H_x = 0,$$

$$H_y = 4\pi \int_0^{+\infty} I_x dz = 2\sqrt{\frac{\pi}{\sigma\omega}} \cos\left(\omega t - \frac{\pi}{4}\right). \tag{5}$$

We shall stress this first result, because it is the key to the proposed method: On the surface of the ground, the magnetic field $\mathcal{H}$ and electric field $\mathcal{E}(E_x = I_x/\sigma)$ are orthogonal. The quotient of the amplitude of the electric field by that of the magnetic field has the value $1/\sqrt{2\sigma T}$. The phase of the magnetic field is retarded by an angle of $\pi/4$ with respect to that of the electric field.

It is well understood that the above result is valid for a telluric sheet flowing in any direction, provided one always chooses the left hand side as positive in measuring the magnetic field. If, for instance, the component of the electric field along oy is of the form

$$E_y = \frac{1}{\sigma} \cos \omega t, \tag{6}$$

it will be necessary to write

$$H_x = -2\sqrt{\frac{\pi}{\sigma\omega}} \cos\left(\omega t - \frac{\pi}{4}\right), \tag{7}$$

with a change of sign relative to the similar formula (5), since the x axis indicates the right hand side when the current flows along the y axis.

The integral in the second member of relation (5) represents the total intensity of the telluric current through a rectangle, vertical and unlimited, going

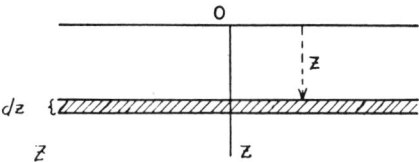

FIG. 3. Section showing horizontal uniform sheet of current.

from the surface, perpendicular to ox, and of unit width. The magnetic field H measures this total intensity within a factor of 4π.

This observation is of great practical importance. *It remains strictly valid for any layered earth, and maintains approximate validity in many cases interesting in exploration.*

Remarks

Assume a horizontal, uniform, extremely thin sheet of direct current of density I, flowing at the depth z between two horizontal planes with sides z and $z+dz$ (Fig. 3). It is well known and easy to show that the magnetic field produced by this horizontal sheet on the surface of the ground is horizontal, that it is directed to the left hand side and that its value is $2\pi I dz$.

For a sheet of direct current, flowing parallel to ox, from the surface of the ground down to depth z, and whose density I_x would be any function of z, one would have

$$H_y = 2\pi \int_0^z I_x(z)dz. \tag{8}$$

Because telluric currents have an extraordinary low frequency, since the length of the wave is enormous relative to p, one might be tempted to apply to them relation (8), assuming their behavior to be that of a direct current, which would lead one to write

$$H_y = 2\pi \int_0^{+\infty} I_x(z)dz, \tag{9}$$

whereas the accurate formula (5) includes the factor 4π, and not the factor 2π.

Units and magnitudes

We measure the magnetic field in γ, the electric field in millivolts/km and the period in seconds. On the other hand, prospectors usually consider the resistivity ρ rather than conductivity σ. They measure resistivities in ohm-meters.

$$1\ \gamma = 10^{-5}\ \text{em cgs}$$
$$1\ \text{mv/km} = 1\ \text{em cgs}$$
$$1\ \text{km} = 10^5\ \text{em cgs} \qquad\qquad 10)$$
$$1\ \Omega\text{m} = 10^{11}\ \text{em cgs}$$

With the new system of units, one obtains:

$$p = \frac{1}{2\pi}\sqrt{10\rho T}, \qquad \rho = 0.2\,T\left(\frac{E}{H}\right)^2. \qquad (11)$$

In order to become familiar with the order of dimensions, it is useful to consult the two tables of numbers which follow. Table 1 gives the values of p for different values of ρ and T. Table 2 gives, also as function of ρ and of T, the values of H corresponding to an electric field of 1 mv/km.

TABLE 1

DEPTHS OF PENETRATION GIVEN IN KM

$\rho\downarrow\backslash T\rightarrow$	1 sec	3 sec	10 sec	30 sec	1 min	2 min	5 min	10 min	30 min
0.2	0.225	0.390	0.712	1.23	1.74	2.47	3.90	5.51	9.54
1	0.503	0.872	1.59	2.76	3.90	5.51	8.72	12.3	21.4
5	1.13	1.95	3.56	6.16	8.72	12.3	19.5	27.6	47.7
10	1.59	2.76	5.03	8.72	12.3	17.4	27.6	39.0	67.5
50	3.56	6.16	11.3	19.5	27.6	39.0	61.6	87.2	151
250	7.95	13.8	25.2	43.6	61.6	87.2	138	195	338
1,000	15.9	27.6	50.3	87.2	123	174	276	390	675
5,000	35.6	61.6	113	195	276	390	616	872	1510

TABLE 2

AMPLITUDES OF THE MAGNETIC FIELD GIVEN IN γ WHEN E IS 1 MV/KM

$\rho\downarrow\backslash T\rightarrow$	1 sec	3 sec	10 sec	30 sec	1 min	2 min	5 min	10 min	30 min
0.2	1	1.73	3.16	5.48	7.75	11.0	17.3	24.5	42.4
1	0.447	0.775	1.41	2.45	3.46	4.90	7.75	11.0	19.0
5	0.2	0.346	0.632	1.10	1.55	2.19	3.46	4.90	8.49
10	0.141	0.245	0.447	0.775	1.10	1.55	2.45	3.46	6
50	0.0632	0.110	0.2	0.346	0.490	0.693	1.10	1.55	2.68
250	0.0283	0.049	0.0894	0.155	0.219	0.310	0.490	0.693	1.2
1,000	0.0141	0.0245	0.0447	0.0775	0.110	0.155	0.245	0.346	0.6
5,000	0.00632	0.0110	0.0200	0.0346	0.0490	0.0693	0.110	0.155	0.268

From now on, one will notice the extent to which the depths of penetration are exactly adapted to the needs of petroleum prospecting. One will also notice very large limits between which the ratio of the amplitudes of the electric and magnetic fields may vary, which is, of course, essential when one wants to establish a "precise" method of prospecting in which this ratio is to be measured.

RELATION BETWEEN THE ELECTRIC AND THE MAGNETIC FIELD FOR A NON-HARMONIC TELLURIC SHEET

If the components of the telluric current no longer vary with time according to a sinusoidal law but instead vary in an absolutely arbitrary way, as in natural telluric sheets, the relations obtained above are easily generalized by means of operational calculus. I shall limit myself to give the result, which does not seem to have any great practical interest in connection with prospecting.

$$
\begin{aligned}
E_y(t) &= -\frac{1}{2\pi\sqrt{\sigma}} \int_{-\infty}^{t} H_x'(u) \frac{du}{\sqrt{t-u}} \\
&= -\frac{1}{2\pi\sqrt{\sigma}} \int_{0}^{+\infty} H_x'(t-u) \frac{du}{\sqrt{u}} \cdot
\end{aligned}
\tag{12}
$$

In this expression, $H_x'(t)$ designates the derivative of $H_x(t)$ with respect to t.

GENERALIZATION FOR ANY HORIZONTALLY STRATIFIED SECTION

If the earth is formed by a number of horizontal strata of arbitrary thicknesses and resistivities, we shall start from the equations of Maxwell and we shall preferably use imaginary notation, stipulating that all the alternating quantities depend on time through a factor $e^{-i\omega t}$. From now on, this factor will be understood rather than expressed explicitly.

If the harmonic sheet, assumed uniform, flows along ox, the components of the Hertz vector Π along oy and oz are null. Furthermore, Π_z depends only on z (and on t).

The equations of Maxwell are satisfied if

$$
\nabla^2\Pi_x + 4\pi\sigma\omega i\Pi_x = 0.
\tag{13}
$$

The electric field $\mathcal{E}$ and magnetic field $\mathcal{H}$ are expressed in a general way by

$$
\begin{aligned}
\mathcal{H} &= 4\pi\sigma \text{ curl } \Pi, \\
\mathcal{E} &= \text{grad div } \Pi - \nabla^2\Pi,
\end{aligned}
\tag{14}
$$

and, specifically, in the actual problem by

$$
\begin{aligned}
H_y &= 4\pi\sigma\frac{\partial\Pi_x}{\partial z}, \qquad H_x = H_z = 0 \\
E_x &= 4\pi\sigma\omega i\Pi_x, \qquad E_y = E_z = 0.
\end{aligned}
\tag{15}
$$

Because, in this case, E_x is proportional to Π_x, we can choose E_x as Hertz vector, so that

$$\frac{\partial^2 E_x}{\partial z^2} + 4\pi\sigma\omega i E_x = 0,$$

$$H_y = -\frac{i}{\omega}\frac{\partial E_x}{\partial z}.$$

(16)

Furthermore, we must assure the continuity of E_x and H_y when crossing the different surfaces of separation.

In order to meet condition (16), E_x must be in the form of

$$E_x = A e^{a\sqrt{\sigma}\,z} + B e^{-a\sqrt{\sigma}\,z},$$

(17)

A and B designating two arbitrary constants and a being defined as

$$a = 2\pi\sqrt{\frac{2}{T}}\, e^{-i\pi/4} = \frac{2\pi}{\sqrt{T}}\,(1 - i).$$

(18)

Let us number from 1 to n the successive formations starting at the surface of the ground. The nth and last one is the lowest stratum. It will be necessary in this layer to put down $A = 0$, because the first term becomes infinite at the same time as z. Furthermore, any solution can always be multiplied by a constant complex arbitrary factor. In other words, the problem is only definite as far as the relative amplitudes and the differences of phase are concerned. For this reason, we can assign an arbitrary value to one of the $2n$ constants A and B. We shall assume that it is constant B corresponding to the bottom stratum which is equal to unity.

In all we have $2(n-1)$ arbitrary constants to meet the same number of conditions at the limits. These conditions are the equality of the two fields at each of the $n-1$ surfaces of separation.

The method of calculation being the same no matter what the value of n, we shall only consider the cases of $n = 2$ and $n = 3$.

It is obvious that these calculations, which do not present any other complications than the resolution of simple algebraic equations of the first degree, are done exclusively by means of addition, multiplication and division and do not necessitate resorting to integrals or series.

SOURCE OF CURRENTS

The above theory does not concern itself with the origin of the currents involved. Whether the source of these currents are internal to the crust of the earth or whether they are ionospheric, whether these sources are natural (actual telluric currents) or whether they are artificial (vagrant currents), the electromagnetic phenomena inside the earth are the same in every case.

In fact, the reasoning depends only on the requirement that the telluric current sheet be sufficiently uniform. But this uniformity is a matter of experience. Telluric prospecting proves that in large sedimentary basins this uniformity extends, in an approximate way, over a considerable expanse, often some ten km in width. Such uniformity should be expected all the more if one only considers the very restricted field that enters into a magneto-telluric comparison. Vagrant currents, because of the relative proximity of the sources which produce them, and because of the poor degree of uniformity of the sheets associated with such artificial currents, are feared by the telluric prospectors. On the contrary, they are looked on as a blessing by magneto-telluric prospectors, because they offer sufficient uniformity to meet the requirements of the new method, and they usefully enlarge the spectrum of frequencies.

Readers of *Geophysics*, as well as this writer, are mainly concerned with what is underneath their feet and are little interested in what goes on above their

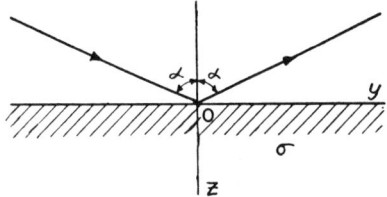

FIG. 4. Plane electromagnetic wave incident upon earth's surface.

heads. However, it may be useful to consider for a few moments longer the nature of the electro-magnetic phenomena as a whole involving the atmosphere.

In the air, where we put down $\sigma = 0$, equation (16) becomes $\partial^2 E_x / \partial z^2 = 0$. E_x appears as a linear function of z, H_y as a constant:

$$E_x(z) = E_x(0) + i\omega z H_y(0); \qquad E_y = E_z = 0.$$
$$H_y(z) = H_y(0); \qquad H_x = H_z = 0. \tag{19}$$

A solution of this kind may surprise the reader. One knows, in particular, that the vertical component of the magnetic field of the earth undergoes quick variations whose correlation with those of the horizontal components of the same field or of the telluric field is evident. But the actual solution shows us that H_z is null.

Let us not forget that, in the expression (13) of the equations of Maxwell, we have, from the start, considered as infinite the speed V of electro-magnetic waves in the ground, as well as the speed c of those waves in empty space. For the real phenomenon of propagation we have substituted from the start a fictitious stationary phenomenon. The approximation was quite sufficient for the calculations we had in mind, but it did not permit an accurate picture of the nature of the physical phenomena involved.

Let us suppose that in the atmosphere, a plane wave spreading in the plane *oyz* hits the surface of the ground at an angle of incidence α (Fig. 4). In order that the conditions at the limits might be met at the surface of the ground, it is, first of all, necessary that the expressions for the characteristic vectors of the three waves (incident, reflected and refracted) include, respectively, the following factors:

Incident Wave:

$$e^{-i\omega\left(t-\frac{y\sin\alpha + z\cos\alpha}{c}\right)}$$

Reflected Wave:

$$e^{-i\omega\left(t-\frac{y\sin\alpha - z\cos\alpha}{c}\right)}$$

Refracted Wave:

$$e^{-i\omega\left(t-\frac{y\sin\alpha + Kz}{c}\right)}$$

The constant K is chosen to satisfy the equation

$$\nabla^2\Pi + \Pi\left(4\pi\sigma\omega i + \frac{\omega^2}{V^2}\right) = 0. \tag{20}$$

It is thus necessary that

$$K^2 = \frac{c^2}{V^2} - \sin^2\alpha + i\,\frac{4\pi\sigma c^2}{\omega}. \tag{21}$$

But, whereas $(c^2/V^2) - \sin^2\alpha$ is at its maximum equal to unity, it happens that the coefficient of i is enormous. For instance, for $\rho = 10\Omega m$ and for $T = 30$ sec, it is equal to 5.4×10^{10} so that in practice, and as an excellent approximation, one may write

$$K^2 = 2\sigma c^2 T e^{i\pi/2},$$

$$K = c\sqrt{2\sigma T}\,e^{i\pi/4}, \tag{22}$$

bearing in mind the fact that the coefficient of i in the imaginary part of K must be positive. Accordingly, we justify in the first place the form itself of the expressions (17) which we have adopted initially as a starting point. After that we notice that an infinity of possible waves in the atmosphere can correspond to a given wave in the ground. Not only is α left completely arbitrary since it does not appear in (22), but the state of polarization of the incident wave remains also totally arbitrary. One is entitled to imagine all kinds of miscellaneous phenomena in the atmosphere, and no particular condition is imposed that the vertical component of the magnetic field must be null or negligible.

SPECIFIC STUDY OF THE TWO LAYER PROBLEM

Let us suppose σ_1 to be the conductivity of the upper formation, and σ_2 that of the lower formation, h being the thickness of the upper one (Fig. 5).

Following the general method sketched above, the general expression for

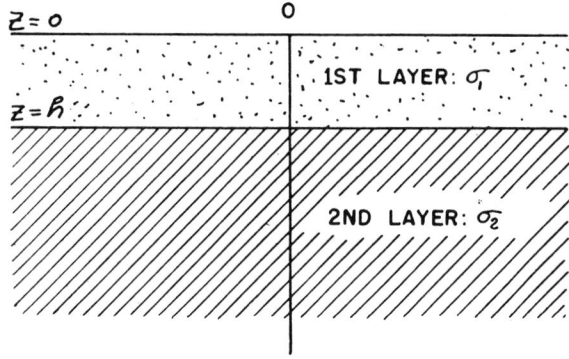

Fig. 5. Two-layer earth section.

the fields will be as follows:

1. In the first formation

$$E_x = A e^{a\sqrt{\sigma_1}\, z} + B e^{-a\sqrt{\sigma_1}\, z} \tag{23}$$
$$H_y = e^{i\pi/4}\sqrt{2\sigma_1 T}\left[-A e^{a\sqrt{\sigma_1}\, z} + B e^{-a\sqrt{\sigma_1}\, z} \right].$$

2. In the second formation

$$E_x = e^{-a\sqrt{\sigma_2}\, z} \tag{24}$$
$$H_y = e^{i\pi/4}\sqrt{2\sigma_2 T}\, e^{-a\sqrt{\sigma_2}\, z}.$$

The continuity of E_x and H_y for $z = h$ involves accordingly the two conditions

$$A e^{a\sqrt{\sigma_1}\, h} + B e^{-a\sqrt{\sigma_1}\, h} = e^{-a\sqrt{\sigma_2}\, h}$$
$$-A\sqrt{\sigma_1}\, e^{a\sqrt{\sigma_1}\, h} + B\sqrt{\sigma_1}\, e^{-a\sqrt{\sigma_1}\, h} = \sqrt{\sigma_2}\, e^{-a\sqrt{\sigma_2}\, h} \tag{25}$$

where

$$A = \frac{\sqrt{\sigma_1} - \sqrt{\sigma_2}}{2\sqrt{\sigma_1}}\, e^{-ah(\sqrt{\sigma_1}+\sqrt{\sigma_2})},$$
$$B = \frac{\sqrt{\sigma_1} + \sqrt{\sigma_2}}{2\sqrt{\sigma_1}}\, e^{ah(\sqrt{\sigma_1}-\sqrt{\sigma_2})}. \tag{26}$$

The result is an expression for the fields on the surface of the ground. In this expression we shall advantageously introduce the depth of penetration p_1 relative to the first formation and we shall be able to set aside a factor common to E_x and H_y, since we are only interested in the relation between those fields. One has then

$$E_x = M e^{-i\phi}$$
$$H_y = \sqrt{2\sigma_1 T}\, N e^{i(\pi/4-\psi)}, \tag{27}$$

in which:

$$M \cos \phi = \left(\frac{1}{p_1} \cosh \frac{h}{p_1} + \frac{1}{p_2} \sinh \frac{h}{p_1} \right) \cos \frac{h}{p_1}, \qquad (28)$$

$$M \sin \phi = \left(\frac{1}{p_1} \sinh \frac{h}{p_1} + \frac{1}{p_2} \cosh \frac{h}{p_1} \right) \sin \frac{h}{p_1},$$

$$N \cos \psi = \left(\frac{1}{p_1} \sinh \frac{h}{p_1} + \frac{1}{p_2} \cosh \frac{h}{p_1} \right) \cos \frac{h}{p_1},$$

$$\qquad (29)$$

$$N \sin \psi = \left(\frac{1}{p_1} \cosh \frac{h}{p_1} + \frac{1}{p_2} \sinh \frac{h}{p_1} \right) \sin \frac{h}{p_1}.$$

Whereupon:

$$\frac{E_x}{H_y} = \frac{1}{\sqrt{2\sigma_1 T}} \frac{M}{N} e^{-i(\pi/4 + \phi - \psi)}. \qquad (30)$$

The formulas given above relating to the case of a single formation are at once found again if one starts from those more general expressions and puts down: $\sigma_1 = \sigma_2 = \sigma$ and $p_1 = p_2 = p$ whereupon

$$M = N = \frac{1}{p} e^{h/p},$$

$$\phi = \psi = \frac{h}{p}, \qquad (31)$$

$$\frac{E_x}{H_y} = \frac{1}{\sqrt{2\sigma T}} e^{-i\pi/4},$$

conforming to the previous result.

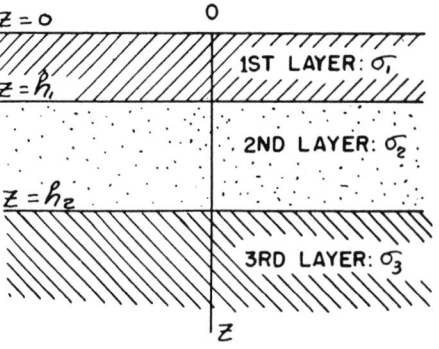

FIG. 6. Three-layer earth section.

FORMULAS FOR THREE FORMATIONS

In the case of three formations of conductivities σ_1, σ_2, and σ_3, when the second one starts at depth h_1 and the third one at depth h_2 (Fig. 6), one uses the following formulas. The ratio between the fields is always in the form

$$\frac{E_x}{H_y} = \frac{1}{\sqrt{2\sigma_1 T}} \frac{M}{N} e^{-i(\pi/4 + \phi - \psi)}, \tag{32}$$

by putting down

$$h_1\left(\frac{1}{p_1} + \frac{1}{p_2}\right) - \frac{h_2}{p_2} = u,$$

$$h_1\left(\frac{1}{p_1} - \frac{1}{p_2}\right) + \frac{h_2}{p_2} = v, \tag{33}$$

$$M \cos \phi = \left(\frac{1}{p_1} - \frac{1}{p_2}\right)\left(\frac{1}{p_2}\cosh u - \frac{1}{p_3}\sinh u\right)\cos u$$

$$+ \left(\frac{1}{p_1} + \frac{1}{p_2}\right)\left(\frac{1}{p_2}\cosh v + \frac{1}{p_3}\sinh v\right)\cos v, \tag{34}$$

$$M \sin \phi = \left(\frac{1}{p_1} - \frac{1}{p_2}\right)\left(\frac{1}{p_2}\sinh u - \frac{1}{p_3}\cosh u\right)\sin u$$

$$+ \left(\frac{1}{p_1} + \frac{1}{p_2}\right)\left(\frac{1}{p_2}\sinh v + \frac{1}{p_3}\cosh v\right)\sin v, \tag{35}$$

$$N \cos \psi = \left(\frac{1}{p_1} - \frac{1}{p_2}\right)\left(\frac{1}{p_2}\sinh u - \frac{1}{p_3}\cosh u\right)\cos u$$

$$+ \left(\frac{1}{p_1} + \frac{1}{p_2}\right)\left(\frac{1}{p_2}\sinh v + \frac{1}{p_3}\cosh v\right)\cos v, \tag{36}$$

$$N \sin \psi = \left(\frac{1}{p_1} - \frac{1}{p_2}\right)\left(\frac{1}{p_2}\cosh u - \frac{1}{p_3}\sinh u\right)\sin u$$

$$+ \left(\frac{1}{p_1} + \frac{1}{p_2}\right)\left(\frac{1}{p_2}\cosh v + \frac{1}{p_3}\sinh v\right)\sin v. \tag{37}$$

APPARENT RESISTIVITY FOR THE CASE OF TWO FORMATIONS

If the comparison of E_x and H_y is made on a ground which is known to be electrically homogeneous, the relation between those two fields allows one to know the true conductivity (or, if one prefers, its reciprocal, the resistivity), of the formation. If the magneto-telluric comparison takes place on any formation, stratified or not, whose structure is not in general known, it will usually happen that the phase of H_y with respect to E_x will not be a retardation of $\pi/4$. This will be the first indication that it is heterogeneous. However, no matter what this phase separation might be, we can agree that the modulus of the ratio is equal to $1/\sqrt{2\sigma_a T}$, in which σ_a would be the conductivity of a homogeneous formation which would give the modulus of the ratio between the fields whose value has been experimentally observed. The quantity σ_a is, by definition, the apparent conductivity and its reciprocal ρ_a is the apparent resistivity.

The apparent resistivity is usually a kind of average of the resistivities one meets in a thickness of ground such that density of the current is not to be neglected with respect to its value along the surface. However, it may happen in exceptional cases that the apparent resistivity might be very slightly less than the smallest of the resistivities of the formations, or on the other hand, very slightly greater than the highest of the resistivities. Actually, one knows that a similar phenomenon occurs for the apparent resistivities that are obtained in the prospecting techniques which use a quadripole of measurement supplied by direct current.

In the case of two formations, the apparent resistivity is easily calculated by means of the formulas established above. In accordance with this definition, one has

$$\frac{1}{\sqrt{2\sigma_a T}} = \frac{M}{N} \frac{1}{\sqrt{2\sigma_1 T}}, \tag{38}$$

or

$$\rho_a = \rho_1 \left(\frac{M}{N}\right)^2; \tag{39}$$

and

$$\frac{\rho_a}{\rho_1} = 1 + \frac{4 \cos \dfrac{2h}{p_1}}{m + \dfrac{1}{m} - 2 \cos \dfrac{2h}{p_1}}, \tag{40}$$

if one writes

$$m = \frac{\sqrt{\dfrac{\rho_2}{\rho_1}} + 1}{\sqrt{\dfrac{\rho_2}{\rho_1}} - 1} e^{2h/p_1}. \tag{41}$$

The fundamental properties of the apparent resistivity as defined in the technique of electrical sounding, with respect to a certain length of injection line of current, appear again at this point in the apparent resistivity defined now in regard to a certain period or to a certain penetration depth p. Indeed, one immediately establishes that:

1. if $p_1 = 0$, $\rho_a = \rho_1$,

2. if $p_1 = \infty$, $\rho_a = \rho_2$.

EXPRESSION FOR THE RETARDATION OF PHASE
IN THE CASE OF TWO FORMATIONS

The other parameter to consider in order to secure interpretation is the phase retardation of the magnetic field with respect to the electric field. In the case of two formations, it is expressed by

$$\theta = \frac{\pi}{4} + \phi - \psi, \tag{42}$$

with

$$\tan \phi = \frac{m-1}{m+1} \tan \frac{h}{p_1},$$

$$\tan \psi = \frac{m+1}{m-1} \tan \frac{h}{p_1}, \tag{43}$$

$$\tan (\phi - \psi) = -\frac{2m}{m^2-1} \sin \frac{2h}{p_1}, \qquad \left(-\frac{\pi}{4} \leqq \phi - \psi \leqq \frac{\pi}{4}\right),$$

m having the meaning given previously (equation 41).

SPECIFIC CASE OF A SECTION WITH TWO LAYERS, ONE BEING AN EXTREMELY
RESISTIVE OR EXTREMELY CONDUCTIVE SUBSTRATUM

In these specific cases, the above formulas become:

1. Extremely resistive substratum:

$$\left. \begin{aligned} \left|\frac{E_x}{H_y}\right| &= \frac{1}{\sqrt{2\sigma_1 T}} \sqrt{1 + \frac{2 \cos 2h/p_1}{\cosh 2h/p_1 - \cos 2h/p_1}} \\ \theta &= \frac{\pi}{4} - \arctan \left(2 \frac{e^{2h/p_1}}{e^{4h/p_1} - 1} \sin 2h/p_1\right) \end{aligned} \right\}, \tag{44}$$

The result becomes particularly simple if h is very much smaller than p:

$$\left|\frac{E_x}{H_y}\right| = \frac{1}{2\sqrt{\sigma_1 T}} \frac{p_1}{h}; \qquad \theta = 0. \tag{45}$$

159

2. Extremely conducting substratum:

$$\left|\frac{E_x}{H_y}\right| = \frac{1}{\sqrt{2\sigma_1 T}}\sqrt{1 - \frac{2\cos 2h/p_1}{\cosh 2h/p_1 + \cos 2h/p_1}}$$
$$\theta = \frac{\pi}{4} + \arctan\left(2\frac{e^{2h/p_1}}{e^{4h/p_1} - 1}\sin 2h/p_1\right) \tag{46}$$

The result becomes particularly simple if h is very much smaller than p_1:

$$\left|\frac{E_x}{H_y}\right| = \frac{1}{\sqrt{\sigma_1 T}}\frac{h}{p_1}, \qquad \theta = \frac{\pi}{2}. \tag{47}$$

LAW OF SIMILITUDE OF THE MAGNETO-TELLURIC SOUNDINGS

It is known that the interpretation of the ordinary electrical soundings is made much easier by the use of logarithmic scales in the construction of theoretical templates on the one hand, and of experimental diagrams on the other. This use of logarithmic scales is based on the laws of similitude (geometric similitude, electric similitude) which are applicable to electrical soundings.

Laws of similitude of the same kind also govern magneto-telluric soundings and will play an important part in their interpretation. Before we explain how to represent the results we have just obtained in the form of master curves, it is necessary to establish these laws of similitude.

Let us consider two structures, as complex as desired, stratified or not, being geometrically similar, the ratio of similitude being K_L. To make it plainer, let us specify that the corresponding parameters of the two structures will be represented by the same letters, respectively primed and unprimed. In this way, L' and L designating corresponding lengths, we shall put down

$$L' = K_L L. \tag{48}$$

At two similar points of the two structures, the resistivities are ρ' and ρ and we postulate electrical similitude

$$\rho' = K_\rho \rho. \tag{49}$$

Finally, if the periods of the electro-magnetic phenomena are T' and T, we require

$$T' = K_T T. \tag{50}$$

If $\Pi'(x', y', z')$ represents a Hertz vector, which is a solution of Maxwell's equations and of the boundary conditions for the primed structure, let us find out the conditions under which

$$\Pi(x, y, z) = \Pi'(x', y', z') \tag{51}$$

is also a solution for the unprimed structure.

It is necessary to consider equation (51) so that $\Pi(x, y, z)$ designates a function of x, y, z obtained when one respectively replaces in $\Pi'(x', y', z')$ the coordinates x', y', z' by $K_L x, K_L y, K_L z$, which, in other words, makes the same Hertz vector correspond at two similar points of the two structures.

When one has

$$\nabla^2 \Pi' = \frac{1}{K_L{}^2} \nabla^2 \Pi; \qquad \sigma' = \frac{\sigma}{K_\rho}; \qquad \omega' = \frac{\omega}{K_T}, \qquad (52)$$

the general equation

$$\nabla^2 \Pi + 4\pi\sigma\omega i\Pi = 0$$

becomes

$$\nabla^2 \Pi' + 4\pi\sigma'\omega' i\Pi' = 0,$$

if

$$K_L{}^2 = K_\rho K_T. \qquad (53)$$

We shall impose this condition.

Besides, one has

$$\mathcal{3C}' = \frac{\mathcal{3C}}{K_\rho K_L}, \qquad \mathcal{E}' = \frac{\mathcal{E}}{K_L{}^2}; \qquad (54)$$

so that the conditions of continuity supposed to be met in one of the structures are also met in the other one.

The ratio E'/H' of an electrical component to a magnetic component is equal to the corresponding ratio with a factor of proportionality, *which is real*. The phase separation between those components is, consequently, the same in both structures.

On the other hand, the ratio ρ_a'/ρ_a of the apparent resistivities has the value

$$\frac{\rho_a'}{\rho_a} = \left(\frac{E'}{E}\right)^2 \left(\frac{H}{H'}\right)^2 \frac{T'}{T} = \frac{1}{K_L{}^4} \cdot K_\rho{}^2 \cdot K_L{}^2 \cdot K_T = K_\rho, \qquad (55)$$

if we take (53) into consideration. In other words, when one goes from one structure to the other, the apparent resistivities are modified in the same ratio as the real resistivities, which moreover might seem obvious enough on the basis of the principles we have considered.

To sum up the preceding, when one knows the apparent resistivity relative to a certain structure and a certain period T, one deduces at once from this *one* apparent resistivity relative to another structure deduced from the first one by geometrical similitude (ratio K_L) and by electrical similitude (ratio K_ρ). The new apparent resistivity is equal to the former one multiplied by the ratio of electrical similitude and it is relative to a period such that

$$K_T = \frac{K_L{}^2}{K_\rho} \, . \tag{56}$$

CONSTRUCTION AND DESCRIPTION OF MASTER CURVES FOR TWO FORMATIONS

A magneto-telluric sounding (in order to abbreviate we shall from now on say MT sounding) will be represented by means of two curves, namely those indicating ρ_a and θ as functions of T. In the preparation of master sets of curves for the case of two formations, it is necessary to consider three arbitrary parameters, namely two resistivities and one thickness, each of which may vary from zero to infinity.

The value of the law of similitude lies in the fact that, in order to represent the whole of the MT-soundings, for two formations, it is sufficient to limit one's self to the specific case of $\rho_1 = 1$ and $h = 1$. In this way there only remains one single arbitrary parameter, namely the resistivity ρ_2 of the substratum, so that the totality of MT-soundings is represented by means of two systems of curves.

Indeed, when, in a more general way, the resistivities of the two present formations will be $\rho_1' \neq 1$ and ρ_2' and when the thickness of the first formation will be $h' \neq 1$, in order to obtain the curve $\rho_a' = \rho_a'(T')$ it will be sufficient to multiply

1. by ρ_1' the ordinates of that one of the curves $\rho_a = \rho_a(T)$ characterized by the ratio ρ_2'/ρ_1' equal in magnitude to the value of the parameter ρ_2.

2. by h'^2/ρ_1' the abscissas of this same curve.

Furthermore, in order to obtain the curve $\theta' = \theta'(T')$, it will be sufficient to multiply by the same factor h'^2/ρ_1' the abscissas of that of the curves $\theta = \theta(T)$ characterized by the value ρ_2'/ρ_1'. There will be no reason to modify the ordinates.

Rather than to carry out these multiplications, it is obviously much easier to choose for each of the systems ρ_a and θ the logarithmic abscissas representing the logarithm of $\sqrt{T}$. Furthermore, for the system ρ_a, the ordinates will represent the logarithm of ρ_a. The two sets of curves reproduced here were constructed in this way, with scales as indicated in Figures 7 and 8.

With the help of these logarithmic master curves, the expansions of the abscissas and of the ordinates described at the beginning of this section will amount from now on to a simple translation. A translation will be carried out parallel to the axis of the abscissas for curve ρ_a, as well as for curve θ, *and this translation will be of the same amplitude in both cases.* Furthermore, in the case of curve ρ_a a second translation will be carried out parallel to the axis of the ordinates.

The whole of the curves of system ρ_a, corresponding to the changing values of ρ_2, have an infinity of points in common, defined by

$$\rho_a = 1, \qquad \cos \frac{2}{p_1} = 0.$$

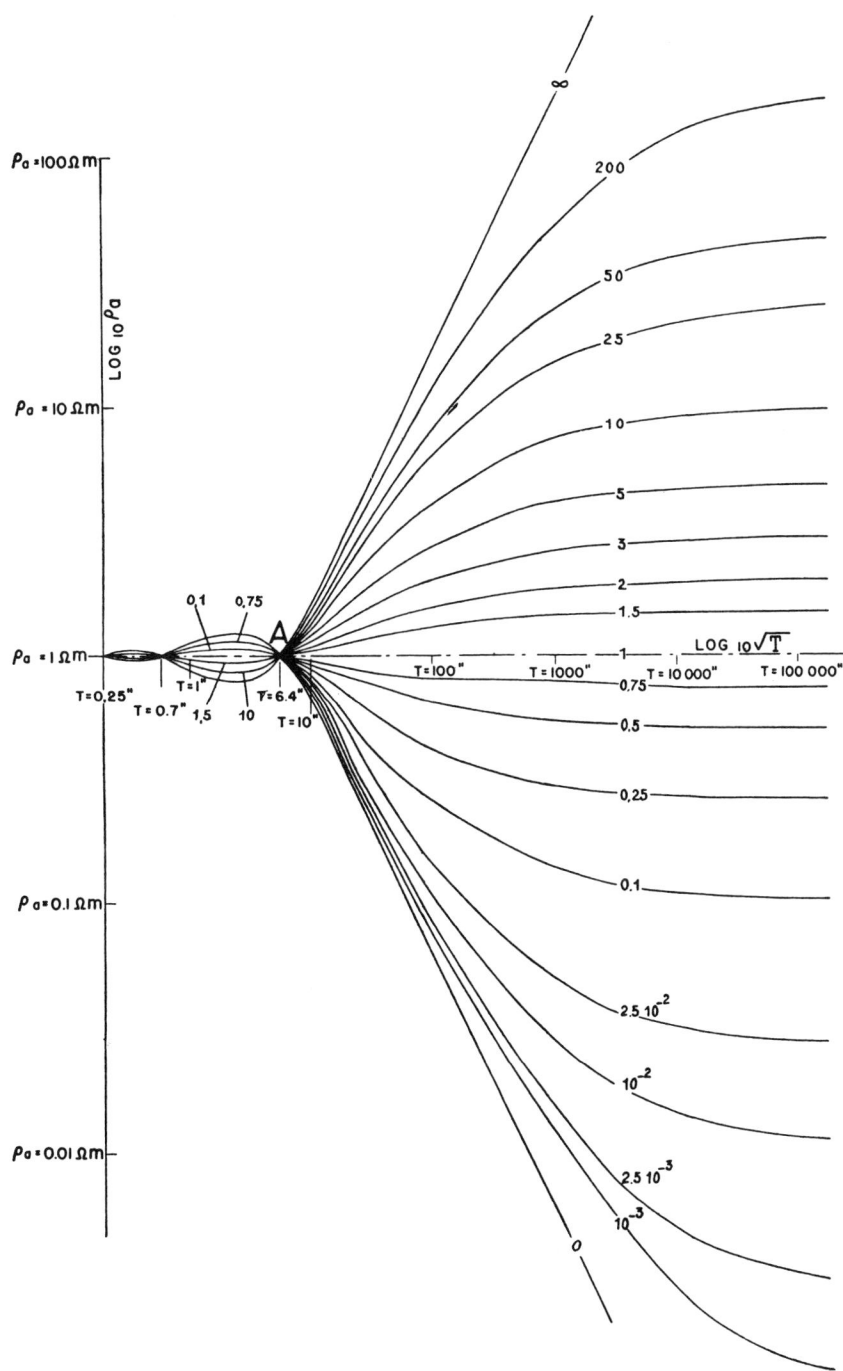

FIG. 7. Master curves of apparent resistivity for magneto-telluric soundings over a two-layer earth. Apparent resistivity plotted as a function of period of the telluric component for various resistivity contrasts. Numbers on the curves show the resistivity of the lower medium in ohm-meters. Resistivity of the upper layer is always 1 ohm-meter.

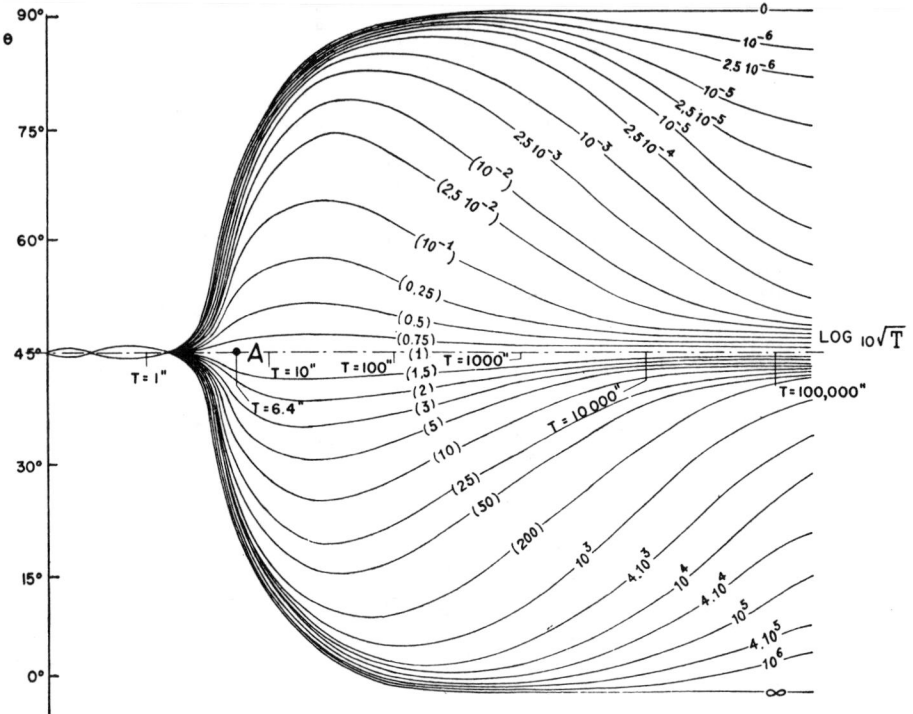

F_IG_. 8. Master curves of phase differences versus the period of the telluric component for various resistivity contrasts in a two-layer earth. Numbers on the curves show the resistivity of the lower medium in ohm-meters. Resistivity of the upper layer is always 1 ohm-meter.

Whereupon

$$\frac{2}{p_1} = (2n + 1)\frac{\pi}{2}; \qquad \sqrt{T} = \frac{8}{2n + 1},$$

n being an integer.

Of their common points, the one which is situated the most to the right, and which is marked A on the chart, is consequently defined by

$$\rho_a = 1, \qquad \sqrt{T} = 8.$$

The curves of system θ also have an infinite number of points in common which are defined by:

$$\theta = \frac{\pi}{4}, \qquad \sqrt{T} = \frac{4}{n},$$

n being an integer.

The coordinates of the point which is situated the farthest to the right, are consequently

$$\theta = \frac{\pi}{4}, \qquad \sqrt{T} = 4.$$

In order to make use of the master curves easier, we have marked on the set for θ the point A, having the coordinates

$$\theta = \frac{\pi}{4}, \qquad \sqrt{T} = 8,$$

which means the point having the same abscissa as point A of the curves for ρ_a.

An examination of system ρ_a shows that the apparent resistivity, equal to unity for $T = 0$ approaches ρ_2 when T becomes infinite. The general configuration of system ρ_a is, consequently, the same as that of the abacus for two formations in classical electrical soundings, which we shall designate from now on as E-soundings in order to abbreviate. Let us notice that when T approaches zero, the apparent resistivity only approaches unity by indefinite oscillation on both sides of its limit. In this way, it is sometimes possible to observe apparent resistivities which are *very slightly greater* than the greatest real resistivities of the formations present, or which are, on the other hand, very slightly smaller than the smallest of those resistivities. This phenomenon, a little paradoxical, is also observed, as one knows, in E-soundings, but only starting with three formations.

The examination of system θ shows that θ is equal to $\pi/4$, as well for $T = 0$ as for T infinite. This set of curves, which has no equivalent in E-soundings, is evidently going to provide one of the most useful means of control in MT-soundings.

PRACTICAL USE OF MASTER CURVES FOR TWO FORMATIONS FOR THE INTERPRETATION OF MT-SOUNDINGS

All the calculations and *theoretical* formulas developed in this memorandum imply the use of electro-magnetic units, which may be of any sort providing they are consistent: cgs for instance. We have said previously which electro-magnetic units we should use in the expression of the experimental results (Formula 10). Those units are very practical, but they are neither classical nor self-consistent.

Therefore, it is necessary to specify now that we no longer want to consider our theoretical master curves as relating to the cases of two formations with resistivities 1 and ρ_2. The resistivities in question are $1\Omega m$ and $\rho_2\Omega m$. The depth of the stratum is not 1 but 1 Km. The abscissa of point A is not 8 but $(8/\sqrt{10})$ $(\sec)^{1/2}$.

This being established, when we represent graphically the *experimental* results of a real MT-sounding we shall plot as our abscissas the logarithms of the square root of the period expressed in seconds. The ordinates of the curve will be logarithms of the numerical value of the apparent resistivities expressed in Ωm.

In addition to this, we shall adopt the same scales as for the theoretical curves. It is convenient to draw the experimental curves on commercial tracing paper on which cross-section lines are printed. The master curves, on the contrary, are drawn on *plain* Bristol board.

In order to know if the two experimental curves ρ_a and θ are characteristic of a subsurface involving two formations, and in order to know the thickness of the first one, or in other words to carry out an interpretation, one must try, by suitable translations, to bring the two experimental curves into coincidence, on the one hand with curve ρ_a, on the other hand with curve θ, of the theoretical set of curves.

If we are to be entitled to consider the result as satisfactory, it is necessary to insure that the two theoretical curves with which we compare the respective experimental curves correspond to the same value of the parameter ρ_2. Furthermore, the two translations which are to be executed parallel to the axis of the abscissas must be identical. From then on, we shall be able to calculate the resistivities ρ_1' and ρ_2' of the two formations at the same time as the depth h' of the second one.

Point A of family ρ_a, as seen through the transparent tracing paper on which we plot the experimental data, has itself an ordinate whose numerical value is the logarithm of $\rho_1' \Omega m$. Likewise, the asymptote of the theoretical curve ρ_a, considered as sufficient, has *on the tracing paper* an ordinate whose numerical value is the logarithm of $\rho_2' \Omega m$. In other words, the value of ρ_1' and ρ_2' can be read at once on the tracing paper if one does not care for a precision of expression which, in this case, has the inconvenience of making things which are very plain look extremely complicated.

The depth h' remains to be determined. Point A of the one or the other abacus, seen through transparent tracing paper, has an abscissa whose numerical value is $X(\sec)^{1/2}$. Conformably to the laws of similitude, one finds, consequently,

$$K_T = \frac{10}{64} X^2; \qquad K_\rho = \rho_1'.$$

Whereupon

$$K_L{}^2 = \frac{10}{64} X^2 \rho_1'; \qquad h' = \frac{X}{8} \sqrt{10\rho_1'} \text{ km.}$$

INTERPRETATION IN THE CASE OF ANY STRATIFIED EARTH. RESOLVING POWER OF MT-SOUNDINGS

Let us now supose that one has to deal with three formations, of resistivities ρ_1, ρ_2, and ρ_3. The depth of the second one is h_1 and that of the third formation or substratum is h_2. If the ratio h_2/h_1 is sufficiently great, the influence of the sub-

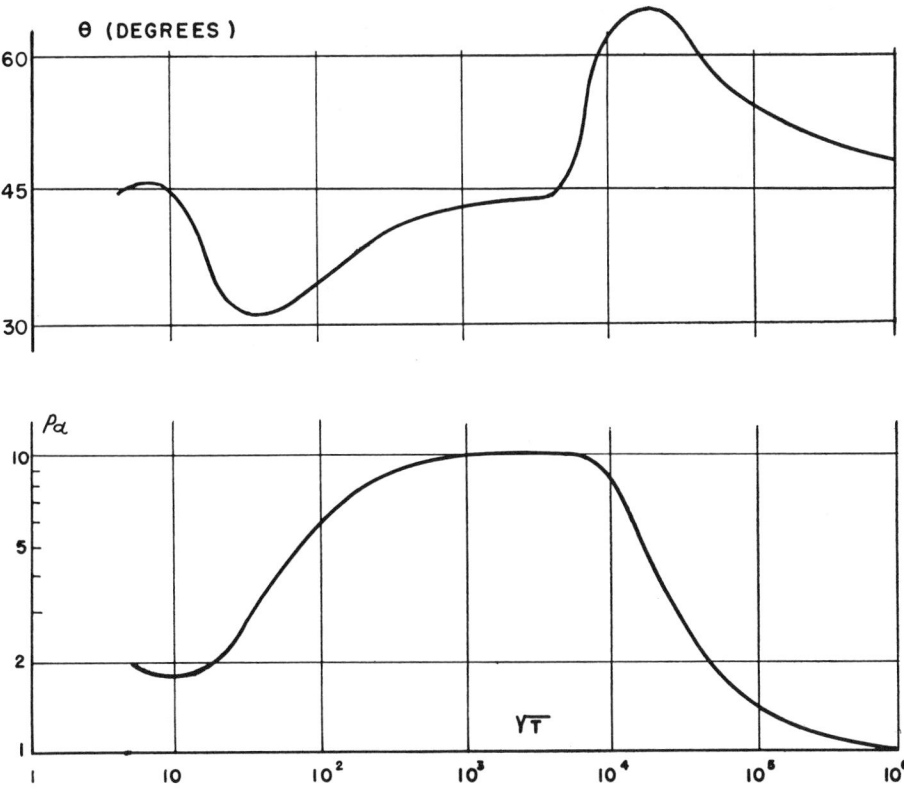

FIG. 9. Computed curves for hypothetical magneto-telluric sounding over three layers in which thickness of second layer is 900 times that of first and in which ρ_1, ρ_2, and ρ_3 are in the ratio of 2:10:1.

stratum starts to be appreciable only for such large periods that the apparent resistivity is already practically equal to ρ_2, while θ has already regained, to a close approximation, its initial value, $\pi/4$. In other words, the influence of the third formation only starts to make itself felt for such periods that the influence of the first formation may be neglected. In order to determine the termination of a graph for three formations of this kind, one is simply led to construct two graphs (ρ_a or θ) for two formations, one after the other. In the second of the graphs for two formations, the formation which is from now on to be known as the first one has the resistivity ρ_2 and the thickness h_2, while the formation from now on to be known as the second one possesses the resistivity ρ_3.

An example of this kind is furnished by Figure 9, in which the ratio h_2/h_1 is supposed to have the value of 900, while the resistivities ρ_1, ρ_2, and ρ_3 are proportional to the numbers 2, 10, and 1.

This highly favorable circumstance in which the master curves for two formations at once allow the interpretation of a sounding carried out over a section

167

involving more than two formations does not occur if one is dealing with strata of insufficient thickness, either for *E*-soundings or for MT-soundings.

Let us imagine, for instance, a subsoil of three formations, such that h_2/h_1 is equal to 10, while the resistivities ρ_1, ρ_2, and ρ_3 are proportional to 9, 1, and ∞. Figure 10 represents the corresponding *E*-sounding, while Figure 11 represents the two curves for the MT-sounding.

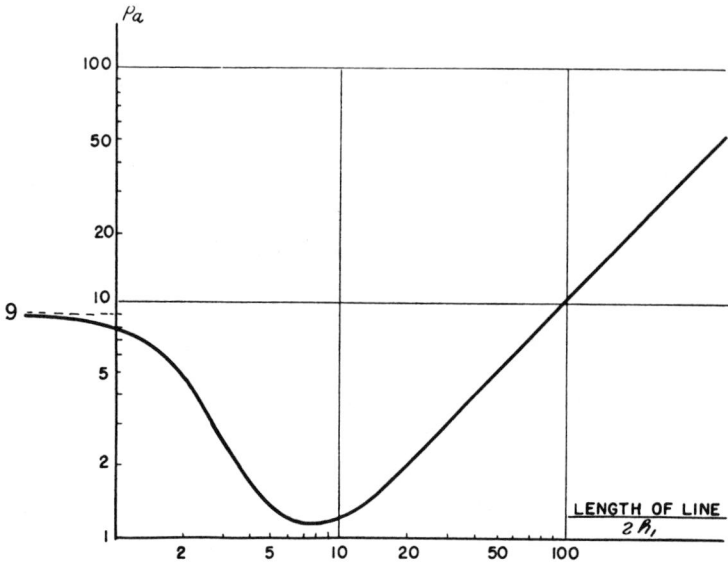

Fig. 10. Computed curve for hypothetical resistivity survey of conventional type over three layers in which thickness of second layer is 10 times that of first and in which ρ_1, ρ_2, and ρ_3 are in the ratio of 9:1: and ∞.

On each of those diagrams, the apparent resistivity, equal to 9 for the short lengths of line (*E*-sounding) or the small periods (MT-soundings), decreases at first when one increases the length of the line or the period, reaches a minimum, and increases indefinitely afterwards. This minimum is not equal to 1, either on the *E*-sounding or on the MT-sounding. One will notice however, that while it is practically equal to 1 in the case of the MT-sounding,* it is only equal to 1.25 in the case of the *E*-sounding. In order to obtain, in the case of the *E*-sounding, with the same resistivities, a minimum practically equal to 1, it would be necessary that the ratio be at least 25.

We shall conclude from this, at first, that the MT-sounding separates the individual effects of the different strata of the subsoil better than the *E*-sounding, and that its resolving power is almost two and a half times higher. Also bearing in mind the additional information furnished by the phase curves, it is consequently

---

* And even slightly less than unity, because of the somewhat paradoxical phenomenon pointed out when we described the master curve for two formations.

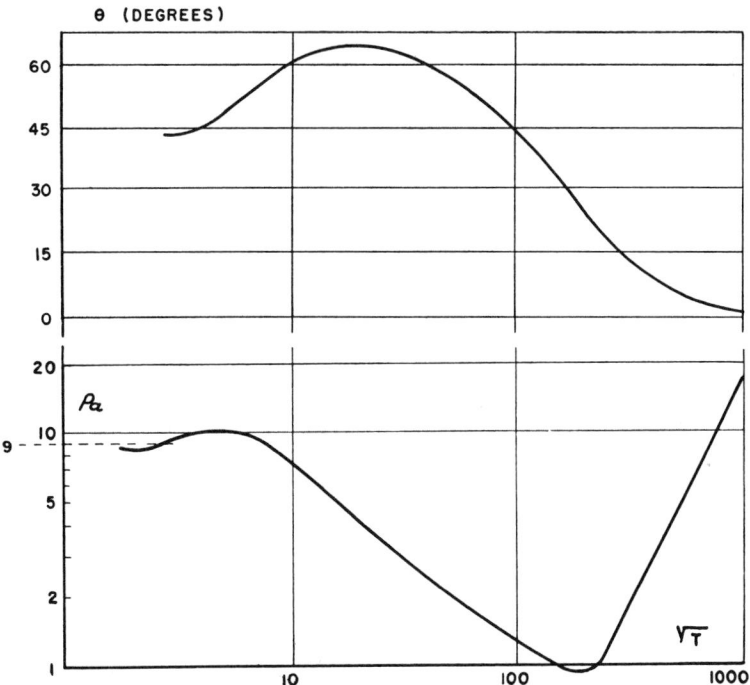

FIG. 11. Computed curves for hypothetical magneto-telluric sounding over three layer configuration of Figure 10.

already very obvious that the MT-sounding allows one to arrive at more precise conclusions than the *E*-sounding, even if one is satisfied with semi-qualitative information.

But still, in the case of MT-soundings, when the problem calls for it, there is nothing to keep us from submitting the semi-qualitative hypothesis we are referring to here to the test of exact calculation. When one has suspected the existence of a certain number of strata, when one has been able to estimate approximately the order of magnitude of their thicknesses and of their resistivities, one can perform the complete calculation of the results that one would obtain if the subsoil presented exactly the supposed structure. If there is disagreement between calculation and experience, one will alter the values formerly assumed for the resistivities and the thicknesses so as to obtain an entirely satisfactory result by a method of successive approximations.

In other words, the MT-sounding can be analyzed by the same method of interpretation one can apply in gravimetry and in magnetism which is so satisfactory for the prospector, but without fear of the disastrous consequences of the fundamental ambiguity which characterizes those last two methods.

Indeed, the calculation in question does not involve integrals nor series, as we have seen. It can be readily carried out when the general formulas for four, five, or more formations have been established in advance in algebraic form, as we have demonstrated here in the case of two and three formations.

However, it is possible to do much better and to save much time by use of an almost exclusively graphical method which is based on the results which will be obtained in the following paragraph.

APPARENT RESISTIVITIES AND PHASES AT THE DIFFERENT LEVELS INSIDE A HOMOGENEOUS FORMATION

Within a stratified section the complex relation E_x/H_y has a specific value at each level of depth z. We are going to obtain a formula particularly important

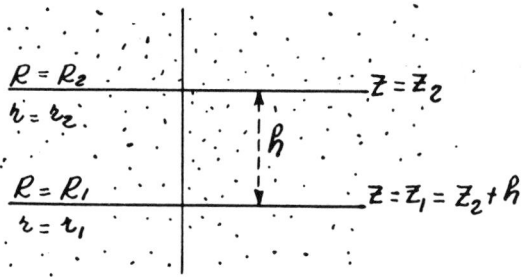

FIG. 12. Geometry for computing relationships between two levels in same medium.

in practice by considering two levels z_1 and z_2 at a distance h, inside the *same* formation of conductivity σ (Figure 12). We shall put down

$$R = - \frac{ia\sqrt{\sigma}}{\omega} \frac{E_x}{H_y} = - i \tan r. \tag{57}$$

The complex numbers R and r are functions of z which represent respectively the values R_1 and r_1 for $z=z_1$ and R_2 and r_2 for $z=z_2$.

A and B designating two constants, we have learned that the expressions for the fields (formulas 17, 18) are of the form

$$E_x = Ae^{a\sqrt{\sigma}z} + Be^{-a\sqrt{\sigma}z},$$
$$H_y = - \frac{i}{\omega} a\sqrt{\sigma}(Ae^{a\sqrt{\sigma}z} - Be^{-a\sqrt{\sigma}z}) \Bigg\} . \tag{58}$$

One deduces from these that:

$$R = \frac{1 + \lambda e^{-2a\sqrt{\sigma}z}}{1 - \lambda e^{-2a\sqrt{\sigma}z}} ; \qquad \lambda = \frac{B}{A} = C^r, \tag{59}$$

or in another form

$$\frac{R-\mathrm{I}}{R+\mathrm{I}}\,e^{2a\sqrt{\sigma}\,z} = \lambda = C^r. \tag{60}$$

Consequently, one calculates at once R_2 as a function of R_1:

$$R_2 = \frac{R_1(\mathrm{I} + e^{2a\sqrt{\sigma}\,h}) + (\mathrm{I} - e^{2a\sqrt{\sigma}\,h})}{R_1(\mathrm{I} - e^{2a\sqrt{\sigma}\,h}) + (\mathrm{I} + e^{2a\sqrt{\sigma}\,h})}, \tag{61}$$

and, afterwards, r_2 as a function of r_1:

$$\tan r_2 = \frac{\tan r_1 - \tan (ia\sqrt{\sigma}\,h)}{\mathrm{I} + \tan r_1 \tan (ia\sqrt{\sigma}\,h)} = \tan (r_1 - ia\sqrt{\sigma}\,h), \tag{62}$$

from which finally

$$r_2 = r_1 - ia\sqrt{\sigma}\,h = r_1 + \sqrt{2}\,(h/p)e^{-3i\pi/4}. \tag{63}$$

The indeterminacy of the argument r does not concern us, since we are only interested in the value of R.

It is easy now to go back to the apparent resistivities ρ_a and to the phases θ defined by

$$\frac{E_x}{H_y} = \sqrt{\frac{\rho_a}{2T}}\,e^{-i\theta}. \tag{64}$$

Consequently one has

$$\sqrt{\frac{\rho_a}{\rho}}\,e^{-i(\theta+\pi/4)} = \tan r, \tag{65}$$

and finally

$$\frac{(\rho_a)_2}{(\rho_a)_1}\,e^{-2i(\theta_2-\theta_1)} = \left(\frac{\tan r_2}{\tan r_1}\right)^2. \tag{66}$$

Since the calculations of the prospectors are not usually carried out to 20 decimal places, a simple chart of the complex values of the tangents of a complex argument allows one to calculate an MT-sounding very quickly for $n+\mathrm{I}$ formations starting from a sounding for n formations when the $(n+\mathrm{I})$st formation is situated on top of the nth one.

REMARKS

1. In the calculation of a theoretical MT-sounding by an operation of successive approximations, the geophysicist, by constructing his theoretical section through the stacking of strata laid down one on top of the other, proceeds exactly in the same way as nature did when the real strata of the ground were laid down by successive processes of sedimentation.

2. At two stations over a sedimentary basin, the section only differs, in principle and as a first approximation, through the addition—or through the subtraction—of a certain number of superficial

strata. Consequently, it will often be convenient as a first working hypothesis to calculate the complex ratio of the complex quotients E_x/H_y, obtained experimentally at the two stations, a ratio whose interpretation involves only the thickness of the superficial layer, which is different for the two stations.

3. When, for one reason or another, one knows with certainty the resistivities of the ground to a certain depth, it may be easy to omit, through calculation, the influence of this known part of the ground and limit the interpretation only to the unknown subjacent portion.

4. This circumstance occurs in particular when one performs a MT-sounding over a body of water for which the depth and conductivity are known. The former calculation allows one in such a case to correct the MT-sounding for the influence of the sea; in other words, it allows one to obtain, through a very accurate calculation, the diagrams for the MT-sounding that one could have determined experimentally if the water were to have been drained away.

VARIATIONS APPLICABLE TO MT-SOUNDINGS PERFORMED UNDER THE SEA

The measurement of the electrical field at sea does not present any specific technical difficulty. The line of measurement is maintained on the surface of the water through the use of floaters, in the same way as fishermen do with their nets. Moreover, there is no difficulty whatsoever in carrying out a correct galvanometric recording on board of a ship tossed about by the waves. One need only be suspicious if one observes phenomena which have a period the same either as that of the marine currents or as that of the swell, since the electrical currents induced by motion of the conducting water in the magnetic field of the earth do not meet the requirements of the theory we have set forth.

The measurement of the magnetic field offers more serious technical difficulties if one is not willing or able to install a self-recording magnetometer on a series of piles forming a foundation or in an immersed box on the bottom of the sea.

One way of avoiding the difficulty consists in registering the magnetic field *on the ground* and the electric field *in the sea*. The daily experience of prospectors who use the telluric method, has shown, indeed, that in sedimentary beds, the line of the telluric current keeps an almost constant direction over expanses as large as 20–70 km. Besides, this direction would be strictly uniform in a precisely stratified earth.

Now the telluric current, even if it is a variable current, is, approximately conservative because of its very low frequency. As I have already pointed out, the magnetic field is very approximately the same at two stations not too distant from each other on the same straight line ox, since it represents, except for a factor 4π, the total intensity of the telluric current through a stratum of unit width starting from the surface normal to ox.

The argument essentially implies that the two stations are situated on the same straight line ox, perpendicular to the magnetic component one is considering. Consequently, it is very advisable to adhere to this condition if possible. However, experience shows that, in practice, this requirement is not always strictly binding.

These observations will not come as a surprise to observatory geophysicists. Through experience, they are well convinced of the fact that the meaning of their magnetic data does not depend particularly on the electrical resistivity of the geological strata in the vicinity of their observatory. But, for over a hundred years, since the first observation of telluric current, it has been found that in telluric registrations, on the contrary, one has had to be greatly concerned with the local geologic structure.

Another way to avoid the difficulty consists in observing that in the homogeneous medium formed by the sea water, where a telluric sheet is flowing parallel to o*x*, the relation between *any* two electro-magnetic dimensions, depending linearly on the Hertz vector, is expressed as a function of parameter λ only (Equation 59). The ratio E_x/H_y is also expressed as a function of λ. In other

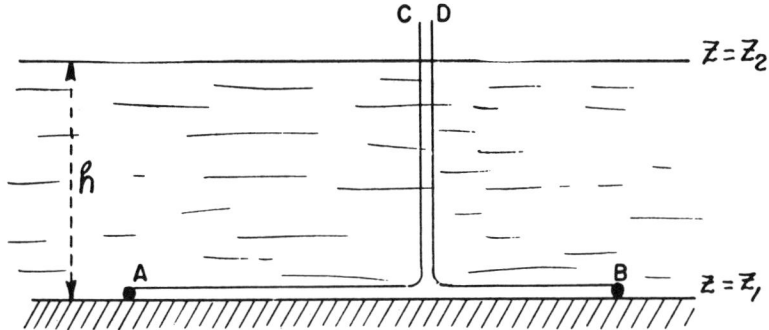

Fig. 13. Configuration of electrodes on water bottom for submarine MT measurements.

words, the study of the relation of any two electro-magnetic quantities is absolutely equivalent to that of the ratio E_x/H_y.

One can, for instance, substitute for the measurement of Hy, the measurement of the electromotive force induced in a large vertical ring parallel to o*x*, this ring being constructed much more easily on the sea than on the ground. Yet one knows that, if the vertical height of this ring is small so that the magnetic component H_y inside it is almost uniform, the measurement of the induced electromotive force is a classical way of measuring H_y.

It may be easier to substitute for the measurement of the magnetic field H_y that of a second electric field. Let us go back to Figure 11, supposing this time that level z_1 represents the horizontal sea bottom, level z_2 the surface of the water (or, in a more general way, any level between the bottom and the surface of the sea). It is easy to measure the field E_x on the bottom of the sea by means of two immersed electrodes A and B connected with recording equipment on the boat by the two lines AC and BD (Fig. 13).

It can be shown that

$$\sqrt{\rho_a}\, e^{-i\theta} = \frac{2\pi p}{\sqrt{T}}\, e^{-i\pi/4} \frac{\dfrac{(E_x)_1}{(E_x)_2} \sinh\left(\sqrt{2}\, e^{-i\pi/4}\, \dfrac{h}{p}\right)}{1 - \dfrac{(E_x)_1}{(E_x)_2} \cosh\left(\sqrt{2}\, e^{-i\pi/4}\, \dfrac{h}{p}\right)}, \tag{67}$$

in which ρ_a and θ have reference to the apparent resistivity and to the phase relative to level z_1, the level of the bottom of the sea; that is to say, the parameters of an MT-sounding that could be performed on the sea bottom if drained.

CONCLUSION. FIELDS OF PRACTICAL APPLICATION FOR THE MAGNETO-TELLURIC METHOD

It follows from the above that the ideal way to apply the magneto-telluric method consists in performing an MT-sounding as described. When the subsoil is approximatively tabular, the harmonic analysis of the telluric and magnetic diagrams makes it possible to conduct a careful quantitative interpretation which gives us the thickness and the resistivity of the various strata.

The periods higher than one second are exactly adapted to the study of large sedimentary beds and their petroliferous structure. Besides, their recording does not involve serious technical difficulties.

The study of the shortest periods, less than one second, seems technically difficult at the present stage of the art. However, it is less urgent in the light of present needs in geophysical prospecting. It should eventually allow us to adapt the magneto-telluric method to various applications requiring detail of the kind involved in civil engineering studies, in mineral prospecting, and in the search for underground water.

We want to draw attention to the fact that an isolated MT-sounding carried on in the center of a large unknown area can present information similar to that given by a wildcat well in a large scale reconnaissance. For instance, the measurement of the number of kilometers thickness of sediment in the center of a large basin presents a problem which cannot be solved even partially by any geophysical method up until now. The magneto-telluric method should be able to solve the problem by use of only a single station.

The discarding of the base station, which is indispensable in the telluric method, gives the operator more freedom of movement and improves the organization of his survey. He is no longer compelled to proceed slowly. He can afford to operate in a more rational way by setting up his initial stations at some distance from each other. Later on, he can locate stations with a closer spacing, but only to the extent required by continuity.

Consequently, one can lay out the survey of a large sedimentary basin by performing at the start a small number of MT-soundings far removed from one another, but with a great depth of investigation. In the second step, one will intercalate stations closer together, and at these he will perform MT-soundings with a more moderate depth of penetration. Finally, the continuity between the stations will be assured either by soundings with a relatively small depth of investigation, or, once in a while, by simple, quick determinations of the apparent resistivity summarily evaluated through a very simplified analysis of the magneto-telluric data. It is unnecessary to add that the magneto-telluric method will be particularly appreciated every time that a deep petroliferous structure appears in complete disharmony with the structure on the surface. In this case it is essential for the prospector to penetrate to a great depth of investigation.

If it should happen that the earth is not even approximately stratified, the quantitative interpretation of the MT-soundings must be just about ruled out. A knowledge of the apparent resistivities and of their variation as a function of the period and of the direction of the line provides, nevertheless, certain indications which can be diagnostic in special cases, even though they are to a large degree qualitative.

ACKNOWLEDGMENTS

I am very glad to express my heartiest thanks to Mrs. F. B. Riek, Jr., to Dr. Robert Van Nostrand, Dr. Milton B. Dobrin and Dr. R. L. Caldwell, Field Research Laboratories, Magnolia Petroleum Company, and to Mr. E. Selzer, physicist at the Institut de Physique du Globe of Paris.

Mrs. Riek, a professional translator, undertook the translation of my French text into English for the Magnolia Petroleum Company. Drs. Van Nostrand, Dobrin, and Caldwell edited the translation with competence and judgment. My friend Mr. Selzer gave me valuable help in correction of the proofs.

The theoretical work reported in this paper was done some time ago and has been mentioned in applications for patents which have been made in several countries to protect the new prospecting method involved. Because of the potential practical applications I have had to postpone any publication related to magneto-telluric phenomena for many years.

Meanwhile, the Russian scientist Tikhonov, and the Japanese scientists Kato, Kikuchi and Rikitake had also recognized the existence of such an effect. To my knowledge, they have not pointed out the possibility disclosed by my work of applying these results to practical geophysical exploration. They have, however, paid attention to their possible use for investigating the electrical conductivities of very deep regions in the Earth's crust.

It is therefore a real pleasure for me to give these scientists proper credit and to list the papers that, to my knowledge, they have published on this subject:

Kato, Y. and Kikuchi, T. (1950) *Sci. Rep. Tôhoku Univ. Ser. V, Geophysics,* Vol. 2, p. 139.

Rikitake, T. (1950) *Bull. Earthquake Res., Inst. Tokyo Univ.,* Vol. 28, p. 45, 219. Also (1951) Vol. 29, pp. 61, 271.

Tikhonov, A. N. (1950) *Dokl. Akad. Nauk S.S.S.R.* 73, 295.

GEOPHYSICS, VOL. 37, NO. 6 (DECEMBER 1972), P. 1005-1021, 3 FIGS.

PROCESSING AND INTERPRETATION OF MAGNETOTELLURIC SOUNDINGS†

G. KUNETZ*

A few methods in the processing and interpretation of magnetotelluric soundings over a stratified earth are investigated, with emphasis on the less commonly used time-domain procedures. Analytical expressions of the theoretical transfer function between the magnetic- and electric-field variations, both in frequency and time domain, are derived. Their properties are studied, and recursive algorithms are given for their numerical computation. On the other hand, a procedure is outlined which leads directly in the time domain to the experimental values of this transfer function. It is similar to the methods used in seismic analysis for signal determination and makes use of the auto- and crosscorrelation functions of the measured field variations. Finally, methods of interpretation, based either on a visual or on an automatic comparison of these theoretical and experimental transfer functions, are proposed. For the case of automatic interpretation, complementary geologic data should be used where possible to take care of the lack of uniqueness of the solution.

INTRODUCTION

The purpose of this paper is to set out the main features of a few methods which may be useful in the processing and the interpretation of magnetotelluric measurements. These methods, however, have not yet been tested on actual field examples.

In all the following developments a plane electromagnetic wave of quasi-static frequencies, propagating perpendicularly to a stratified earth, is assumed. Hence, all the necessary information is, in principle, contained in the simultaneous time variations of a pair of mutually perpendicular horizontal-field components, one of the electric field $E(t)$ and one of the magnetic field $H(t)$, or in their complex spectra $E(\omega)$ and $H(\omega)$.

Three main points will be considered (see Figure 1).

The first point consists of the determination of convenient theoretical expressions, especially in the time domain, of the transfer function leading from the magnetic field (or its time derivative) to the corresponding electric field. To this end a recurrence procedure and a series expansion are developed, which lead to an expression of the frequency-domain transfer function $Z(\omega)$, well adapted to a transposition into the time domain, giving $U_{th}(t)$.

The second point deals with a method for the computation of the experimental values of this transfer function from the measured magnetic and electric-field variations. A statistical approach, similar to those used in seismic analysis for signal determination, involving the autocorrelation and crosscorrelation functions of the electric and magnetic fields, gives a least-mean-square estimate of the experimental values $U_{exp}(t)$ of this transfer function.

The third point is devoted to methods of interpretation, either in the frequency or in the time domain, principally based on an optimal fit of the theoretical and experimental transfer functions. First, however, the application in the time domain of the classical curve-matching techniques [comparison of experimental apparent resistivity curves derived from $U_{exp}(t)$ to apparent resistivity master curves derived from $U_{th}(t)$] is shown.

† Paper presented at the 40th Annual International SEG Meeting, November 10, 1970, New Orleans, Louisiana. Manuscript received by the Editor March 15, 1971; revised manuscript received June 12, 1972.
* Compagnie Générale de Géophysique, Paris, France.

1005

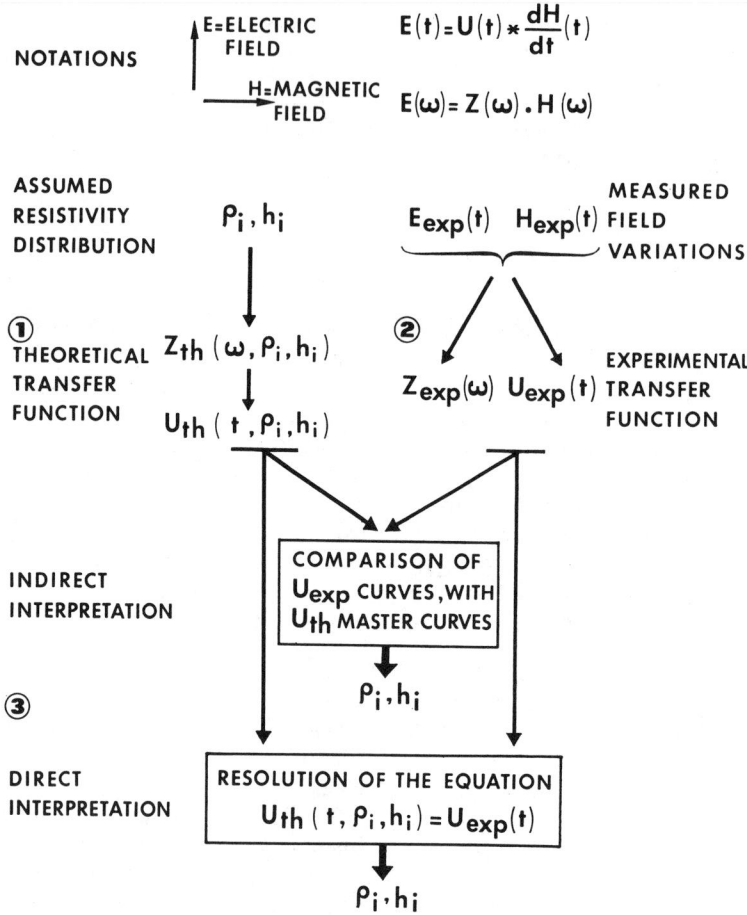

FIG. 1. Overview of the discussed methods.

Direct, and more or less automatic, interpretation techniques are thereafter considered, making use of the equation $U_{\text{th}}(t, \rho_i, h_i) = U_{\text{exp}}(t)$ (or the same equation in the complex frequency domain) and trying to find the resistivity distribution (ρ_i, h_i) satisfying the above equation as well as possible. Because of the lack of uniqueness of the solution of this problem, additional data, derived from geological or other geophysical sources, must be introduced, either at the beginning of the computation or at the final stage of the interpretation.

In the deduction of the proposed methods, several results or techniques borrowed from other fields of geophysical prospecting are used, such as seismic or direct-current electrical work. They are more or less classical in their own domain, and

it did not seem necessary to elaborate on them. However some of the less-known algorithms or formulas are given in the appendices.

EXPRESSIONS AND PROPERTIES OF THE THEORETICAL TRANSFER FUNCTION IN THE FREQUENCY AND IN THE TIME DOMAIN

Recurrence formula and series expansion in the frequency domain

If we consider a stratified earth of n layers with thicknesses

$$d_1, d_2, \cdots, d_{n-1}, d_n = \infty,$$

and resistivities

$$\rho_1, \rho_2, \cdots, \rho_n,$$

and put

$$E(\omega) = Z(\omega) \cdot H(\omega),$$

and

$$W(\omega) = \frac{ik_1}{\omega} Z(\omega),$$

with

$$k_1 = \sqrt{-\frac{4\pi i \omega}{\rho_1}},$$

several procedures can be used for the determination of $W(\omega)$ (Berditchevski, 1968 and Kunetz, 1969). The best known procedure makes use of a sequence of auxiliary functions R_p, related by

$$R_p = \frac{L_p + R_{p+1} \cdot u_{p+1}}{1 + L_p \cdot R_{p+1} \cdot u_{p+1}},$$

where

$$L_p = \frac{\sqrt{\rho_{p+1}} - \sqrt{\rho_p}}{\sqrt{\rho_{p+1}} + \sqrt{\rho_p}},$$

and

$$u_p = \exp\left[-2 \frac{d_p}{\sqrt{\rho_p}} \sqrt{-4\pi i \omega}\right].$$

Starting from $R_{n-1} = L_{n-1}$, it leads to R_1, from which the function W is deduced by

$$W = 1 + 2 \frac{R_1 u_1}{1 - R_1 \cdot u_1}.$$

This recurrence procedure, which leads to the expression of W, shows also that this transfer function is minimum phase. First it follows from

$$|u_p| \leq 1 \quad -1 \leq L_p \leq +1$$

and

$$-1 \leq R_{n-1} = L_{n-1} \leq +1$$

that one has, for any p,

$$|R_p| \leq 1,$$

and, in particular, $|R_1| \leq 1$, and $|R_1 u_1| \leq 1$. Hence, both $1 + R_1 u_1$ and $1 - R_1 u_1$ are minimum phase, and so is $W = (1 + R_1 u_1)/(1 - R_1 u_1)$.

The constraint on the phase of W is even stronger; this phase has to remain in the interval

$[-(\pi/2), (\pi/2)]$.[1] As a consequence, the real part of $W(\omega)$ has to be positive. This, however, is only a necessary, but not a sufficient, condition to be satisfied by W in the case of a horizontal stratification.

For $n = 2$ and $n = 3$ the following well-known expressions of W are found:

$$W_{(2)} = 1 + 2 \frac{u_1 L_1}{1 - u_1 L_1}$$

and

$$W_{(3)} = 1 + 2 \frac{u_1 L_1 + u_1 u_2 L_2}{1 - u_1 L_1 + u_2 L_1 L_2 - u_1 u_2 L_2}.$$

This recurrence formula, as well as the above particular expressions of W, are identical in form to those arrived at for the determination of the Stefanesco kernel function in the case of direct-current distribution in a stratified earth (Baranov and Tassencourt, 1959), provided the definitions of the parameters L_p and u_p are modified as follows:

$$(L_p) = \frac{\rho_{p+1} - \rho_p}{\rho_{p+1} + \rho_p}; \quad (u_p) = e^{-2\lambda d_p}.$$

λ is the variable of integration in the direct-current case. In this latter case it is well known that, assuming that all the thicknesses d_p are multiples of a finite thickness h, and, writing $(u) = e^{-2\lambda h}$, the function (W) can be developed in a series:

$$(W) = 1 + 2 \sum_{m=1}^{\infty} (q_m)(u)^m,$$

where the (q) are the so-called electrical images. In the same way, if all the ratios $d_p/\sqrt{\rho_p}$ are multiples of a finite ratio a, the function W can be developed in a series:

---

[1] A still more restrictive condition on the phase φ_W of W is obtained by observing that as

$$|u_1| = e^{-\alpha\sqrt{\omega}} \text{ and } |R_1| \leq 1$$

$$\left(\text{where } \alpha = 2 \frac{d_1}{\sqrt{\rho_1}} \sqrt{2\pi}\right), \quad |R_1 u_1| \leq e^{-\alpha\sqrt{\omega}}.$$

Hence,

$$|t_g \varphi_w(\omega)| \leq \frac{1}{\sinh \alpha\sqrt{\omega}}.$$

$$W = 1 + 2 \sum_{m=1}^{\infty} q_m u^m,$$

where

$$u = e^{-2a\sqrt{-4\pi i\omega}}.$$

The above assumptions mean that the earth may be subdivided into "elementary" layers, either of constant thickness (direct current) or of constant value $h/\sqrt{\rho}$ (magnetotellurics). These elementary layers define a new set of "reflection coefficients" L_m (some of which will be zero, when adjacent elementary layers have the same resistivity).

As a consequence of the identical form of the two sets of formulas, relative to direct current and magnetotellurics, the image values q_m are linked to the reflection coefficient L_m by exactly the same relationship in both cases. Moreover, it has been shown (Baranov and Kunetz, 1958) that this relationship is also the same as the one which, in the case of a plane acoustic wave in an elastic stratified medium, links the amplitude of the reflections at the surface (including multiples, i.e., the synthetic seismogram with multiples) to the reflection coefficients (the elementary layers corresponding, in the case of seismic work, to constant traveltime $\tau_i = h_i/V_i = $ constant).

The algorithm which allows the calculation of the images q_m from the reflection coefficients L_m presents no difficulties (see Appendix A1).

The inverse algorithm, which gives the reflection coefficients L_m, starting from the images q_m, is just as simple in principle (see Appendix A2). However, for an arbitrary set of numbers q_m, the reflection coefficients L_m will generally have values greater than one, which would have no physical meaning, as they correspond to negative resistivities. It has been demonstrated, in the case of acoustic waves (Kunetz and d'Erceville, 1962), that a necessary and sufficient condition for obtaining $|L_m| \leq 1$ and, hence, positive resistivity values is that the images q_m satisfy the condition

$$\phi(\theta) = 1 + 2 \sum_{m=1}^{\infty} q_m \cos m\,\theta \geq 0,$$

for any θ.

This condition will be used in the section devoted to interpretation.

The transfer function in the time domain

The above series expansion of $W(\omega)$ will be used now to obtain a time-domain expression of the operator linking the electric and magnetic fields. More precisely, the operator $U(t)$ sought will be one which, when convolved with dH/dt, leads to $E(t)$:[2]

$$E(t) = U(t) * \frac{dH}{dt}.$$

The Fourier transform of dH/dt is $-i\omega H(\omega)$. We write $E(\omega)$ as

$$E(\omega) = -\frac{Z(\omega)}{i\omega}\left[-i\omega H(\omega)\right].$$

It is clear, therefore, that $U(t)$ will be the Fourier transform of

$$U(\omega) = -\frac{Z(\omega)}{i\omega} \equiv \frac{W(\omega)}{k_1}.$$

The series expansion of $U(\omega)$ is

$$U(\omega) = \sqrt{\frac{i\rho_1}{4\pi\omega}}\left[1 + 2\sum_{m=1}^{\infty} q_m e^{-2am\sqrt{-4\pi i\omega}}\right].$$

Its Fourier transform, term by term, is

$$U(t) = \frac{1}{2\pi}\sqrt{\frac{\rho_1}{t}}\left[1 + 2\sum_{m=1}^{\infty} q_m e^{-(4\pi a^2/t)m^2}\right].$$

The function $U(t)$ thus determined[3] is such that

$$E(t) = U(t) * \frac{dH}{dt}$$

$$\equiv \int_0^{\infty} U(\tau)\frac{dH}{dt}(t-\tau)d\tau.$$

The above expression of $U(t)$ is well adapted for the numerical calculation, starting from the thicknesses and the resistivities of the layers. This

---

[2] It would also be possible to determine the operator $Y(t)$ which convolved by $E(t)$ would lead to $H(t)$:

$$H(t) = Y(t) * E(t).$$

$Y(t)$ would be the Fourier transform of

$$\tilde{Y}(\omega) = \frac{4\pi}{\rho_1}\cdot\frac{1}{k_1}\cdot\frac{1}{W(\omega)}.$$

[3] The parameters in the above expressions are in cgs em units. In particular, the value of the coefficient $4\pi a^2$, which has the dimensions of time, can be given in "geophysical" units as:

$$4\pi a^2 = t_{0(\text{seconds})} = \frac{4\pi h_1^2 \text{ (kilometers)}}{10\rho_1 \text{ (ohm-meters)}}.$$

would lead to master curves which could be compared to the experimental values of the time-domain transfer function obtained from the measured field variations $E(t)$ and $H(t)$.

The expression between brackets in $U(t)$ can be written

$$S(t) = 1 + 2 \sum_1^\infty q_m e^{-m^2(t_0/t)},$$

with

$$t_0 = \frac{4\pi h_1^2}{\rho_1}.$$

For $t \to 0$, $S(t) \to 1$, it is known, in direct current theory, that for $t \to \infty$,

$$S(t) \to 1 + 2 \sum_1^\infty q_m = \sqrt{\frac{\rho_n}{\rho_1}}.$$

Hence,

$$S^2(t) = \frac{4\pi^2}{\rho_1} t U^2(t)$$

may be considered as a reduced apparent resistivity ρ_{app}/ρ_1.

As an example, a few theoretical apparent resistivity curves have been calculated and are represented in Figure 2 in a bilogarithmic scale. They are relative to three-layer cases in which the ratio $d/\sqrt{\rho}$ of the middle layer is nine times the corresponding ratio in the first layer. This entails that, in our subdivision in elementary layers, the middle layer has to be split into nine identical elementary layers, each having a thickness h_2, satisfying the relation

$$\frac{h_2}{\sqrt{\rho_2}} = \frac{h_1}{\sqrt{\rho_1}}.$$

Ten resistivity ratios have been adopted, leading to

$$L_1 = \frac{\sqrt{\rho_2} - \sqrt{\rho_1}}{\sqrt{\rho_2} + \sqrt{\rho_1}}$$

$$= .7 \ .6 \ .5 \ .4 \ .3 \quad -.3 \ -.4 \ -.5 \ -.6 \ -.7$$

and, thus, to the series of reflection coefficients to be used:

$$L_1 = L_1, L_2 = 0 \cdots L_9 = 0, L_{10} = -L_1.$$

It may be of interest to compare some of these curves to the corresponding curves in the frequency domain. The amplitude and phase curves relative to the two cases,

$$\frac{D_2}{D_1} = 36, \qquad \frac{\rho_2}{\rho_1} = 16,$$

and

$$\frac{D_2}{D_1} = 2.25, \qquad \frac{\rho_2}{\rho_1} = \frac{1}{16},$$

are represented on Figure 3 together with the corresponding time-domain curves.

It appears that the time-domain and the spectral-phase curves show the characteristic features of the succession of layers for smaller values of t (for the time-domain curve), or of the period T (for the phase curve), than does the spectral-amplitude curve. This, however, does not prove an advantage with respect to the depth of investigation of the use of time-domain or phase curves. Such curves may be more difficult to obtain, with comparable precision, from the same length of field records.

DETERMINATION OF THE EXPERIMENTAL VALUES OF THE TRANSFER FUNCTION

Several methods are used for the determination of this transfer function in the frequency domain. Therefore, in this paper, only a few lesser-known possibilities will be considered, mainly for the determination of this function in the time domain.

Direct determination in the time-domain—Use of the Wiener-Levinson algorithm

Because of the presence of noise on the experimental data and the very simplified assumptions on the nature of the electromagnetic field and that of the medium, the theoretical relation

$$E(t) = U(t) * H'(t)$$

cannot be satisfied exactly with an operator $U(t)$ starting at the time $t = 0$. Therefore, the experimental operator $U_{exp}(t)$ may be defined as the one to be applied to the measured $H'(t)$ in order to minimize its difference with $E(t)$:

$$\int_{-\infty}^{\infty} \left[E(t) - \int_0^\infty U(x) \cdot H'(t-x) dx \right]^2 dt$$

$$= \text{minimum}.$$

This leads to the Wiener-Hopf-type integral equation

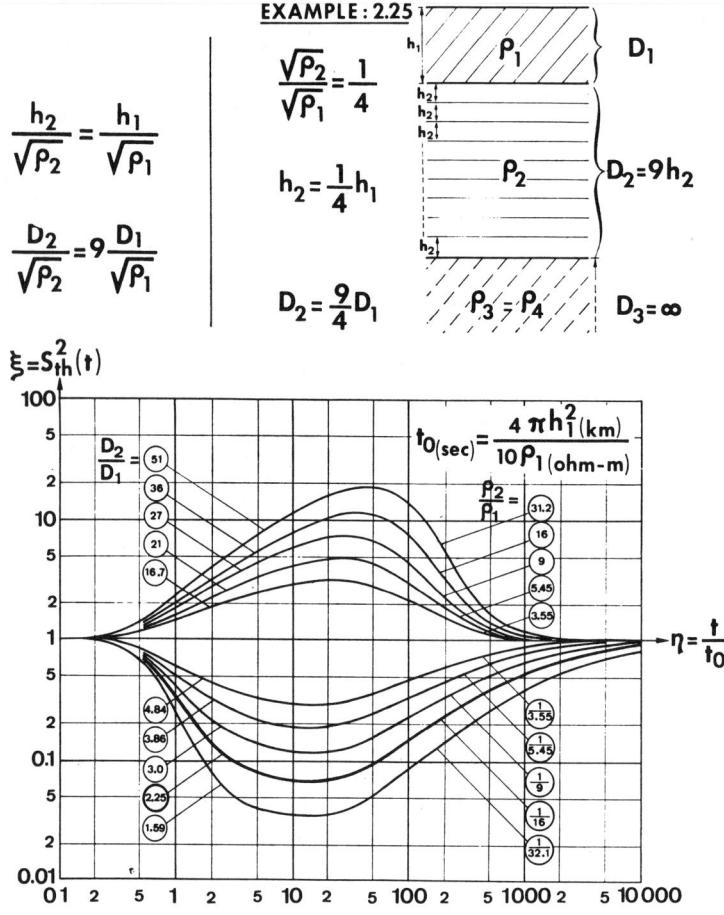

FIG. 2. Apparent resistivity master curves in the time domain.

$$\int_0^\infty U(x)\left[\int_{-\infty}^\infty H'(t)H'(t-x+\tau)dt\right]dx = \int_{-\infty}^\infty E(t)H'(t-\tau)dt,$$

which can be written

$$\int_0^\infty U(x)A(\tau-x)dx = F(\tau).$$

$A(\tau)$ is the autocorrelation function of $H'(t)$, and $F(t)$ is the crosscorrelation of $E(t)$ by $H'(t)$.[4]

[4] However, this way of determining $U(t)$ is optimal only if the noise mainly affects the electric component $E(t)$. Otherwise, it yields a biased estimate of $U(t)$ with systematically too low energy. This is apparent from the fact that the noise disappears, in principle, from the crosscorrelation function $F(\tau)$ on the right-hand side of the integral equation giving $U(t)$, whereas it remains in the autocorrelation $A(\tau)$ of $H'(t)$, on the left-hand side.

The assumption of a noise predominant on the magnetic component would lead to the determination of the operator $Y(t)$ (footnote 2) from the integral equation:

The corresponding system of linear equations

$$U_0A_0+U_1A_1+U_2A_2+\cdots U_pA_p = F_0,$$
$$U_0A_1+U_1A_0+U_2A_1+\cdots U_pA_{p-1}=F_1,$$
$$U_0A_2+U_1A_1+U_2A_0+\cdots U_pA_{p-2}=F_2,$$

and

$$U_0A_p+U_1A_{p-1}+U_2A_{p-2}+\cdots U_pA_0=F_p$$

$$\int_0^\infty Y(x)\left[\int_{-\infty}^\infty E(t)E(t-x+\tau)dt\right]dx = \int_{-\infty}^\infty H(t)E(t-\tau)dt$$

On the other hand, the methods using the ratio of the power spectra (or of the autocorrelation functions), with subsequent determination of the phase (Determination of the transfer function starting from its amplitude spectrum—Use of Bode's formula), are based on the hypothesis of equal signal-to-noise ratio on the electric and the magnetic components.

is well known in seismic processing and can be solved, thanks to the special form of the left-hand-side matrix (Toeplizian matrix), even for a very large number of unknowns by algorithms derived from the one given by Wiener-Levinson (Wiener, 1949).

Instead of solving the preceeding equations for the unknown $U(t)$, it may be convenient to solve them for the function

$$V(t) = U(t) - \frac{1}{2\pi} \sqrt{\frac{\rho_1}{t}} \cdot$$

Provided a good estimation of ρ_1 coherent with the measured data is available, the value of the integral

$$\frac{1}{2\pi} \sqrt{\rho_1} \int_0^\infty \frac{1}{\sqrt{x}} A(\tau - x)dx$$

may be calculated and added to the right-hand side of the above equations.

On the other hand, the particular form of the linear system given above holds only if the functions involved are sampled at constant time intervals. Such a sampling does not seem to be well adapted to the nature of the function $U(t)$, for which a logarithmic scale of the abscissas is more

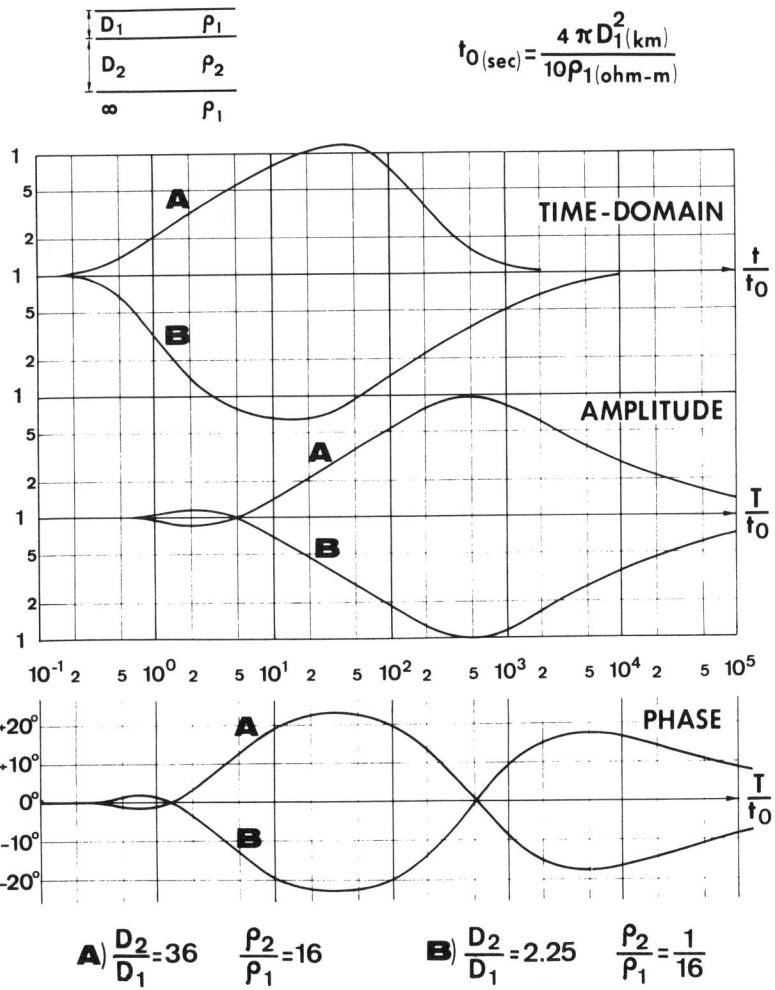

FIG. 3. Comparison of apparent resistivity master curves derived from the amplitude spectrum, the phase spectrum, and the time-domain response.

suitable. However, the use of a variable sampling interval for $U(t)$ makes the resolution of the above integral equation somewhat more difficult, because neither $A(t)$ nor $F(t)$ can be sampled at large intervals (see Appendix B); and, moreover, the resulting linear system cannot be solved by the Wiener-Levinson algorithm. This latter difficulty is offset by the fact that the use of a variable sampling interval may considerably reduce the number of equations and hence permit the resolution of the system, even without the aid of any special algorithm. This procedure is somewhat easier to apply in the frequency domain, as will be shown.

Determination of the transfer function starting from its amplitude spectrum—Use of Bode's formula

It is also possible to calculate the time-domain values of the transfer function by using only its amplitude spectrum. This latter function is the one usually deduced from the field measurements.

Whatever means are used to obtain this amplitude spectrum, its values are always given for frequencies (or periods) in geometric progression.[5] This allows a correct description of the whole useful spectrum with a reasonably small number of values. Moreover, it is possible, using only these unequally spaced samples, to obtain a satisfactory determination of the time-domain transfer function.

This is done by performing two successive transformations. 1) Taking advantage of its minimum-phase property, the phase $\varphi(\omega)$ of the function

$$W(\omega) = \frac{ik_1}{\omega} Z(\omega) \equiv \frac{ik_1}{\omega} \frac{E(\omega)}{H(\omega)}$$

can be deduced from its known amplitude:

$$|W(\omega)| = \sqrt{\frac{4\pi}{\rho_1} \cdot \frac{1}{\sqrt{\omega}}} \frac{|E(\omega)|}{|H(\omega)|}.$$

When the sample abscissas are in geometric progression, Bode's well-known formula,

$$\varphi(\omega) = \frac{2}{\pi} \omega \int_0^\infty \frac{\text{Log} |W(x)|}{x^2 - \omega^2} dx,$$

[5] However, in order to avoid aliasing, the value given for a period T_j has to be the ratio of the total energies in the band between $T = \sqrt{T_j T_{j-1}}$ and $T = \sqrt{T_j T_{j+1}}$.

may be reduced to a convolution by putting $x = e^y$ and $\omega = e^z$ (Appendix C1).

The phase thus obtained may not satisfy the necessary conditions for a horizontal stratification discussed in the section on recurrence formula and series expansion in the frequency domain. It is then possible to look for a phase which satisfies these conditions and leads, by Bode's formula, to an amplitude $|W(\omega)|$ as close as possible to the experimentally found values. If this is done in the least-mean-square sense, one is led to a problem of quadratic programming, which will be discussed later.

2) Once the phase and amplitude of $W(\omega)$ are known, the real and imaginary part of the function

$$U(\omega) = \frac{1}{k_1} W(\omega) = \sqrt{\frac{i\rho_1}{4\pi\omega}} W(\omega)$$

can be calculated for the same values of the abscissas ω.

From $U(\omega)$ a Fourier integral will lead to the time-domain operator $U(t)$. The corresponding integration may be performed making use only of the unequally sampled values of $U(\omega)$ determined above. Moreover it is convenient to calculate the resulting time-domain operator $U(t)$ for arguments t in geometric progression, and, in this case, the integration is again reduced to a convolution (see Appendix C2).

METHODS OF INTERPRETATION

Due to the integrating character of all electrical methods and to the limited accuracy of the measurements, it is not possible to deduce a unique resistivity distribution from the measured electric and magnetic field variations without using additional information.

In the case of the usual indirect interpretation procedures, based on curve matching, this additional information is introduced by the choice of the master curves to which the field curves are compared.

On the other hand, if no other information is available, direct interpretation can merely aim at the calculation of one or several among the many resistivity distributions compatible with the measured data. Additional information may be used to choose between the computed solutions or to modify them according to well-known equivalence principles. If available in a convenient form (for instance a hypothesis on the most probable resistivity distribution), it also can be introduced in the computing procedure.

Indirect interpretation by curve matching

Comparison in the time domain.—Comparison of the experimental transfer function—usually of its power spectrum—to precalculated master curves is the most commonly used interpretational technique. There may be some advantage in applying this technique in the time domain, thus taking into consideration amplitude as well as phase characteristics of the transfer function. Theoretical master curves, similar to those of Figure 2, can be calculated easily by the formulas given above and in Appendix A1, and experimental time-domain values can be calculated by the procedures described in the previous section. More details on this curve matching in the time domain have been given in an earlier work (Kunetz, 1969).

Comparison in the frequency domain.—If experimental values of amplitude and phase are available, they can, of course, be compared to pairs of master curves in amplitude and phase. However, this may not be very practical, and it may be preferable to take advantage of the theoretical relation between phase and amplitude to improve the estimate of this latter parameter.

Let the experimental values of the phase and of the logarithm of the amplitude be

$$b_j = \text{phase of } W(\omega_j)$$

and

$$a_j = \log W(\omega_j)$$

for frequencies (ω_j) in geometric progression, and let x_j be the unknown "improved" estimate of a_j. As already indicated in the section on Determination of the transfer function—Use of Bode's formula, the theoretical phase y_j corresponding to x_j could be calculated by a convolution:

$$y_j = \sum_i x_i s_{j-i}.$$

Hence, an estimate of the unknowns x_j can be obtained by minimizing the expression

$$\sum_j \left\{ (x_j - a_j)^2 + \lambda^2 [(\sum_i x_i s_{j-i}) - b_j]^2 \right\},$$

where λ^2 is a weighting factor whose value will depend on the comparative precision of the amplitudes and of the phases.

This leads again to a Wiener-Hopf-type linear system of the same kind as the one previously encountered in the section on Direct determination in the time-domain—Use of the Wiener-Levinson algorithm.

Direct interpretation in the time domain

Use of the transfer function.—Thanks to the knowledge of the analytical expression of the transfer function $U(t)$ and of the condition which the "images" q_m have to satisfy, it is possible to try a direct approach to interpretation, i.e., the determination of the resistivity distribution as a function of depth.

First, the analytical expression of $U(t)$,

$$U(t) = \frac{1}{2\pi} \sqrt{\frac{\rho_1}{t}} \left[1 + 2 \sum_{m=1}^{\infty} q_m e^{-(4\pi a^2/t)m^2} \right],$$

must be modified by introducing the Fourier transform of q_m, an even, periodical, and nonnegative function $\phi(\theta)$, linked to q_m by the relation:

$$q_m = \frac{1}{\pi} \int_0^{\pi} \phi(\theta) \cos m\theta d\theta.$$

This leads to

$$U(t) = \frac{1}{2\pi} \sqrt{\frac{\rho_1}{\pi t_0}}$$
$$\cdot \int_0^{\pi} \phi(\theta) \left[\sum_{k=-\infty}^{\infty} e^{-((\theta - 2k\pi)^2 t)/4t_0} \right] d\theta.$$

If the values of the transfer function U have been determined from the measured field variations $E(t)$ and $H(t)$, the unknown function $\phi(\theta)$ must satisfy the above integral equation, with the additional condition $\phi(\theta) \geq 0$. Because of the errors on the experimental values of the transfer function (as well as to the nonconformity of the actual physical conditions with the assumptions made), no exact solution will generally exist. Hence, an optimal solution in the least-square sense will be sought.

The determination of $\phi(\theta)$ is the "heart of the matter" and it will be dealt with in more detail later. Once $\phi(\theta)$ is known, the images q_m are calculated as the coefficients of the Fourier series of

$\phi(\theta)$. The reflection coefficients L_m are deduced from the images by the algorithm given in Appendix A2, and, hence, the resistivities and finally the thicknesses are obtained. It should be emphasized that the above algorithm leads to physically acceptable values of $L_m(|L_m| < 1)$ only because the function $\phi(\theta)$ from which it starts has no negative values.

For the determination of the function $\phi(\theta)$, the following method may be used.[6]

The integral equation to be satisfied may be written

$$\int_0^\pi \phi(\theta)G(\theta,t)d\theta - U_{\exp}(t) = 0,$$

with

$$\phi(\theta) \geq 0$$

and

$$G(\theta, t) = \frac{1}{2\pi}\sqrt{\frac{\rho_1}{\pi t_0}}\sum_{k=-\infty}^{+\infty} e^{-((\theta-2k\pi)^2 t)/4t_0}.$$

The function $(h_1/\rho_1)G(\theta_2 t/t_0)$ is dimensionless and independent of the parameters ρ and h, which are specific for a given problem.

The above equation must be considered for only the discrete values of t, $t = t_i$, for which $U_{\exp}(t)$ has been determined; for these values of t, the mean square of the left-hand side of this equation must be minimized. Putting

$$U_{\exp}(t_i) = U_i$$

and

$$G(\theta, t_i) = G_i(\theta),$$

the above condition can be written

$$\sum_i \left[U_i - \int_0^\pi \phi(\theta)G_i(\theta)d\theta \right]^2 = \min.$$

The integral in this expression will be replaced by a finite sum, assuming that $\phi(\theta)$ may be considered, in small enough intervals, as constant:

$$\phi(\theta) = \phi_j \quad \text{for } \theta_j < \theta < \theta_{j+1}.$$

[6] This method is the transposition of the method given previously for direct-current electrical sounding interpretation (Kunetz and Rocroi, 1970) and which has been applied successfully in the latter case to a large number of field results.

For the choice of the intervals (θ_j, θ_{j+1}) it is useful to consider the characteristics of the function $\phi(\theta)$. Starting from given resistivities and thicknesses, this function is easy to calculate, either as the Fourier transform of the images q_m, or by a simple iterative algorithm (Kunetz, 1969). It appears that the sampling rate must be increased with increasing resistivity contrasts and an increasing number of layers. In the case of direct-current interpretation, it has always proved satisfactory to represent $\phi(\theta)$ at less than 100 (generally unequally spaced) values of θ.

Putting

$$J_{i,j} = \int_\theta^{\theta_{j+1}} G_i(\theta)d\theta,$$

the minimum condition becomes

$$\sum_i \left[U_i - \sum_j \phi_j J_{i,j} \right]^2 = \min,$$

and must be satisfied by nonnegative values of the unknowns ϕ_j.

This type of problem is known as "quadratic programming," and there are several algorithms which lead to its resolution (Boot, 1967). It should be noted, in view of some of the following developments, that quadratic programming, when applied to the resolution of a least-square problem, in which the only constraints are positive values for the unknowns, starts with the same linear system (the same normal equations) as that solved when there are no constraints at all. The method of resolution is, of course, different.

Instead of minimizing the mean square of the absolute differences between the values of the theoretical and the experimental transfer functions, as in the expression given above, it may be convenient to minimize their relative differences. Moreover, to take care of the possibly unequal precision of the data, weighting factors p_i may be introduced. Consequently, the quantity to be minimized would be written

$$Q = \sum_i p_i \left(1 - \frac{1}{U_i}\sum_j \phi_j J_{i,j}\right)^2.$$

The way additional information is taken into account, when available, is an important point. If an estimate of probable values of resistivities and thicknesses can be made, it may be introduced into the expression to be minimized in order to arrive at a solution as close as possible to these approximate values.

This may be done by calculating the function $\phi^0(\theta)$, corresponding to the estimated resistivity distribution (by one of the methods indicated above), and then minimizing the sum

$$Q + \mu^2 R$$

with

$$R = \sum_j (\phi_j - \phi_j^0)^2.$$

μ^2 is a weighting factor which depends on how close one wants to keep to the initial estimate.

Interpretation without an explicit transfer function.—The method described in the preceding section is performed in two stages; the procedure of each stage is separately optimized.

First, from the field variations E and H, the "best" experimental values of the transfer function are determined:

$$[E(t) - U_{exp}(t) * H'(t)]^2 = \text{min.}$$
$$\Downarrow$$
$$U_{exp}(t)$$

Second, from the comparison of the thusly determined experimental values with the theoretical transfer function, the "best" resistivity distribution is inferred:

$$[U_{th}(t, \rho_i, h_i) - U_{exp}(t)]^2 = \text{min.}$$
$$\Downarrow$$
$$\rho_i, h_i$$

In principle, a better result should be expected by attempting to deduce the best resistivity distribution directly from the measured field variations, without the intermediate determination of $U_{exp}(t)$. The corresponding least-mean-square condition would be

$$\int_{-\infty}^{\infty} \left[E(t) - \int_0^{\infty} U_{th}(\tau, \rho_i, h_i) \right.$$
$$\left. \cdot H'(t - \tau) d\tau \right]^2 dt = \text{min,}$$

which can be written, introducing the nonnegative function $\phi(\theta)$,

$$\int_{-\infty}^{\infty} \left[E(t) - \int_0^{\infty} H'(t - \tau) \right.$$
$$\left. \cdot \int_0^{\pi} \phi(\theta) G(\theta, \tau) d\theta d\tau \right]^2 dt = \text{min.}$$

or

$$\int_{-\infty}^{\infty} \left\{ E(t) - \int_0^{\pi} \phi(\theta) \right.$$
$$\left. \cdot \left[\int_0^{\infty} H'(t - \tau) G(\theta, \tau) d\tau \right] d\theta \right\}^2 dt = \text{min.}$$

The expression between brackets [ ] is a function of t and θ. This minimum condition is of the same form as that encountered in the preceding section, and, hence, the solution could be obtained, in principle, by the same methods. However the discretization with respect to the time t would be much more difficult here, as $E(t)$, as well as $H'(t)$, are infinitely long, quasi-stationary functions of time. Hence, the above expression must be transformed in order to allow practical calculation. Such a transformation is given in Appendix D. It involves the explicit determination of the normal equation coefficients to be introduced in the quadratic programming. The experimental data enter into these coefficients through the Laplace transform of the autocorrelation function $A(u)$ of $H'(t)$

$$B(y) = \int_0^{\infty} A(u) e^{-uy} du$$

and the Laplace transform of the cross correlation $F(\tau)$ of $E(t)$ by $H'(t)$,

$$C(y) = \int_0^{\infty} F(\tau) e^{-\tau y} d\tau.$$

This procedure appears, nevertheless, substantially more complicated than the one which makes use of a predetermined transfer function U.

Direct interpretation in the frequency domain

This approach is of interest only if there has been a preliminary determination of the experimental values of the complex transfer function $U(\omega)$.

Introducing

$$q_m = \frac{1}{\pi} \int_0^\pi \phi(\theta) \cos m\theta d\theta$$

into the theoretical expression of $U(\omega)$ gives

$$U(\omega) = \sqrt{\frac{i\rho_1}{4\pi\omega}} \left\{ 1 + \frac{2}{\pi} \int_0^\pi \phi(\theta) \right.$$
$$\left. \cdot \left[\sum_{m=1}^\infty e^{-2am\sqrt{-4\pi i\omega}} \cos m\theta \right] d\theta \right\} .$$

This can be transformed into

$$U(\omega) = \frac{4h_1}{\pi} \int_0^\pi \phi(\theta)$$
$$\cdot \left[\sum_{k=-\infty}^{+\infty} \frac{(\theta - 2k\pi)^2 + 4\omega t_0 i}{(\theta - 2k\pi)^4 + 16\omega^2 t_0^2} \right] d\theta.$$

This is of the form:

$$U(\omega) = \int_0^\pi \phi(\theta) B(\theta, \omega) d\theta$$
$$+ i \int_0^\pi \phi(\theta) C(\theta, \omega) d\theta.$$

If $B_{exp}(\omega)$ and $C_{exp}(\omega)$ are the real and imaginary parts of the transfer function deduced from the experimental data, the minimum condition will be

$$\int_{\omega_1}^{\omega_2} \left\{ \left[B_{exp}(\omega) - \int_0^\pi \phi(\theta) B(\theta, \omega) d\theta \right]^2 \right.$$
$$\left. + \left[C_{exp}(\omega) - \int_0^\pi \phi(\theta) C(\theta, \omega) d\theta \right]^2 \right\} d\omega = \min,$$

with

$$\phi(\theta) \geqq 0,$$

where ω_1 and ω_2 are the lower and upper boundaries of the frequencies for which the experimental transfer function $U_{exp}(\omega)$ has been determined. This is again a problem which can be solved, for the unknown function $\phi(\theta)$, by quadratic programming.

When no weighting factors are applied, the expressions of the coefficients $M_{i,j}$ of the normal equations

$$\sum_i \phi_i M_{i,j} - N_j$$

depend on the specific problem considered only through the frequency limits ω_1, ω_2, whereas the experimental values of $U(\omega)$ appear in the right-hand-side coefficients N_j (see Appendix E).

Instead of minimizing the mean-square difference between the theoretical and experimental values of the function $U(\omega)$, it may be more convenient to apply this condition to the function

$$V(\omega) = \sqrt{\frac{4\pi\omega}{i\rho_1} \cdot U(\omega) - 1},$$

as this latter function has finite total energy in the interval $0 < \omega < \infty$.

CONCLUSIONS

The following methods have been described in this paper:

1. Analytical and numerical determination in the complex frequency domain and in the time domain of the transfer function between corresponding components of the electric and magnetic fields for an assumed resistivity distribution.

2. Statistical approach to the determination of the "best" experimental values of this transfer function, especially in the time domain, from the measured variations of the electric and magnetic fields.

3. Interpretation, i.e., determination of possible resistivity distributions, through visual or automatic comparison of the experimental transfer function to its theoretical form or expression.

Although these methods apply to successive phases of the processing of magnetotelluric measurements, it must be emphasized that they are quite independent of each other: the proposed methods of interpretation can be applied by whatever means the theoretical and experimental transfer functions have been obtained, and, conversely, the methods and expressions for the determination of these functions can be used for any interpretational procedure chosen.

The proposed methods need to be tested on actual field results in order to appreciate their advantages and limitations. For these applications, however, a good knowledge of the amplitude and phase responses of the measuring layout and instruments is necessary, as well as a fair agreement between the assumed and actual geologic conditions. It is hoped that those who have at their disposal such experimental material will find an interest in testing and improving some of the above described procedures.

As for the direct interpretation methods, it must be remembered that a resistivity distribution given by such a method always must be considered as only a guide toward the solution of the geologic problem and never as its final solution.

REFERENCES

Baranov, V., and Kunetz, G., 1958, Distribution du potentiel dans un milieu stratifié: C.R. Académie des Sciences, 247, p. 2170–2171.
Baranov, V., and Tassencourt, J., 1959, Calculation of resistivity master-curves with the electronic computer: 16th EAEG meeting, Munich.
Berditchevski, M. N., 1968, Elektritcheskaia razvedka metodom magnito-telluritcheskogo profilirovania: Moscow, Edition Nedra.
Boot, J. C. G., 1967, Quadratic programming: Amsterdam, North Holland Publ. Co.
Kunetz, G., 1969, Traitement et interprétation des sondages magnéto-telluriques: Revue de l'Institut Français du Pétrole et Annales des Combustibles Liquides, v. 24, n. 6.
Kunetz, G., and Rocroi, J. P., 1970, Traitement automatique des sondages électriques: Geophys. Prosp. v. 18, no. 2.
Kunetz, G., and d'Erceville, I., 1962, Sur certaines propriétés d'une onde acoustique plane de compression dans un milieu stratifié.:Ann. Géophys.,v. 18, no. 4, p. 351.
Wiener, N., 1949, Extrapolation, interpolation and smoothing of stationary time-series: New York, John Wiley and Sons, Inc.

REFERENCES FOR GENERAL READING

Baranov, V., and Kunetz, G. 1960, Film synthétique avec réflexions multiples. Théorie et calcul pratique: Geophys. Prosp., v. 8, no. 2, p. 315–325.
Cagniard, L., 1953, Principe de la méthode magnéto-tellurique: Ann. Géophys., v. 9, fasc. 2, p. 95–125.
Chauveau, J., 1967, Sur l'analogie des calculus d'un sismogramme et d'un sondage magnéto-tellurique théorique: Ann. Géophys. ,v. 23, fasc. 1, p. 49.
Chetaev, D. N., 1966, Inverse problem in the theory of magneto-telluric prospecting: Izv. Akad. Nauk, S.S.S.R., Fiz. Zemli, no. 9, p. 105–107.
Fournier, H., 1967, Proposition d'une méthode pour déterminer la structure du premier millier de kilomètres de la terre d'après la résistivité apparente: Saint-Gall. General Assembly IUGG, IAGA.
Lavergne, M., and Chaize, L., 1970, Signal et bruit en magnéto-tellurique: Geophys. Prosp., v. 18, no. 1. p. 64.
Rankin, D., and Nabetani, S., 1968, The electromagnetic impulse response of the earth and its resistivity

structure: J. Min. Coll. Akita Univ., Ser. A, v. 4, no. 3, p. 23–28.
Stefanesco, S. S., Schlumberger, C., and Schlumberger, M., 1932, Etudes théoriques sur la prospection électrique du sous-sol: Bucarest, fasc. 2.
Vozoff, K., Hasegawa, H., and Ellis, R. M., 1963, Results and limitations of magnetotelluric surveys in simple geologic situations: Geophysics, v. 28, no. no. 5, p. 778–792.
Waeselynck, M. Y. R., 1966, Méthode magnétotellurique—dépouillement des résultats par corrélation: Onde electrique, v. 46, 475, p. 1121–1124.
Wu, E. T., 1968, The inverse problem of magnetotelluric sounding: Geophysics, v. 33, no. 6, p. 972–979.
Yungul, S., 1961, Magnetotelluric sounding three-layer interpretation curves: Geophysics, v. 26, p. 465–473.

APPENDIX A

ALGORITHMS LINKING ELECTRICAL IMAGES AND REFLECTION COEFFICIENTS

1) For the determination of the "images" q_m from the "reflection coefficients" L_m, an auxiliary set of functions Z_j^m is used. The iterative formulas are as follow:

$$\alpha_m = (1 - L_1^2) \cdot (1 - L_2^2) \cdots (1 - L_{m-1}^2),$$

$$q_m = \alpha_m L_m - (q_{m-1} Z_1^{m-1} + q_{m-2} Z_2^{m-1} + \cdots + q_1 Z_{m-1}^{m-1}),$$

$$Z_j^m = \begin{cases} Z_j^{m-1} - L_m Z_{m-j}^{m-1} & \cdots j \leq m \\ 0 & \cdots j > m, \end{cases}$$

and

$$Z_0^0 = 1 \qquad \alpha_1 = 1.$$

2) The above procedure, in principle, can be reversed to give the reflection coefficients L_m from the images q_m. With the same notations as above, this gives:

$$L_m = \frac{1}{\alpha_m} (q_m + q_{m-1} Z_1^{m-1} + \cdots + q_1 Z_{m-1}^{m-1}).$$

The remaining formulas are unchanged.

However, as stated, this reversed procedure is valid only if the images q_m satisfy the condition given in the text under Recurrence formula and series expansion in the frequency domain. For the first few values of m the direct or inverse procedure gives:

m	1	2	3
Results	$\alpha_1 = 1$	$\alpha_2 = 1 - L_1^2$	$\alpha_3 = (1 - L_1^2)(1 - L_2^2)$
Direct	$q_1 = \alpha_1 L_1$	$q_2 = \alpha_2 L_2 - q_1 Z_1^1$	$q_3 = \alpha_3 L_3 - (q_2 Z_1^2 + q_1 Z_2^2)$
Inverse	$L_1 = \dfrac{1}{\alpha_1} q_1$	$L_2 = \dfrac{1}{\alpha_2}(q_2 + q_1 Z_1^1)$	$L_3 = \dfrac{1}{\alpha_3}(q_3 + q_2 Z_1^2 + q_1 Z_2^2)$

j			
0	$Z_0^1 = Z_0^0 - 0 = 1$	$Z_0^2 = Z_0^1 - 0 = 1$	$Z_0^3 = Z_0^2 - 0 = 1$
1	$Z_1^1 = 0 - L_1 Z_0^0$	$Z_1^2 = Z_1^1 - L_2 Z_1^1$	$Z_1^3 = Z_1^2 - L_3 Z_2^2$
2		$Z_2^2 = 0 - L_2 Z_0^1$	$Z_2^3 = Z_2^2 - L_3 Z_1^2$
3			$Z_3^3 = 0 - L_3 Z_0^2$
.			
.			
.			

APPENDIX B

USE OF UNEQUAL SAMPLING INTERVALS FOR THE DETERMINATION OF $U(t)$

If the values of the unknown function $U(t)$ are looked for at points of abscissas t_k, the discretization will lead to a linear-equation system of the form

$$\sum_k a_{j,k} U_k = b_j,$$

with

$$b_j = \int_{t_j}^{t_{j+1}} F(\tau)\,d\tau$$

and

$$a_{j,k} = \int_{t_j}^{t_{j+1}} d\theta \int_{t_k}^{t_{k+1}} A(\theta - \varphi)\,d\varphi.$$

This last double integral can be reduced, of course, to a single integration. It may be convenient to choose

$$t_0 = 0 \quad \text{and} \quad t_{k+1} = \Delta \cdot \alpha^k$$

$$(k = 0, 1, 2, 3, \cdots).$$

In the special case of $\alpha = 2$, the values b_j and $a_{j,k}$ can be approximated by the expressions

$$(b_j) = \sum_{p=2^j-1}^{2^{j+1}-2} F_p,$$

and

$$a_{j,k} = \sum_{q=0}^{q=2i-1} \sum_{p=2^k-2^j-q}^{p=2^{k+1}-2^j-q-1} A_p,$$

where F_p and A_p are the values of $F(t)$ and $A(t)$ sampled at constant intervals:

$$F_p = F(p \cdot \Delta)$$

and

$$A_p = A(p \cdot \Delta).$$

APPENDIX C

BODE'S FORMULA AND FOURIER TRANSFORM WITH SAMPLING INTERVALS IN GEOMETRIC PROGRESSION

1) Bode's formula states:

$$\varphi(\omega) = \frac{2}{\pi}\,\omega \int_0^\infty \frac{\log |W(x)|}{x^2 - \omega^2}\,dx.$$

The integral is taken in Cauchy's sense. Putting $\omega = e^y$ and $x = e^z$, and

$$\varphi(e^y) = b(y)$$

$$\log |W(e^z)| = a(z),$$

it becomes

$$b(y) = \frac{2}{\pi} e^y \int_{-\infty}^{\infty} \frac{a(z)}{e^{2z} - e^{2y}} e^z dz$$

$$= \frac{1}{\pi} \int_{-\infty}^{\infty} \frac{a(z)}{\sinh(z - y)} dz.$$

When, for practical calculation, this convolution integral is digitized, the zero of the denominator, for $z = y$, must be taken into account. In the less-sophisticated way of performing the digitization, this leads to the convolution:

$$b_j = \sum_i a_i s_{j-i},$$

with

$$s_0 = 0,$$

$$s_1 = -\frac{1}{\pi} \left(\frac{\Delta}{\sinh \Delta} + \frac{1}{2} \right),$$

$$s_j = -\frac{1}{\pi} \frac{j \cdot \Delta}{\sinh j\Delta},$$

$$s_{-1} = -s_1,$$

and

$$s_{-j} = -s_j,$$

where

$$b_j = b(j \cdot \Delta)$$

and

$$a_i = a(i \cdot \Delta).$$

2) In the Fourier cosine transform, for instance,

$$g(t) = \int_0^{\infty} f(\omega) \cos \omega t \, d\omega.$$

Put $t = e^{\tau}$ and $\omega = e^{\theta}$. This gives

$$g(e^{\tau}) = \int_{-\infty}^{\infty} f(e^{\theta}) \cos(e^{\theta + \tau}) e^{\theta} d\theta.$$

This is of the form:

$$h(\tau) = \int_{-\infty}^{\infty} a(\theta) b(\theta + \tau) d\theta.$$

This integral, when discretized with equally spaced values of θ and τ, leads to a convolution. One of the terms of this convolution is, moreover, independent of the specific problem considered and can be calculated once for all.

APPENDIX D

NORMAL EQUATION COEFFICIENTS FOR THE DIRECT TIME-DOMAIN INTERPRETATION

The minimum condition can be written:

$$\int_{-\infty}^{\infty} \left[E(t) - \int_0^{\tau} \phi(\theta) L(t, \theta) d\theta \right]^2 dt = \min,$$

with

$$L(t, \theta) = \int_0^{\infty} H'(t - \tau) G(\theta, \tau) d\tau.$$

The corresponding normal equations would be

$$\int_0^{\tau} \phi(\theta) M(\theta, \omega) d\theta = N(\omega),$$

with

$$M(\theta, \omega) = \int_{-\infty}^{\infty} L(t, \theta) L(t, \omega) dt,$$

and

$$N(\omega) = \int_{-\infty}^{\infty} L(t, \omega) E(t) dt.$$

We introduce the expression of $L(t, \theta)$ in $M(\theta, \omega)$ and change the order of integrations:

$$M(\theta, \omega) = \int_0^{\infty} \int_0^{\infty} \left[G(\theta, \tau) G(\theta, x) \right.$$
$$\left. \cdot \int_{-\infty}^{\infty} H'(t - \tau) H'(t - x) dt \right] d\tau dx$$

$$= \int_0^{\infty} \int_0^{\infty} G(\theta, \tau) G(\theta, x)$$
$$\cdot A(\tau - x) d\tau dx$$

$$= \frac{1}{2} \int_0^{\infty} \left\{ A(u) \int_u^{\infty} \left[G\left(\theta, \frac{v + u}{2}\right) \right. \right.$$
$$\cdot G\left(\omega, \frac{v - u}{2}\right) + G\left(\theta, \frac{v - u}{2}\right)$$
$$\left. \left. \cdot G\left(\omega, \frac{v + u}{2}\right) \right] dv \right\} du.$$

The analytical expression of the function G is

$$G(\theta, t) = c \sum_{k=-\infty}^{\infty} e^{-[(\theta - 2k\pi)^2 t]/4t^0}.$$

(It should be emphasized that the summation with respect to k always can be limited to a few values of k, because very small values of t/t_0 do not have to be considered.)

If we introduce this expression of G into the above integral, the integration with respect to v gives:

$$c^2 \sum_k \sum_l \frac{e^{-((\theta - 2k\pi)^2 u)/4t_0} + e^{-((\omega - 2l\pi)^2 u)/4t_0}}{(\theta - 2k\pi)^2 + (\omega - 2l\pi)^2} \cdot 8t_0.$$

Putting

$$B(y) = \int_0^\infty A(u)e^{-yu}du,$$

the final result is

$$M(\theta, \omega) = \frac{c^2}{2} \sum_{k,l} \frac{8t_0}{(\theta - 2k\pi)^2 + (\omega - 2l\pi)^2}$$

$$\cdot \left\{ B\left(\frac{(\theta - 2k\pi)^2}{4t_0}\right) + B\left(\frac{(\omega - 2l\pi)^2}{4t_0}\right) \right\}.$$

It is readily seen that this is merely the folding up, over the square $0 \le \theta \le \pi$ and $0 \le \omega \le \pi$, of the function[7]

$$m(\theta, \omega) = \frac{c^2}{2} \frac{8t_0}{\theta^2 + \omega^2}$$

$$\cdot \left\{ B\left(\frac{\theta^2}{4t_0}\right) + B\left(\frac{\omega^2}{4t_0}\right) \right\}$$

calculated for

$$-\infty < \theta < +\infty \quad -\infty < \omega < +\infty.$$

Thus,

$$M(\theta, \omega) = \sum_{k=-\infty}^{\infty} \sum_{l=-\infty}^{\infty} m[\theta - 2k\pi, \omega - 2l\pi].$$

Similarly,

$$N(\omega) = c \sum_{k=-\infty}^{\infty} C\left[\frac{(\omega - 2k\pi)^2}{4t_0}\right],$$

[7] $m(0,0)$ must be calculated separately as it is of the form 0/0.

where

$$C(y) = \int_0^\infty F(\tau)e^{-y\tau}d\tau.$$

$F(\tau)$ is the crosscorrelation of $E(t)$ by $H'(t)$.

For its practical determination, the unknown function $\phi(\theta)$ must be discretized. If it is assumed that

$$\phi(\theta) = \phi_j \quad \text{for} \quad \theta_j < \theta < \theta_{j+1},$$

the coefficients of the corresponding normal equations

$$\sum_j \phi_j M_{i,j} - N_i$$

become

$$M_{i,j} = \int_{\theta_i}^{\theta_{i+1}} d\theta \int_{\theta_j}^{\theta_{j+1}} M(\theta, \omega)d\omega$$

and

$$N_i = \int_{\theta_i}^{\theta_{i+1}} N(\omega)d\omega.$$

APPENDIX E

NORMAL EQUATION COEFFICIENTS FOR THE DIRECT FREQUENCY-DOMAIN INTERPRETATION

Before discretization, one has

$$M(\theta, \psi) = \frac{16h_1^2}{\pi^2} \int_{\omega_1}^{\omega_2} [B(\theta, \omega)B(\psi, \omega) + C(\theta, \omega)C(\psi, \omega)]d\omega,$$

with

$$B(\theta, \omega) = \sum_{k=-\infty}^{\infty} \frac{(\theta - 2k\pi)^2}{(\theta - 2k\pi)^4 + 16\omega^2 t_0^2},$$

and

$$C(\theta, \omega) = \sum_{k=-\infty}^{\infty} \frac{4\omega t_0}{(\theta - 2k\pi)^4 + 16\omega^2 t_0^2}.$$

Putting

$$m(\theta, \psi) = \frac{16h_1^2}{\pi^2}$$

$$\cdot \int_{\omega}^{\omega_2} \frac{\theta^2 \psi^2 + 16\omega^2 t_0^2}{(\theta^4 + 16\omega^2 t_0^2)(\psi^4 + 16\omega^2 t_0^2)} d\omega,$$

one can write

$$M(\theta, \psi) = \sum_{k=-\infty}^{+\infty} \sum_{l=-\infty}^{+\infty} m(\theta - 2k\pi, \psi - 2l\pi).$$

The integration with respect to ω gives

$$m(\theta, \psi)$$

$$= \frac{4h_1^2}{\pi l_0} \frac{1}{\theta^2 + \psi^2} \left[\tan^{-1} \frac{4l_0\omega_2}{\theta^2} - \tan^{-1} \frac{4l_0\omega_1}{\theta^2} \right. $$

$$\left. + \tan^{-1} \frac{4l_0\omega_2}{\psi^2} - \tan^{-1} \frac{4l_0\omega_1}{\psi^2} \right],$$

and finally the discretization leads to

$$M_{i,j} = \int_{\theta_i}^{\theta_{i+1}} d\theta \int_{\theta_j}^{\theta_{j+1}} M(\theta, \psi) d\psi.$$

Similarly, N_j is given by

$$n(\psi) = \frac{h_1}{4\pi} \int_{\omega_1}^{\omega_2} \left[B_{\exp}(\omega) \frac{\psi^2}{\psi^4 + 16l_0^2\omega^2} \right.$$

$$\left. + C_{\exp}(\omega) \frac{4l_0\omega}{\psi^4 + 16l_0^2\omega^2} \right] d\omega,$$

$$N(\psi) = \sum_{l=-\infty}^{+\infty} n(\psi - 2l\pi),$$

and

$$N_j = \int_{\theta_j}^{\theta_{j+1}} N(\psi) d\psi.$$

For these coefficients the integration with respect to ψ, involved by the discretization, can be performed explicitly before the integration with respect to ω, and thus independently of the specific problem under consideration.

GEOPHYSICS, VOL. 37, NO. 1 (FEBRUARY 1972), P. 98–141, 36 FIGS.

THE MAGNETOTELLURIC METHOD IN THE EXPLORATION OF SEDIMENTARY BASINS†

KEEVA VOZOFF*

The paper describes the theory of the magnetotelluric (MT) method, and some of the experimental, analytical, and interpretive techniques developed for its use in petroleum exploration in the past five years. Particular emphasis is placed on interpretation, since it is the area least amenable to routine treatment. Whereas present interpretation techniques are adequate, interpretation is the area of both the greatest progress and the greatest need for improvement.

Field results are presented from traverses in South Texas bordering on the Gulf of Mexico, and the Anadarko Basin of southwestern Oklahoma. Wide station spacings were used, such as might typify basin evaluations. The South Texas results are compared directly with smoothed induction logs. No useable logs could be found for Oklahoma. Comparisons with known and inferred geology show that the surveys mapped resistivity successfully in the known parts of these basins as well as in portions inaccessible seismically.

The capabilities and economics of the MT method justify its consideration for evaluating large unexplored blocks and "no record" areas.

NOTATION

x, y, z Coordinate axes. Positive north, east, and down.

j Current density, amperes/square meter

V Potential difference, volts

I Current, amperes

R Resistance, ohms

f Frequency, hertz (hz)

ω Angular frequency $=2\pi f$, radians/second

ρ Resistivity, ohm-meters (ohm-m)

σ Conductivity, mho/m

μ Permeability, henry/meter (h/m)

μ_0 Free space permeability, $=4\pi\times10^{-7}$ h/m

E Electric field, volts/meter (practical units—mv/km)

H Magnetic field, amperes/meter [practical units—gammas (γ)]

E_x Component of E in the x direction

l Electrode separation, meters

δ Skin depth, meters

ρ_a Apparent resistivity, ohm-m

Z_{ij} Impedance tensor element relating E_i to H_j, ohms

Z'_{ij} Z_{ij} after rotation through θ (in principal axes)

ρ_{ij} Apparent resistivity corresponding to Z_{ij}

θ_0 Direction of E for largest apparent resistivity, clockwise from north

x', y', z Coordinate system after rotation through θ (principal axes)

$E_\parallel, H_\parallel$ Components parallel to strike

ϕ Tipper azimuth measured clockwise from north

S Skewness

T Tipper

INTRODUCTION

Advances in magnetotelluric techniques and preliminary indications of success in their use

---

† Dr. Vozoff's paper is the second in a series of survey papers sponsored by the Research Committee of the SEG. The series was the subject of an editorial in the December, 1971 issue of GEOPHYSICS (v. 36, p. 1252).

Presented at the 39th Annual International SEG Meeting, September 17, 1969, Calgary, Alberta. Manuscript received by the Editor April 30, 1971; revised manuscript received July 20, 1971.

* Professor of Geophysics, School of Earth Sciences, Macquarie University, North Ryde, N.S.W. 2113, Australia.

have led to an acceleration in interest and application. As a rough estimate, more than a million dollars was spent on magnetotelluric (MT) contract fieldwork and research in 1969. For the first time, fieldwork is being supported by the operating divisions of major oil companies as well as by their research laboratories.

This paper is intended to help fill the increasing need of the petroleum geophysicist to understand the method so he can use it intelligently. Field results are included to illustrate the method's potential and its inherent limitations.

BASIC CONCEPTS

The MT method is a way of determining the electrical conductivity distribution of the subsurface from measurements of natural transient electric and magnetic fields on the surface.

Results interpreted from measurements at a single site are sometimes compared to an induction log, very heavily smoothed, obtained without drilling a well. Results from a line of measuring stations are interpreted to give underlying conductivity distribution and structure. That picture of the subsurface can in turn be related to porosity and salinity, since conductivity depends primarily on those two factors in common sedimentary rocks.

The time variations of the earth's electric and magnetic fields at a site are recorded simultaneously over a wide range of frequencies, usually on digital tape. The variations are analyzed by computer to obtain their spectra and apparent resistivities as a function of frequency are computed from the spectra. Interpretation consists of matching the computed plots of apparent resistivity against frequency to curves calculated for simplified models.

The MT method depends on the penetration of electromagnetic energy into the earth. Depth control comes as a natural consequence of the greater penetration of the lower frequencies. The measurements are absolute. Their interpretation gives true resistivity values and true depths, not just anomalies. Depth interpretation based on MT data is therefore much more definitive than that based on gravity or magnetic data.

THE SIGNALS

The time-varying magnetic (H) signal is the always present "noise" in the earth's magnetic field. When very large, it interferes with magnetic surveys. In the conducting earth, the changing magnetic field induces telluric (eddy) currents and voltages: the latter are the electric (E) signals. They are very similar in appearance to the H signals. On chart records, both sets of variations look irregular and noise-like for the most part, as in Figure 1. At times, in certain frequency bands, the variations may appear sinusoidal (Figure 2) but the sinusoids are not an important part of the signal for MT purposes. Signal amplitudes fall off rapidly with increasing frequency over most of the range of frequencies used. Figure 3 shows typical spectral behavior for E and H as recorded in four overlapping frequency bands. Signal level can increase very rapidly at the onset of magnetic storms, an increase of a factor of 10 being common and even a factor of 100 is not unusual.

Most of the magnetic noise reaching the earth below 1 hz is due to current flow in the ionized layers surrounding the earth. The currents are powered by solar activity, and by the relative motions of the earth, sun, and moon. At frequencies above 1 hz, worldwide electrical thunderstorm activity within the atmosphere is the major contributor. The transient fields due to thunderstorms can be exceedingly large locally, while those associated with tornados are greater still.

EFFECT OF EARTH CONDUCTIVITY ON H

When the magnetic fluctuations reach the surface of the earth, reflection and refraction occur. Although there was considerable disagreement at one time, it is now well established that, as a working theory, the signals can be treated as plane electromagnetic waves. This will not be true under all conditions, but holds for the vast majority of geological situations of interest in petroleum prospecting (Madden and Nelson, 1964; Rikitake, 1966; Vozoff and Ellis, 1966).

Although the majority of the incident energy is reflected, a small portion is transmitted into the earth and slowly travels vertically downward. To the conducting rocks, this energy appears as a magnetic field which is changing with time, and electric fields are induced so that currents, called telluric currents, can flow. These telluric currents are completely analogous to the eddy currents which flow in transformers due to the changing magnetic fields caused by the ac current in the primary windings.

Energy in the downgoing disturbance is quickly dissipated as heat. As a result, the field penetration is relatively small in terms of its wavelength in air. The penetration mechanism in this situation is actually diffusion rather than wave propagation.

Current density in the earth depends on resistivity ρ, as might be expected. Within a rock, the normal relationship between the electric field and the current density at each point is

$$j = E/\rho.$$

This differential form of Ohm's law is really a definition of *resistivity*, and is very similar to the Ohm's law definition of resistance,

$$I = V/R.$$

In mks units, E is in volts/meter; j is in amperes/square meter; ρ is in ohm-meters; and H is in amperes/meter. However, because the fields are so small, the more commonly used practical units are mv/km for E and gammas for H. The practical units will be used in later sections.

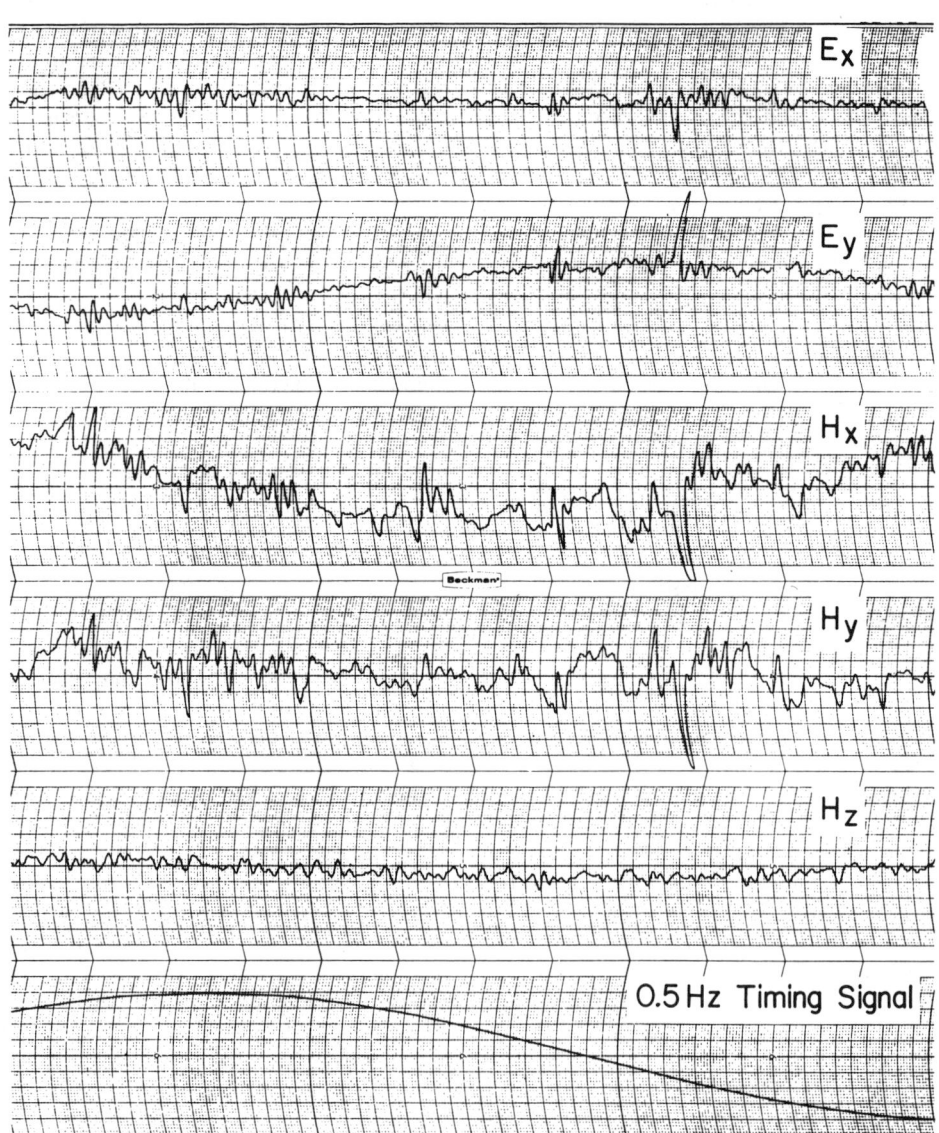

FIG. 1. Typical noise-like telluric signals. Sine wave at bottom shows time scale.

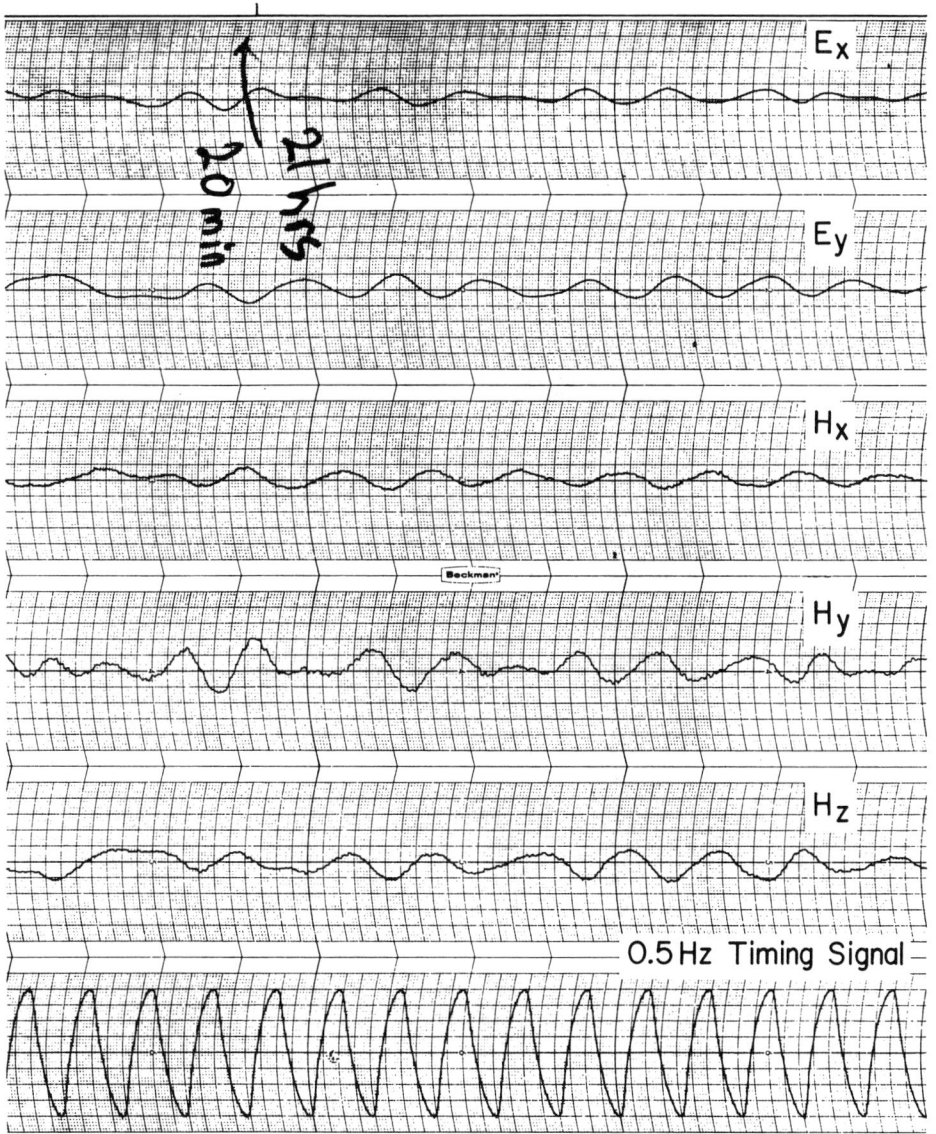

FIG. 2. Large near-sinusoidal telluric signals with superposed noise-like signals. Time scale on bottom trace.

The E measurement is actually a voltage difference measurement between two electrodes. In a uniform earth, the voltage difference V between electrodes a distance l apart would be

$$V = lE.$$

In the MT method it is usually assumed that E is constant over the length of the wire; i.e.,

$$E = V/l.$$

The depth of penetration of the fields into the earth is inversely related to rock conductivity. In a uniform earth E and H weaken exponentially with depth; the more conductive the earth, the less the penetration. The depth at which the fields have fallen off to $(e)^{-1}$ of their values at the surface is called the skin depth δ.

$$\delta = \sqrt{2/\omega\mu\sigma} \ \text{m} \tag{1}$$

$$\approx \tfrac{1}{2}\sqrt{\rho/f} \ \text{km}, \tag{1a}$$

where f is frequency, $\omega = 2\pi f$, and μ is permeability. (μ in the earth is taken equal to μ_0 except in highly magnetic materials.) Frequency enters into the equations because the magnitudes of the induced telluric currents depend on the time rate of change of the magnetic fields.

In a uniform or horizontally layered earth all currents, electric fields, and magnetic fields are practically horizontal, regardless of the direction from which these fields enter the earth. This comes about because of the high conductivity of earth relative to air. It can be thought of in terms of Snell's law in optics, with the velocity in the earth being orders of magnitude smaller than that outside. Furthermore, the currents and electric fields are at right angles to the associated

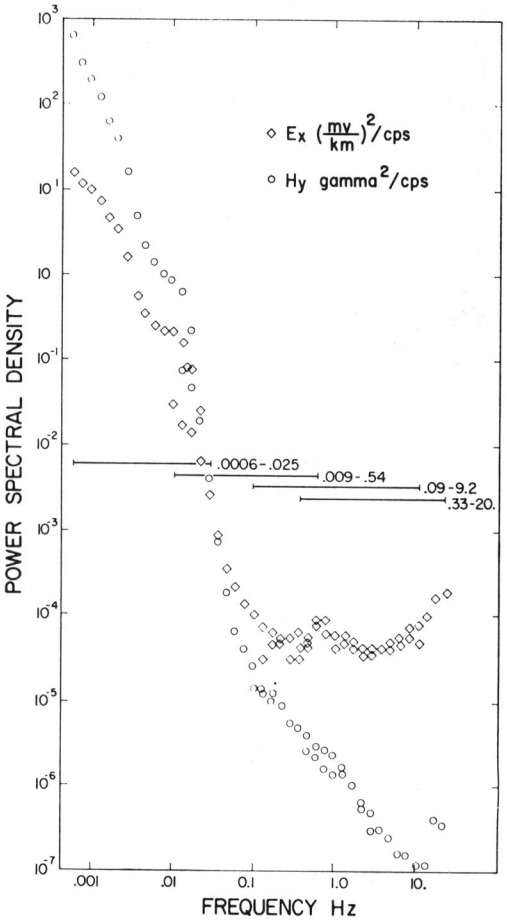

FIG. 3. Selected power spectral densities recorded in the four overlapping frequency bands indicated. Bands were recorded at different times.

magnetic fields at each point. If E is positive to the north, H is positive to the east. That is, viewed from above, E must be rotated 90 degrees clockwise to obtain the direction of positive H.

The mathematical description of the perpendicular E and H fields in a uniform isotropic conductor includes all these features in a concise form:

$$H_y = H_y^0 e^{-i\omega t + (i-1)z/\delta}; \qquad (2)$$

$$E_x = E_x^0 e^{-i\omega t + (i-1)z/\delta}; \qquad (3)$$

$$E_x^0 = (1 - i)\omega\mu\delta H_y^0/2. \qquad (4)$$

The superscript indicates the value at the surface. Particularly interesting is the ratio

$$\frac{E_x^0}{H_y^0} = \frac{(1 - i)\omega\mu\delta}{2} \text{ ohms} \qquad (5)$$

$$= (1 - i)(\omega\mu/2\sigma)^{1/2}.$$

Since E and H are recorded at frequencies which can be accurately measured and since μ varies little from μ_0 in most rocks, the ratio shows the relationship which exists between the conductivity and the measured fields. The equation can be solved for conductivity, giving

$$\sigma^{1/2} = (1 - i)(\omega\mu/2)^{1/2} \frac{H_y^0}{E_x^0}. \qquad (6)$$

Equation (6) is usually rewritten in mks units as

$$\rho = \frac{i}{\omega\mu}\left(\frac{E_x}{H_y}\right)^2 \qquad (6a)$$

and the superscripts are omitted.

In practical units, where E is given in mv/km and H is in gammas, the magnitude of ρ is

$$\rho = \frac{1}{5f}\left(\frac{|E_x|}{|H_y|}\right)^2. \qquad (6b)$$

When ρ (or σ) is calculated from E and H values, it is called an *apparent* resistivity ρ_a (or apparent conductivity σ_a). ρ and ρ_a are related, but they must be clearly distinguished. ρ_a is the resistivity that a uniform earth must have to give the measured value of the impedance Z. ρ is a property of the medium, whereas ρ_a depends on how it is measured. The ratio of E_i to H_j at each frequency is the impedance Z_{ij} for those components at

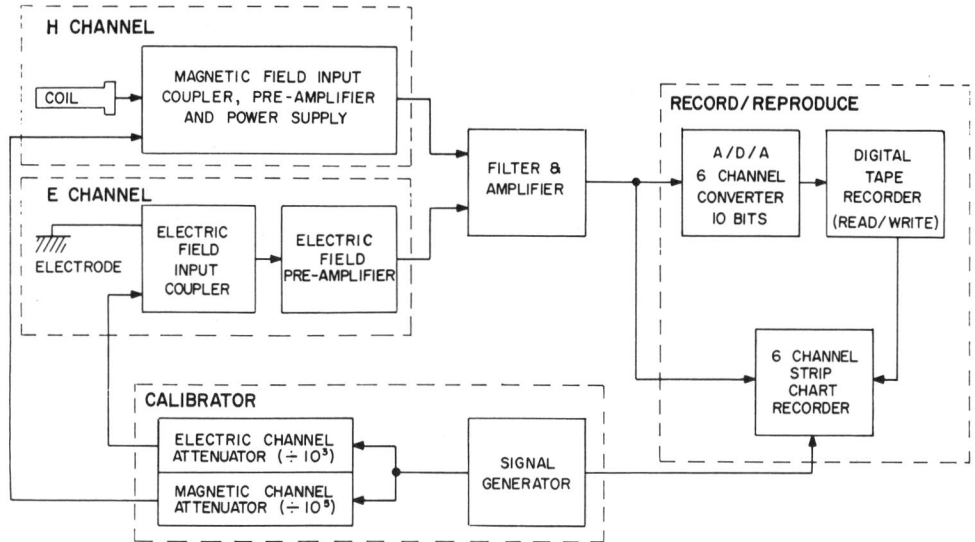

Fɪɢ. 4. Block diagram of a basic magnetotelluric instrumentation system.

that frequency. Since E and H are usually not in-phase, Z_{ij} is taken to be a complex number.

In a uniform earth, ρ_a has to be the same at every frequency, and E leads H in-phase by 45 degrees at all frequencies. [That this is so can be checked by substituting equation (5) into equation (6b)]. Thus, if we plot ρ_a and phase against frequency, we see that both are constants.

In discussing kinds of electrical structure it is useful to define two-dimensional and three-dimensional structures. In the two-dimensional case [$\sigma = \sigma(x, z)$], conductivity varies along one horizontal coordinate and with depth. The other horizontal direction is called the strike. When conductivity varies with both horizontal coordinates and with depth [$\sigma = \sigma(x, y, z)$], the structure is said to be three-dimensional and has no strike. If σ depends only on z, the structure is one-dimensional. In each case, σ at each point can depend on the direction of current flow; if σ does depend on direction, the medium is anisotropic.

If the conductivity changes with depth, ρ_a varies with frequency, since lower frequencies penetrate more deeply. Apparent resistivity can be written and computed *exactly* for any desired combination of horizontal layers, whether isotropic or arbitrarily anisotropic. It can be calculated *approximately* for any two-dimensional model structure. As might be expected, Z for horizontally isotropic and homogeneous layers

does not depend on the directions used as long as E is measured perpendicular to H.

When due to faulting or jointing, σ varies laterally or with direction, the j and E which are induced by a given H depend on their direction relative to strike. In order that these effects can be sorted out, we record complete horizontal E and H fields (two perpendicular components of each) at every site. In addition, the vertical component of H is also recorded, for a total of five recorded signals in all. These are designated H_x, H_y, H_z, E_x, and E_y.

In general, ρ_a at each frequency varies with measurement direction. We assume that there is a strike but that its direction is unknown. Then E_x is due partly to H_y, but also partly to currents induced by H_x, which have been deflected by the structure. The same is true of E_y, so the relations are written

$$E_x = Z_{xx}H_x + Z_{xy}H_y, \tag{7}$$

$$E_y = Z_{yx}H_x + Z_{yy}H_y. \tag{8}$$

For example, Z_{yx} gives the part of E_y which is due to H_x, and so forth. Since E_y and H_x are generally not in-phase, the Z's are complex. E and H component amplitudes are obtained by computer-analyzing the records using methods described in the section on data analysis. Computation of the Z_{ij} for two-dimensional models will be discussed under Interpretation.

INSTRUMENTATION

Over most of the frequency range of 0.0006–10 hz used for most MT work, the signals weaken rapidly with increasing frequency. Figure 3 shows typical (smoothed) power density spectra in this range. The absolute levels can rapidly increase by 30 db or more on the advent of a magnetic storm. They are also often observed to gradually diminish by 10–20 db in a matter of hours. Furthermore the detailed shape of the spectrum changes, with some frequency bands being enhanced relative to the general trend. Heirtzler and Davidson (1967) show a sample of these variations in H, for the component along the main geomagnetic field.

A block diagram of a typical measuring system is shown in Figure 4. The complete equipment set includes three H channels and two E channels. Of the components, nearly all are custom-made except for the signal generator, E field preamplifiers, strip chart recorder, and the digital recording system. A single 1 kva liquid propane-powered motor generator will operate the system.

The H sensors are long, slender induction coils comprised of 30,000 turns of #22 copper wire wound on cores laminated of moly-permalloy strips. The strips are 1 inch wide and 6 ft long, filling a $1\frac{1}{3}$ inch diameter tube. Earth contacts for E sensors are nonpolarizing electrodes made up of cadmium metal immersed in a cadmium chloride solution, in porous pots $1\frac{3}{4}$ inches in diameter and $2\frac{3}{4}$ inches high. (The solution is highly toxic!)

The other highly critical component is the H preamplifier. This is a guarded, differential input, chopper carrier amplifier with very low-noise connections and components, with heat sinks and shock mounts. No common commercial amplifiers have yet been found which can be fed by the high inductance coils while retaining the bandwidth, low noise, and high gain necessary in this application.

The rest of the system was designed for geophysical field use. Its response is accurately known, but its specifications are not otherwise remarkable.

Rather than record the entire frequency range at once, the 10^{-3} to 10 hz band is recorded in

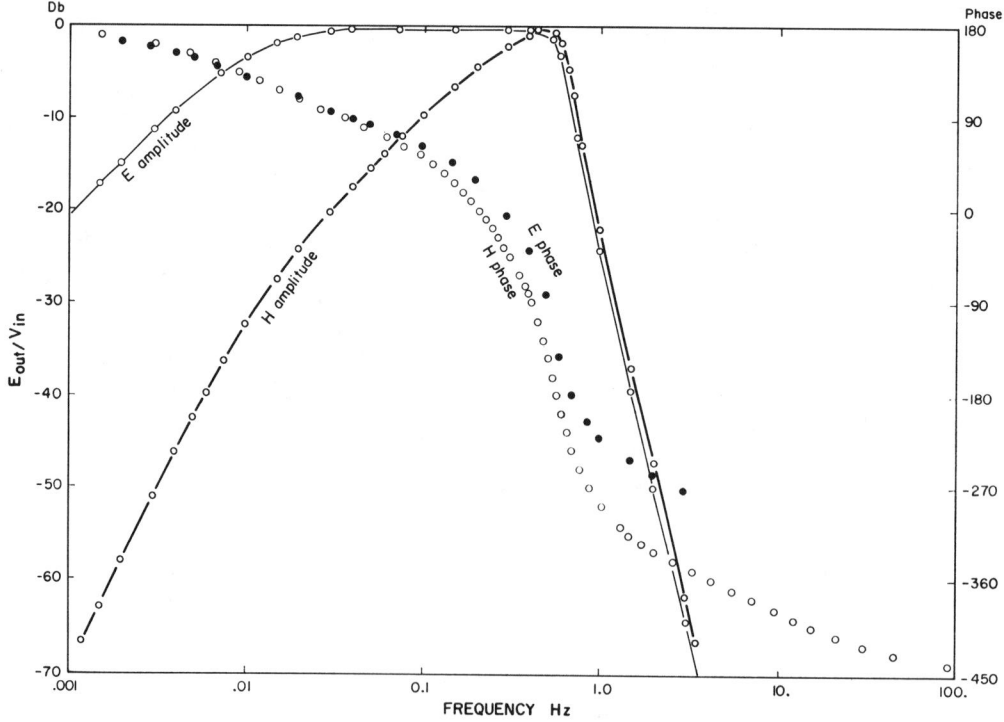

Fig. 5. Responses of commercial E and H systems in one typical recording band.

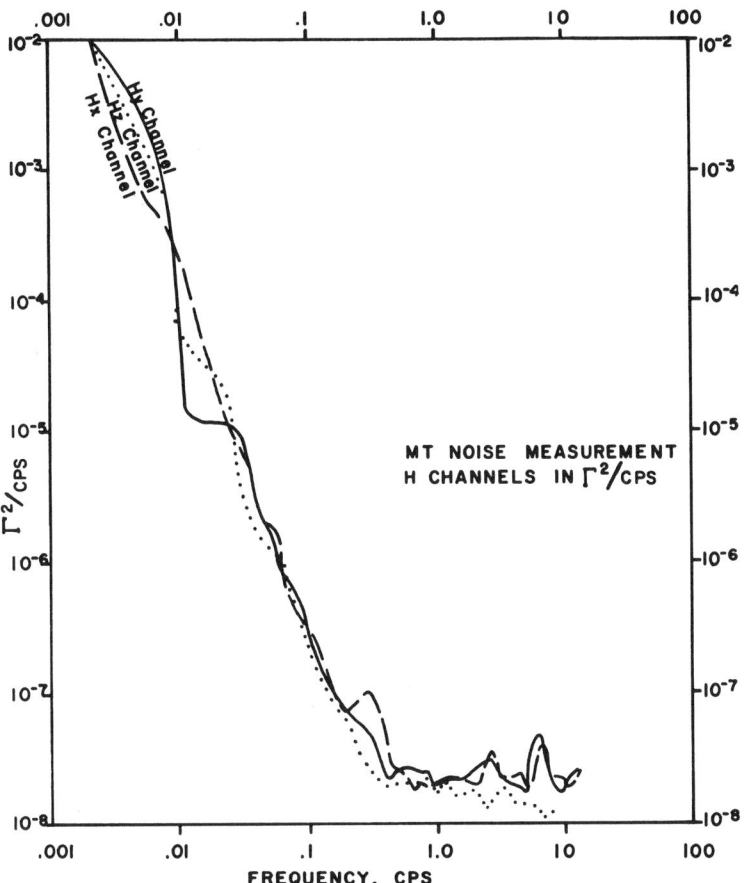

FIG. 6. Noise power spectral density of commercial H system. Obtained using noise measurement procedures of Hopkins (1965).

three narrower, overlapping segments. Commonly these are (nominal) 0.002–0.025, 0.01–0.5, and 0.1–7.5 hz, with $\Delta t = 10$ sec, 0.5 sec, and 1/30 sec, respectively. (Δt is the sampling rate of the digital equipment.) Other filter settings can be used when desired. The actual frequency range analyzed extends from 0.0006 to 9.8 hz. In areas of high near-surface resistivity, a still higher frequency band is recorded, extending from 0.33 to 15.5 hz ($\Delta t = 1/37.5$ sec). The frequency range is divided this way for two important reasons: to make best use of the available dynamic range and for economy in digital recording. That is, dividing the frequency range permits the use of large gains in some parts of the spectrum and lower gains in others, according to the signal levels actually present at recording time. The economic aspect enters, since the total recording duration is a

multiple of the longest period which is being recorded, whereas digitizing rate is at least twice the highest frequency being recorded. If the band is very wide, far more short period data will be acquired than is necessary while waiting for adequate sampling of long periods. The 1- to 2-decade segments used are a compromise.

Total number of samples to be recorded in a data set is selected in this equipment by means of a front panel switch. Most commonly, data sets consist of 4096 multiplexed scans. Each voltage sample is converted to a 10 bit binary number, giving a full scale resolution of 2^{-10} or about 60 db.

Typical normalized system response curves are shown in Figure 5. In terms of the parameters measured, the maximum full scale sensitivities for this equipment are $H = 9 \times 10^{-4}$ gammas at 1 hz and $E = 12.5\ \mu\mathrm{v/km}$ at 1 km electrode spacing.

Noise power spectral densities for this equipment are shown in Figure 6.

Quoted sensitivities for other magnetometers made for exploration use are in the neighborhood of 10^{-1} gammas full scale for optically pumped devices, and 1 gamma full scale for proton precession and fluxgate magnetometers. Noise levels in each case are 10 to 20 percent of full scale. The flat frequency responses of these other magnetometers gives them the advantage at periods longer than 30–100 sec. Superconducting magnetometers have been built with noise levels comparable to those of the induction coil system described here, but they do not yet seem suitable for use in the field (see Nisenoff, 1969, for example).

From the practical point of view, there are several improvements which would be desirable in present state-of-the-art equipment. Induction coils of adequate sensitivity are clumsy and heavy, and the vertical component coil requires that a hole six ft deep be augered. Several smaller coils in combination might be much more convenient to use. A large loop laid flat on the surface is the only alternative at this time to the vertically emplanted coil.

For work in loose or unsteady surface materials (marsh, ice-floes, loose soil near trees), a system of motion stabilization or compensation would be very helpful.

Hopkins (1965) discusses the engineering and design problems associated with MT instrumentation.

FIELD PROCEDURE

Although basically simple, field procedures require a great deal of planning and attention to detail, since they dominate the costs; and the sensitivity of the measurement makes it highly vulnerable to disturbances at the measuring site.

Two pairs of electrodes aligned at right angles must be laid out at each site, as must three mutually perpendicular magnetometers. A setup is shown schematically in Figure 7. The electrodes provide low resistance, low noise electrical connections with the earth, for the E measurements. The input to each of the E-signal channels is the voltage difference between an electrode pair. (Although one usually thinks of the earth as being at zero potential, voltage differences must exist if telluric currents flow because the earth has a finite resistivity, and $j = E\sigma$.) The farther apart a pair of electrodes, the larger the signal voltage measured, so it is usually desirable to put the electrodes as far apart as possible, subject to

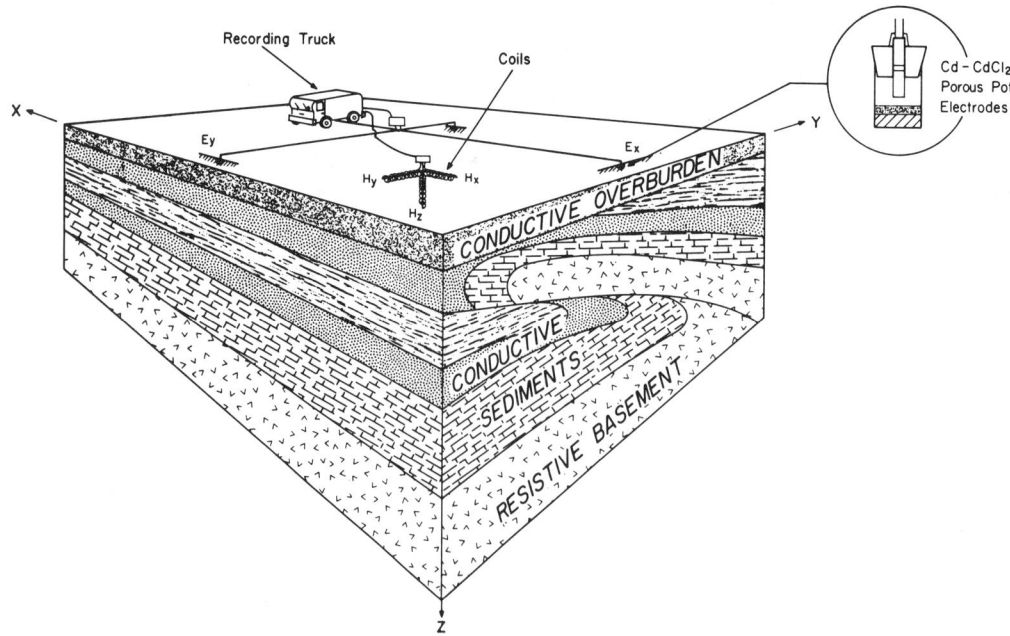

Fig. 7. Field setup.

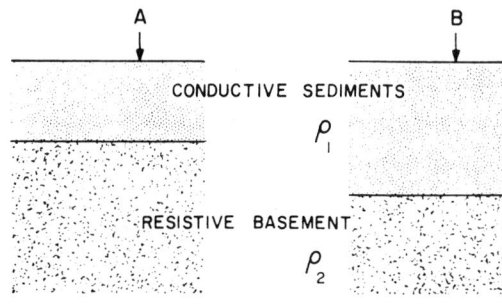

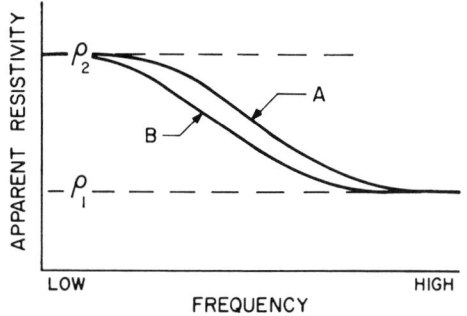

FIG. 8. Diagrammatic two-layer apparent resistivity curves for the models shown.

other factors, such as obstructions, property boundaries, the time needed to lay out connecting wires, and the minimum spacing tolerable between adjacent measurement sites. For routine operations, it is desirable to use fixed wire lengths. Finally, since the wires must not be permitted to move in the earth's main magnetic field as this induces noise, clods of dirt are placed every few feet along the wire to restrain it, a nontrivial task. A 2000-ft spacing is common because it fits conveniently within a quarter section.

The two electrode pairs are intended to measure the two perpendicular components of an electric field vector which exists at each site. However, it is possible for the electric field on the surface to change in both direction and intensity over very short distances, due to large lateral resistivity changes near the surface. Large electrode spacings should be used in this situation to average over as much of the variation as possible, or the resulting data will apply to conditions which are too localized to be of use. For best averaging in these circumstances, it is also important that the two electrode pairs form a cross whose four arms are as nearly equal in length as possible, rather than being arranged to form an **L**

or a **T**, (Swift, 1967), and that the coils be near the center.

Topographic features can cause distortions similar to those caused by resistivity heterogeneities. While these can also be modeled, it is better to avoid them if possible, especially if the relief is more than 10 percent or so of the electrode spacing.

Induction coil magnetometers are even more sensitive to motions than are the wires connecting the electrodes. To prevent their moving or vibrating due to wind, the two horizontal coils are buried in shallow trenches 12 to 18 inches deep. The vertical coil is placed in a hole dug by auger. Coils are leveled to within a fraction of one degree by means of sensitive level bubbles. Horizontal azimuth is adjusted to similar accuracy by a simple transit. Burying the coils has the added advantage of reducing thermal transients and the resulting noise. Coils having permeable cores

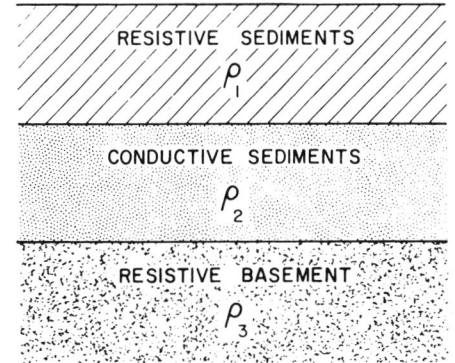

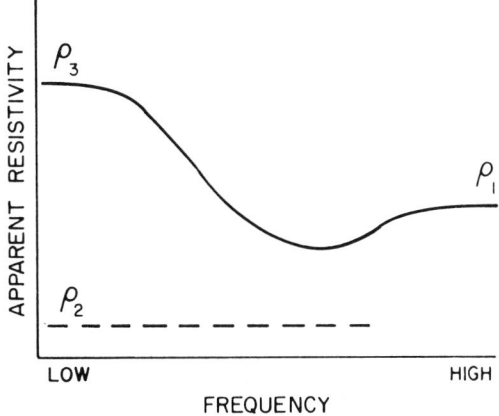

FIG. 9. Diagrammatic three-layer apparent resistivity curve for the model shown.

must be separated from each other by several times their length when they are emplaced, to avoid the effects on each of the local field distortions surrounding the others.

Sites must be chosen with care, to avoid possible sources of disturbance such as cathodic protection circuitry, power and fence lines, unprotected pipelines, and vehicle and pedestrian traffic.

By the use of separate but well-coordinated crews for recording and sensor emplacement, efficient field operations can be achieved. For further efficiency, recordings can be scheduled so as to avoid known or anticipated noise. Customarily, several recordings are obtained in each frequency band, because noise and signal are both variable and largely unpredictable.

DATA ANALYSIS

The purpose of data analysis is to extract reliable values of impedances, apparent resistivities, and the other earth response functions (ERF) from the field records, and to present them in a form convenient for interpretation. Operationally, data analysis consists of (a) manual editing of records, to reject those judged to be contaminated by noise; (b) computer manipulation of tape-recorded data to transform all records into the frequency domain, to derive the ERF which

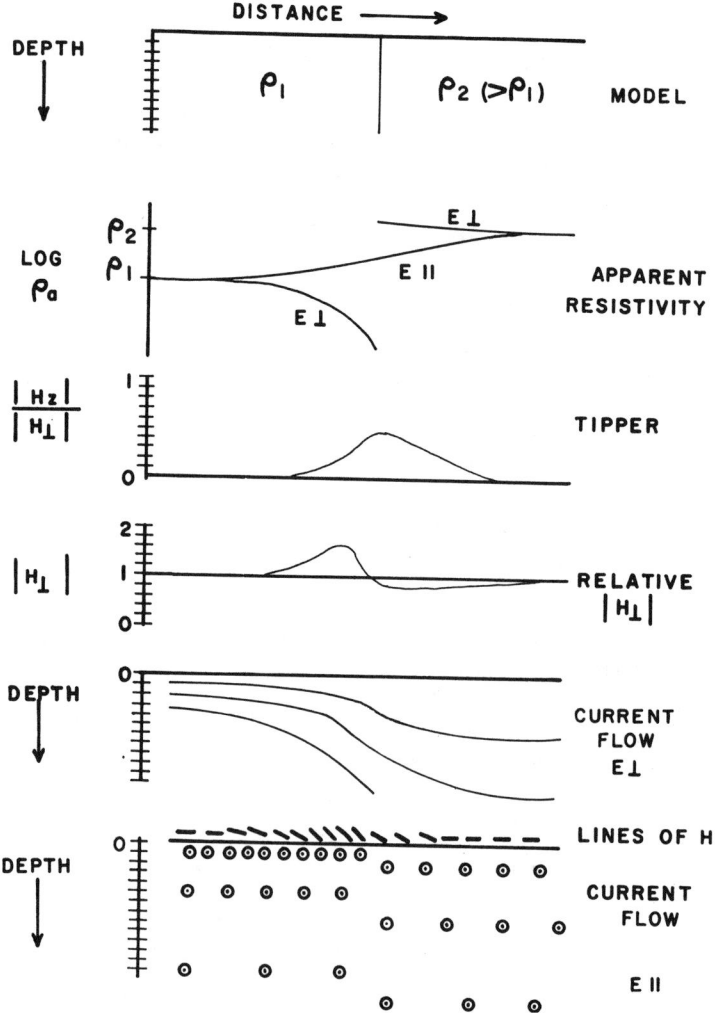

FIG. 10. Diagrammatic response curves for a simple vertical contact at frequency f. Horizontal scales are distance.

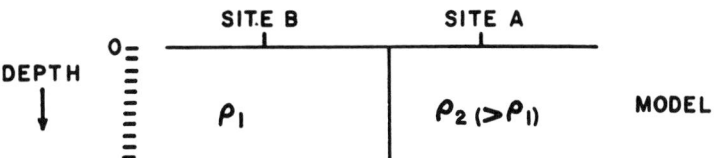

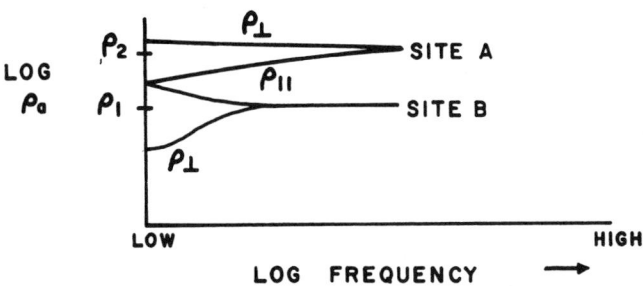

FIG. 11. Diagrammatic response curve for a simple vertical contact. Apparent resistivity versus frequency.

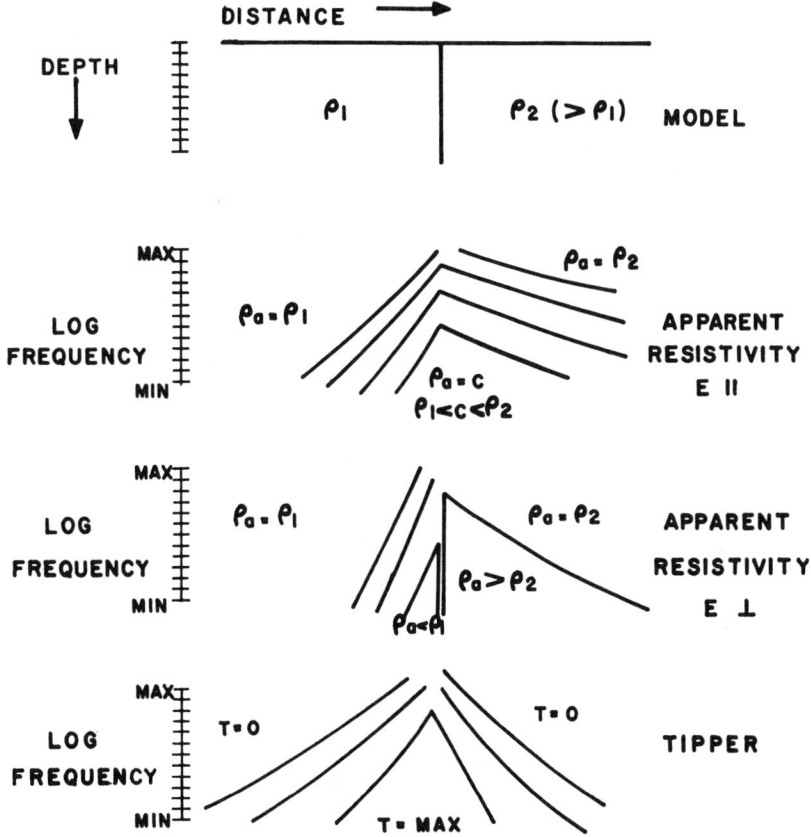

FIG. 12. Diagrammatic pseudosections for a simple vertical contact.

are used for interpretation, to screen each value computed, and to plot in a form convenient for interpretation the values which pass the screening; and (c) manual evaluation of the results so as to reject suspect data and to attempt to extract useful portions of previously rejected data when necessary.

Later, in the interpretation stage, the ERF are compared with those computed for models. These earth response functions are customarily (1) horizontal direction θ_0 of largest apparent resistivity, (2) impedances and apparent resistivities for E in the direction θ_0 and for E perpendicular to θ_0, (3) phases of the two apparent resistivities, (4) the portion of H_z that is linearly related to the horizontal field, (5) the direction of the horizontal magnetic field component most highly coherent with (4), and (6) the phase difference between (4) and (5). Motivation and theory for the first three functions can be found in Madden and Swift (1969), Cantwell (1960), Madden and Nelson (1964), and Swift (1967). Some of the analytical techniques for their computation are summarized and discussed by Sims and Bostick (1969), and Word et al (1969). Motivation and theory for the last three functions are discussed in Vozoff and Swift (1968), Sims and Bostick (1969), and Word et al (1969). Their computation was devised by T. R. Madden and is summarized below. Rankin (1969) and Kunetz (1969) have suggested techniques for and advantages of extracting the ERF in the time domain.

The first phase, manual editing, involves the examination of both recording log books and chart records for evidence that artificial or wind noise is large enough to downgrade a record. Records degraded by noise are not used if their use can be avoided, although it is sometimes necessary to reconsider using their quieter parts when insufficient quiet data exist.

Fourier transformation of each of the five recorded field components is the second step in analysis. This step yields in-phase and quadrature values for each component at as many frequencies as there are samples. The ERF are derived from these numbers.,

It is helpful at this point to describe a few of the properties of the ERF, anticipating the section on interpretation, in order to explain some of the procedures which are used in data analysis. In two-dimensional structures when neither of the coordinate axes is along strike, all four elements of the impedance tensor Z_{ij} are nonzero and have different values. Magnetic field components in the x direction give rise to some currents along x, in addition to the y-directed currents, which would be the only ones if the earth were uniform or horizontally layered. Magnetic y components are likewise associated with both E_x and E_y, so that Z_{xx}, Z_{xy}, Z_{yx}, and Z_{yy} will all have some finite values. Now if the coordinate axes are rotated (either physically or by computation) until one of them is along strike, then currents due to H_x can no longer be deflected into the x direction and those due to H_y flow only in the x direction. In this situation, Z_{xx} and Z_{yy} must be zero. The other pair are nonzero and unequal, since current densities will differ in the two directions, and the E components must also differ. If the coordinates are rotated a further 90 degrees, the same situation is found, except that the Z values are inter-

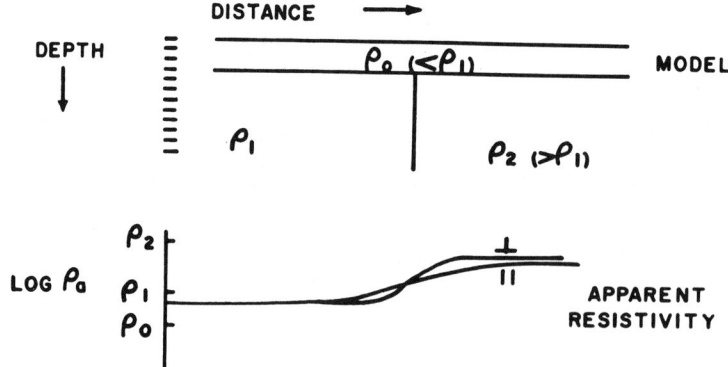

Fig. 13. Effect of burial on the apparent resistivity of a contact. Compare with Figure 10 at the same frequency.

changed. Some of the other properties of the impedance elements are not so apparent. One of these is that

$$Z_{xx} = -Z_{yy} \qquad (9)$$

regardless of the angle between the coordinate axes and strike. Equation (9) was also found by computation to be valid for an arbitrary number of horizontal layers, each of which is arbitrarily anisotropic. Another important property is that

$$Z_{xy} - Z_{yx} = \text{constant} \qquad (10)$$

at all orientations.

In three-dimensional structures, the tensor elements are still well behaved, according to Sims and Bostick (1969). By plausibility arguments, they arrived at

$$Z_{xx} + Z_{yy} = \text{constant} \qquad (11)$$

and

$$Z_{xy} - Z_{yx} = \text{constant}. \qquad (12)$$

The Z_{ij} are first found from the transformed data by solving equations (7) and (8). This involves using two equations in four unknowns. The apparent discrepancy is resolved by taking advantage of the facts that the Z_{ij} change very slowly with frequency and can therefore be computed at far fewer frequencies than there are transform values. That is, the Z_{ij} are calculated as averages over frequency bands with each band including many points of the transform. This has been done in a number of different ways, the most common of which is that described by Madden and Nelson (1964). Each equation is written as

$$\langle E_x A^* \rangle = Z_{xx} \langle H_x A^* \rangle + Z_{xy} \langle H_y A^* \rangle \quad (13)$$

and

$$\langle E_x B^* \rangle = Z_{xx} \langle H_x B^* \rangle + Z_{xy} \langle H_y B^* \rangle, \quad (14)$$

where A^* and B^* are the complex conjugates of any two of H_x, H_y, E_x, and E_y, and CD^* is the cross power of C and D,

$$\langle CD^* \rangle(\omega_1) = \frac{1}{\Delta\omega} \int_{\omega_1-(\Delta\omega/2)}^{\omega_1+(\Delta\omega/2)} CD^* \, d\omega. \quad (15)$$

Six different combinations are possible, so that six different values of each Z_{ij} can be calculated (H_x, H_y; H_x, E_y; etc). Most commonly, the two H components are used, as they are expected to

have a greater degree of independence (smaller cross power) than any other pair. The denominator of each resulting equation is

$$\langle H_x A^* \rangle \langle H_y B^* \rangle - \langle H_x B^* \rangle \langle H_y A^* \rangle. \quad (16)$$

If A and B are strongly dependent, the two terms in the denominator become nearly equal. Although other pairs can and have occasionally been used, the (H_x, H_y) pair is found in practice to give results that are as good or better than any of the others in terms of the numbers of points passing acceptance tests. Exceptions have been observed, particularly when one H channel has been contaminated by artificial noise. The nature of the acceptance tests is discussed below.

Performing the multiplications and solving the equations give the following four expressions to be evaluated:

$$Z_{xx} = \frac{\langle E_x A^* \rangle \langle H_y B^* \rangle - \langle E_x B^* \rangle \langle H_y A^* \rangle}{\langle H_x A^* \rangle \langle H_y B^* \rangle - \langle H_x B^* \rangle \langle H_y A^* \rangle}; \quad (17)$$

$$Z_{xy} = \frac{\langle E_x A^* \rangle \langle H_x B^* \rangle - \langle E_x B^* \rangle \langle H_x A^* \rangle}{\langle H_y A^* \rangle \langle H_x B^* \rangle - \langle H_y B^* \rangle \langle H_x A^* \rangle}; \quad (18)$$

$$Z_{yx} = \frac{\langle E_y A^* \rangle \langle H_y B^* \rangle - \langle E_y B^* \rangle \langle H_y A^* \rangle}{\langle H_x A^* \rangle \langle H_y B^* \rangle - \langle H_x B^* \rangle \langle H_y A^* \rangle}; \quad (19)$$

$$Z_{yy} = \frac{\langle E_y A^* \rangle \langle H_x B^* \rangle - \langle E_y B^* \rangle \langle H_x A^* \rangle}{\langle H_y A^* \rangle \langle H_x B^* \rangle - \langle H_y B^* \rangle \langle H_x A^* \rangle}. \quad (20)$$

Having computed the Z_{ij}, we can then substitute back into equations (7) and (8) to compute values of E_x and E_y. These values, which are predicted from H_x and H_y, have the interesting characteristic that they must be wholly dependent on the horizontal H field. Thus, any differences between the observed E and the predicted E must be due to contamination of either E or H by noise. For this reason, the coherency between predicted and observed E components has proven to be the most sensitive measure of noise available. It is loosely called the predictability. At "normal" sites a predictability of 0.95 or greater is required of an E component as one of the criteria for the corresponding apparent resistivity to appear on the final plots.

Once the Z_{ij} have been found in the original (x, y, z) coordinate system, they can be rotated to any other system (x', y', z') by an angle θ in the clockwise direction. The rotated impedances are

$$2Z'_{xx}(\theta) = (Z_{xx} + Z_{yy}) + (Z_{xx} - Z_{yy})\cos 2\theta$$
$$+ (Z_{xy} + Z_{yx})\sin 2\theta, \qquad (21)$$

$$2Z'_{xy}(\theta) = (Z_{xy} - Z_{yx}) + (Z_{xy} + Z_{yx})\cos 2\theta$$
$$- (Z_{xx} - Z_{yy})\sin 2\theta, \qquad (22)$$

$$2Z'_{yx}(\theta) = -(Z_{xy} - Z_{yx}) + (Z_{xy} + Z_{yx})\cos 2\theta$$
$$- (Z_{xx} - Z_{yy})\sin 2\theta, \qquad (23)$$

$$2Z'_{yy}(\theta) = (Z_{xx} + Z_{yy}) - (Z_{xx} - Z_{yy})\cos 2\theta$$
$$- (Z_{xy} + Z_{yx})\sin 2\theta. \qquad (24)$$

The principal axes of Z are the values of θ at which Z'_{xy} and Z'_{yx} take on their largest and smallest values, respectively. One way of finding these directions is to compute Z for many values of θ and interpolate to find maxima and minima. It is preferable to use an analytical technique if possible, to reduce computation. However, the only such technique which has thus far been developed does not directly maximize either $Z'_{xy}(\theta)$ or $Z'_{yx}(\theta)$. Instead it solves for the angle θ_0 at which

$$|Z'_{xy}(\theta_0)|^2 + |Z'_{yx}(\theta_0)|^2 = \text{maximum.} \qquad (25)$$

Setting the derivative with respect to θ of this sum equal to zero gives (Swift, 1967)

$$\tan 4\theta_0 = \frac{(Z_{xx} - Z_{yy})(Z_{xy} + Z_{yx})^* + (Z_{xx} + Z_{yy})^*(Z_{xy} + Z_{yx})}{|Z_{xx} - Z_{yy}|^2 - |Z_{xy} + Z_{yx}|^2}. \qquad (26)$$

This same value of θ_0 also satisfies

$$|Z'_{xx}(\theta_0)|^2 + |Z'_{yy}(\theta_0)|^2 = \text{minimum,} \qquad (27)$$

so that in the case of two-dimensional structures the scheme finds the true principal axes. In the three-dimensional case, the method picks a slightly more general maximum, i.e.,

$$|Z'_{xy}(\theta) + Z'_{yx}(\theta)| = \text{maximum} \qquad (28)$$

(Sims and Bostick, 1969). The results are seldom shown as impedance values. Instead, the Z'_{ij} are converted to apparent resistivities ρ'_{ij}, with

$$\rho'_{ij} = \frac{1}{5f}|Z'_{ij}|^2. \qquad (29)$$

Apparent resistivity has the phase of Z'_{ij}, that is, the phase difference between E_i and H_j.

Four different Z'_{ij} are extracted at each frequency: Z'_{xx}, Z'_{xy}, Z'_{yx}, and Z'_{yy}. The main purpose

of analysis and plotting is to permit interpretation, which is now possible only for two-dimensional structures. Hence, only ρ'_{xy} and ρ'_{yx} are routinely plotted, since they are the only two which appear in two-dimensional models. The other pair of elements, ρ'_{xx} and ρ'_{yy}, is also useful in interpreting data from the field. Although ρ'_{xx} and ρ'_{yy} are often small compared with the larger of the others, they are never found to be zero. This fact can be used to judge the degree to which the structure at a site departs from true two-dimensionality. If ρ'_{xx} and ρ'_{yy} are both very small, we have either a well-defined strike or horizontal layers.

As noted above, both $(Z_{xx} + Z_{yy})$ and $(Z_{xy} - Z_{yx})$ are independent of θ, as is their ratio. The magnitude of the complex ratio of the quantities is called the skewness,

$$S = \frac{|Z_{xx} + Z_{yy}|}{|Z_{xy} - Z_{yx}|}. \qquad (30)$$

If S is large, structure at the site must appear to be three-dimensional in that frequency range.

An aspect of this method is that it must use a very wide frequency range to be effectively interpreted. Apparent resistivity curves are smooth and regular when they are plotted on log-log scales. Semilog or linear scales give very steep slopes at very low frequencies and uselessly flat slopes at high frequencies. Since Fourier transforms normally appear on a linear frequency scale but interpretation is done on a log-frequency basis, it is helpful to do the integrations of equation (15) with bands whose center frequencies are equispaced on a logarithmic scale. If n such bands are desired in each frequency decade, the ratio of center frequencies of adjacent bands is

$$\log \frac{f_{i+1}}{f_i} = \frac{1}{n}. \qquad (31)$$

Bandwidth, in terms of the number of points of the transforms, must also increase with frequency for consistent smoothing on a log-frequency scale. This is the frequency domain equivalent of the time domain constant Q filters used by Swift

(1967). The bandwidth consideration explains the spacing of points on the plot in this and some other papers on MT.

The other ERF are designed to use the vertical magnetic component H_z to help determine which of the two principal impedance axes is the strike direction. At the same time, the remaining ERF aid the interpreter in understanding the cause of apparent anisotropy, point out distant lateral conductivity changes, and often provide additional warning when three-dimensional structural conditions occur. The concept which guides the use of H_z is the same as that behind the AFMAG and some VLF techniques of electromagnetic propsecting at audio and VLF frequencies.

From the field data we want to find the horizontal direction in which the magnetic field is most highly coherent with H_z. In two-dimensional structures that direction will be constant, perpendicular to strike, for reasons discussed in the next section. It is assumed that, if we wished, the measurement axes could be physically rotated at each frequency to find this direction.

The procedure, due to T. R. Madden (1968, unpublished), is to assume that H_z is linearly related to H_x and H_y and to write at each frequency

$$H_z = A H_x + B H_y, \qquad (32)$$

where A and B are unknown complex coefficients.

$$A = A_r + i A_i, \\ B = B_r + i B_i. \qquad (33)$$

Following the derivation of the Z_{ij}, we integrate

$$\langle H_z H_x^* \rangle = A \langle H_x H_x^* \rangle + B \langle H_y H_x^* \rangle \qquad (34)$$

and

$$\langle H_z H_y^* \rangle = A \langle H_x H_y^* \rangle + B \langle H_y H_y^* \rangle \qquad (35)$$

and solve for A and B. This pair of coefficients can be thought of as operating on the horizontal magnetic field and tipping part of it into the vertical. For that reason, (A, B) is called the "tipper." Its magnitude in each frequency band,

$$|T| = \{|A|^2 + |B|^2\}^{1/2} \\ = (A_r^2 + A_i^2 + B_r^2 + B_i^2)^{1/2}, \qquad (36)$$

shows the relative strength of H_z. Its phase is

$$\tan^{-1} \left[(A_r^2 + B_r^2)/(A_i^2 + B_i^2) \right]^{1/2}. \quad (37)$$

For a two-dimensional structure striking in the direction $(\phi \pm 90)$ degrees from x, A and B will have the same phase if we assume noise-free data, so that A/B is a real number and the tipped horizontal component H_ϕ is at an angle ϕ from the x axis, where

$$\phi = \arctan (B/A). \qquad (38)$$

(It is important in the use of the tipper to know whether the phase difference between H_z and H_ϕ lies near 0 degrees or near 180 degrees, since, for a simple contact, phase at a single site can be used to determine both strike of the contact and direction to the contact from the site.)

In the three-dimensional case, when A and B have different phases, ϕ can be defined in several ways. For example, the definition of ϕ which maximizes the cross power of horizontal and vertical components is

$$\phi_1 = \frac{(A_r^2 + B_r^2) \tan^{-1}(A_r/B_r) + (A_i^2 + B_i^2) \tan^{-1}(A_i/B_i)}{T^2}. \qquad (39)$$

This is a weighted average of rotation angles for the real and imaginary parts of the horizontal. Another rotation criterion is maximum coherency of a horizontal component and H_z. Measures of three-dimensionality are the differences between the principal axis direction obtained by impedance tensor rotation and those obtained from vertical-horizontal field relationships. An indicator of the variations in the latter is the tipper skew, defined as

$$\text{Tipper Skew} = \frac{(A_r^2 + B_r^2) \tan^{-1}(A_r/B_r) - (A_i^2 + B_i^2) \tan^{-1}(A_i/B_i)}{T}. \qquad (40)$$

Tipper skew is zero if both the real and imaginary parts of H_z are most coherent with the *same* horizontal component, as for two-dimensional structures.

From A and B we can also calculate a predicted H_z, and determine its coherency with the measured vertical component by

$$\text{coh}(H_z H_z^{pred}) = \frac{A^*\langle H_z H_x^*\rangle + B^*\langle H_z H_y^*\rangle}{\langle H_z H_z^*\rangle^{1/2}[AA^*\langle H_x H_x^*\rangle + BB^*\langle H_y H_y^*\rangle]^{1/2}}. \qquad (41)$$

Taken together, these quantities describe the relation of H_z to the two horizontal components. They show the importance and locations of lateral conductivity changes and the reliability of H_z for interpretive purposes. For nearly two-dimensional structures, the tipper can be compared directly with the vertical/horizontal field ratios computed for various models. The absence of a significant (i.e., smooth and consistent) tipper permits one to use layered model interpretation with some confidence. The tipper is also a sensitive indicator of certain noise sources.

It has been pointed out on theoretical grounds that tippers could be caused by factors external to the conducting earth. Fortunately for the method, and for several possible reasons, the MT data so far studied by the writer have shown no evidence of tippers of external origin.

INTERPRETATION

The general procedures of interpretation are discussed in this section. Examples of their application to field results will be shown in the following section.

Interpretation consists of three stages. These are (1) a general qualitative overview of the results, (2) fitting layered models at each site, and (3) two-dimensional and three-dimensional interpretation. When properly carried out, this interpretation sequence results in the best possible estimate of conductivity structure within the survey area. In practice, the third stage is often followed by an attempt to assign rock types and a structure to the results, when experience indicates the attempt may be warranted. Interpretation may draw on other sources of information about the area or about similar areas.

Since all geophysical interpretation is based on comparing observed earth response data with model data, this section will begin with a discussion of models. In horizontally layered structures, a single apparent resistivity curve, or its phase, contains all information of significance. Strike and tipper are undefined. Derivation of apparent resistivity and phase curves for the layered model appear in several published references, for example Madden and Swift (1969), Tikhonov (1950), and Wait (1962). Layered

model curves can be calculated as accurately as desired, for as many layers as desired, at computing costs of a few cents each. Suites of curves appear in Cagniard (1953), Yungul (1961), and Srivastava (1967).

Figure 8 compares a pair of two-layer models and their idealized apparent resistivity curves. At a sufficiently *high* frequency, the skin depths are small enough that practically no energy penetrates to the basement. Apparent resistivity is therefore asymptotic to ρ_1 at high frequencies and the upper layer is not penetrated. When frequencies are *low* enough, the top layer has little effect and the apparent resistivity approaches ρ_2. The transition with frequency is gradual, so smooth curves result. The greater layer thickness of model B requires going to a lower frequency to obtain the same amount of basement influence on the curve.

A typical three-layer curve is shown in Figure 9 for a second layer which is more conductive than the other two. As expected, the curve approaches ρ_1 at high frequencies, drops toward ρ_2 at lower frequencies, and then goes to ρ_3. Although ρ_a never reaches ρ_2, the presence of a conductive second layer is obvious as long as it is not too thin. (Ways of interpreting its thickness and resistivity are discussed below.) The extension to cases of four or more layers is clear.

The gradual way in which the effect of each layer appears, as contrasted with the abrupt onset of a reflection in a seismic record, characterizes the weaknesses and the strengths of this method. If, for example, the second layer were thin enough, the apparent resistivity curve would go smoothly from ρ_1 to ρ_3: the second layer would not be seen unless it were extremely conductive. On the other hand, this tendency to average together the minor features permits the method to show up weakly systematic variations which might be lost with a higher resolving power. It can be thought of as emphasizing the grosser (longer wavelength) structural features at the expense of the finer details. This characteristic carries over to the two-dimensional case, so that structural features may show up even though traverses do not cross directly over them, justifying large traverse spacings.

Figure 10 shows the simplest two-dimensional structure, the vertical fault. It displays all of the characteristics of more complicated two-dimensional models, but is one of the few for which any theoretical solutions have been found. Consequently it makes a very instructive example.

The gross differences between this and the horizontally layered models examined earlier are that (1) apparent resistivity and phases vary with rotation of measurement axes and (2) the tipper is nonzero. We consider two sites, one on each side of the fault (see Figure 11). When measurement axes at site A are aligned with the structural axes, the two apparent resistivity curves which result are shown (labeled A) in Figure 11. Measurements made at point B give the two curves labeled B. The subscript for each curve refers to the direction of E relative to strike. H_z and the tipper approach zero with increasing frequency at each location. At low frequencies, the tippers at both sites are asymptotic to the same constant values.

When the values of several quantities are plotted versus location along a traverse crossing the fault, curves such as the apparent resistivity, tipper, and relative $|H_\perp|$ of Figure 10 are obtained. The features to note are the following:

1. ρ_a varies smoothly for currents flowing along to strike, but varies discontinuously, with an overshoot, for currents across strike.
2. ρ_a is asymptotic to the appropriate resistivity value at large distances from the fault.
3. H_z is largest near the fault, decreasing smoothly to zero in both directions.
4. The horizontal H component across strike, $H_\perp$, varies significantly near the fault.

Most of these effects can be understood by examining the current flow patterns near the fault in Figure 10. At a large distance from the interface, currents are crowded closer to the surface on the conductive side. Currents across strike must be continuous through the contact, and they adjust smoothly as shown. Far away from the contact, the values of current density and electric field are appropriate to the uniform medium. Approaching the contact from the conductive side, the current density at the surface decreases; hence, E and ρ_a decrease. Although the current densities on the two sides are identical at the contact, the abrupt resistivity change causes an abrupt change in E, and hence in ρ_a. Because

current is continuous and $E=j\rho$, E changes in the same ratio as ρ; ρ_a, which depends on E^2, changes as the *square* of the change in ρ.

Currents and electric fields parallel to strike are able to adjust without having to cross the boundary. The parallel E component and the corresponding ρ_a are continuous across the contact, approaching the appropriate uniform earth solution smoothly with distance on each side. However, the current densities change abruptly here, as illustrated, causing a magnetic field that curves around the higher density region. This is the origin of the vertical H component. Direction of largest H in the vertical plane is shown by the lines of H above $E_\parallel$ current flow.

More rigorously, there will be a vertical H component only if curl E has a vertical component. If y is the strike direction $(\partial/\partial y=0)$, then for currents across strike $(E_y=0)$ the curl of E has no z component. Currents parallel to strike have a nonzero value for $\partial Ey/\partial x$, so they also have an associated H_z component.

When measurements are made along a traverse line as is natural in two-dimensional situations, a more convenient presentation of apparent resistivities than that of Figures 10 and 11 is essential. One such presentation is the pseudosection, suggested by Neves (1957). Figure 12 shows pseudosections for the fault. The horizontal scales are distance; the vertical scales are frequency with lowest frequencies at bottom. Apparent resistivity value is plotted beneath site location at the appropriate frequency for each site, and the results are contoured. Separate pseudosections have to be constructed for E parallel and H parallel to strike.

As a means of further explaining the logical basis of the second stage of interpretation, we will discuss three additional models. These are the buried steeply dipping contact, the thin dipping conductive dike, and the thin dipping resistive dike.

When a cover of resistivity ρ_0 is laid over the previous model, the result is to smooth and attenuate all of the boundary effects which were seen. The degree of smoothing increases with the thickness and conductivity of the overburden (Figure 13). Apparent resistivity curves at large distances from the contact are those of the appropriate two-layer model. The two pseudosections for E parallel and E perpendicular become very similar. Although the transition in apparent resistivity will still be more abrupt for currents flowing across strike, the actual discontinuity vanishes because both current density and resistivity are continuous. The tipper then becomes a more important clue to strike.

Many faults in older consolidated and metamorphosed rocks have the appearance of narrow conductive dikes. Shale beds, which often are the loci of imbricate faulting, also behave like conductive dikes. A steeply dipping thin conductive dike has a pronounced effect on parallel current and electric field and is therefore effective in generating a tipper. However, the dike's effect on the E field across it does not become large until the dip is very shallow, when the dike starts to behave as a conductive layer. Figure 14 shows the behavior of the dike when it is vertical. The $E_\parallel$ apparent resistivity responds at a substantial distance on either side, giving the dike the appearance of a conductive valley. If this actually were a conductive valley, however, the $E_\perp$ apparent resistivity would also show it. The lack of response in $E_\perp$ is diagnostic of the narrow steeply dipping conductor. The same lack of response holds even for dips as shallow as 45 degrees.

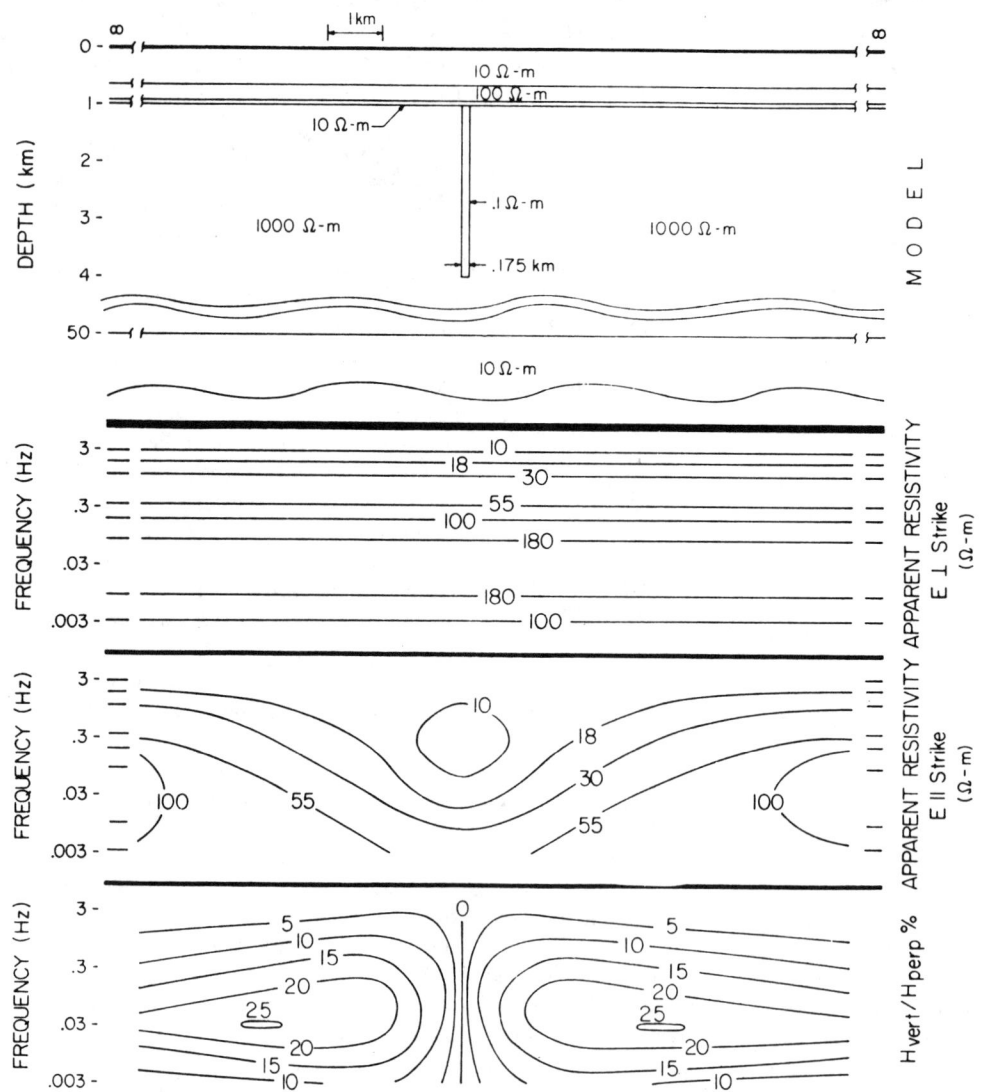

FIG. 14. Computed model of a buried vertical conductive dike.

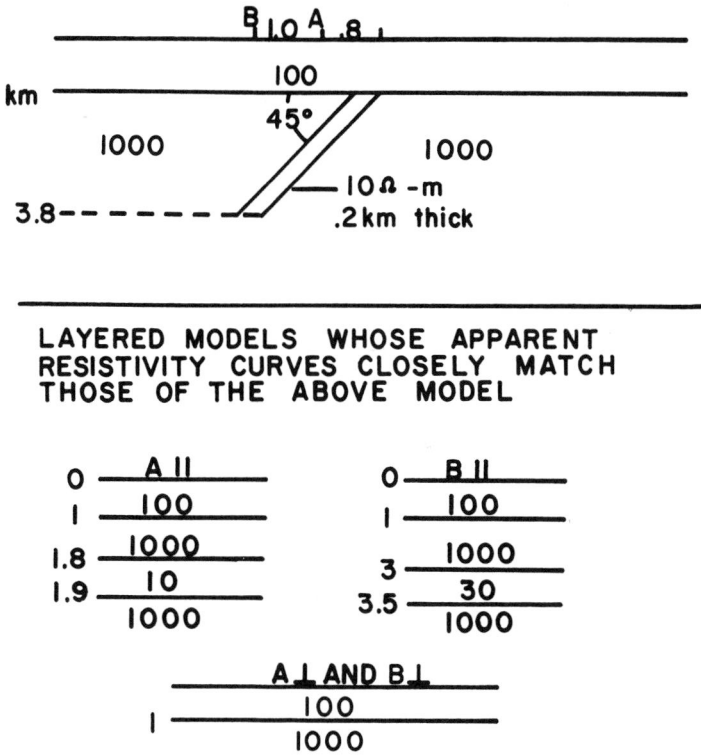

LAYERED MODELS WHOSE APPARENT
RESISTIVITY CURVES CLOSELY MATCH
THOSE OF THE ABOVE MODEL

Fig. 15. Result of fitting layered models to response curves of a two-dimensional structure
at sites A and B.

The result of fitting horizontally layered models at two locations over a conductive dike is shown in Figure 15. Site A is 0.8 km from the trace of the dike, and B is 1.8 km from the trace. The $E_\perp$ components show no evidence of the dike. In the $E_\parallel$ components, the dike has much the same effect on apparent resistivity as does a conductive layer. If the $E_\parallel$ interpretations were used to construct an electrical cross-section, they would show a conductive bed dipping in the right direction. Although the interpreted thickness and conductivity of the bed would not be individually correct, the conductivity-thickness products would be reasonably close to that of the dike (0.01 and 0.015 versus 0.02).

In the extreme case, many thin conductors in a more resistive matrix cause the material to behave as a uniform but anisotropic material. Such a model can be used to study the effects of dip on apparent resistivities. Figure 16 shows the apparent resistivities for an anisotropic model. In this case, the anisotropy was confined to a second layer, the first and third layers being isotropic

and homogeneous. The results further illustrate that the dip must be fairly shallow (here less than 30 degrees) for the thin conductors to have much effect on $E_\perp$.

If the thin dike is more resistive than its surroundings, the situation is reversed: $E_\perp$ apparent resistivity is the component most affected and little effect is seen in the $E_\parallel$ apparent resistivity until the dip is small enough that the dike begins to behave as a resistive layer. This behavior can be seen in the previous (anisotropic layer) model if the 100 ohm-m of the second layer is now considered due to many thin resistive beds in a 10 ohm-m matrix.

The picture just discussed is seen in the immediate vicinity of an isolated resistive dike(Figure 17). Diagnostics are an anomaly in resistivity across strike, with little or no effect on $E_\parallel$ and tipper (not shown). The thin resistive dike seems to represent a fault in young, unconsolidated sediments, as well as salt ridges and intrusive dikes.

It is observed that the $E_\perp$ pseudosections com-

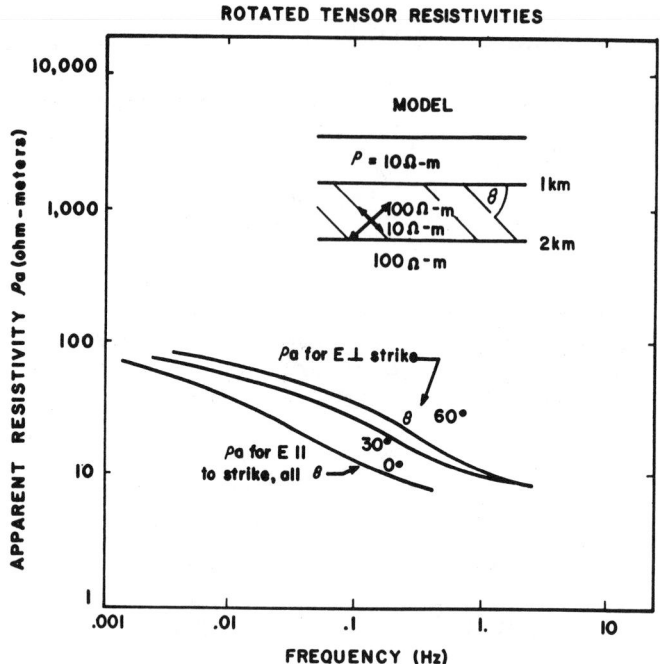

Fig. 16. Changes in apparent resistivity with dip of principal conductivity axis in an anisotropic medium.

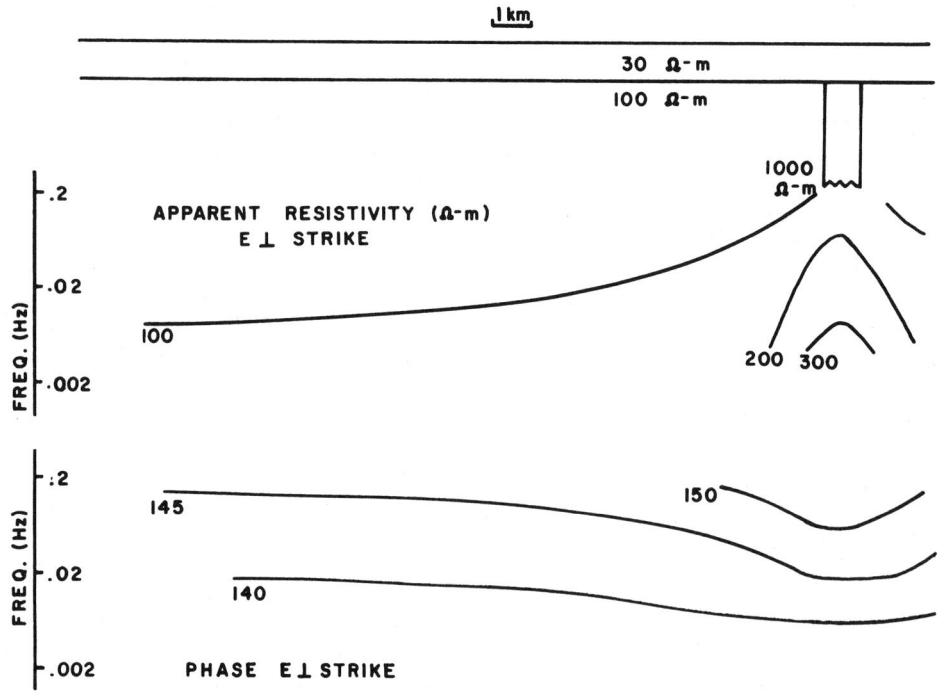

Fig. 17. Computed model of a vertical resistive dike. E‖ resistivity is not distinguishable from that of the two-layer case with the dike absent.

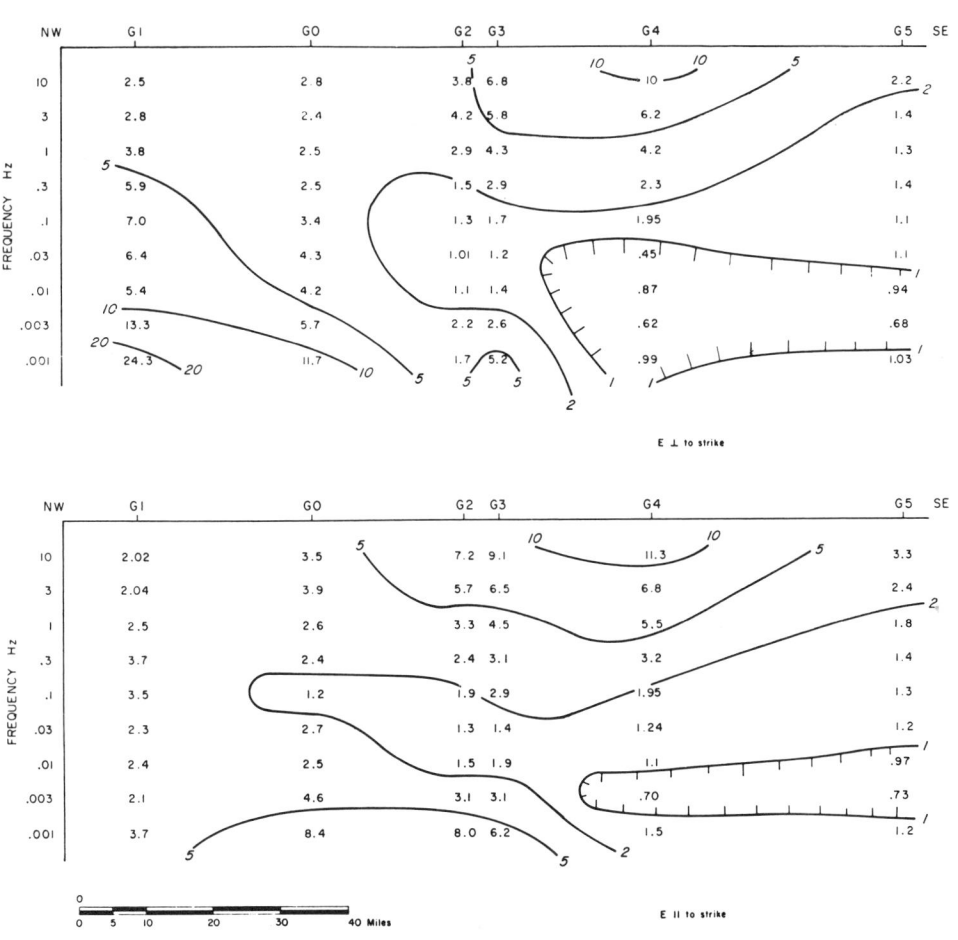

FIG. 18. Pseudosections, South Texas traverse.

monly have a greater range of values and more lateral variation in values than do the $E_{\parallel}$ pseudosections. This is not unexpected, since even in the simplest models the apparent resistivities for $E_{\perp}$ commonly display an overshoot, while a smooth transistion is more typical of those for $E_{\parallel}$. Furthermore, the simplest real earth is far more complex in its variability than any model we might have the incentive to construct, even though there appears to be more regularity to the broad averages for geological units than there is to the fine structure (Keller, 1968; Harthill, 1968). As a result, we are seldom very far from some lateral conductivity change. These two factors explain the larger variance typical of $E_{\perp}$. Any continuity of structure or lithology which may exist is therefore more easily followed in the

$E_{\parallel}$ pseudosection, so it is the one most often used for the first (horizontal layer) interpretation. (In the important case in which a narrow resistive structure extends upward into a more conductive section, $E_{\parallel}$ may not resolve the structure and therefore $E_{\perp}$ will yield a better depth estimate on the structure.) If many contours are found between adjacent sites on the $E_{\perp}$ pseudosection, it is taken to indicate a "break" between the sites, and the interpretation is usually confirmed by tipper behavior.

In some areas, the two pseudosections are nearly indistinguishable, indicating no more than gradual changes between sites. In other areas, the two are quite different and change rapidly with location, indicating sharp lateral breaks. Figure 18 shows gradual changes at the southeastern end, and apparent anisotropy at the northwest.

Obviously, the smoothness of pseudosections depends in part on the spacing between sites, since even with shallow dips the resistivity-depth curve may change considerably between sites 10 or more miles apart, as in a reconnaissance survey. Here again the tipper can be helpful, since it is found to be negligibly small in the absence of sharp lateral breaks.

If the layered model fitted at each site is ex-

tended half way to each adjoining site, we obtain the first-try two-dimensional model. The parameters of this model can then be modified systematically until the pseudosections match the field data. In practice, only a few such trials are usually made because of the large number of variables involved, and the fact that the geology usually departs enough from being two-dimensional that the end result will not be quantitative no matter how good the fit.

Waeselynck (1967) outlined an approach to

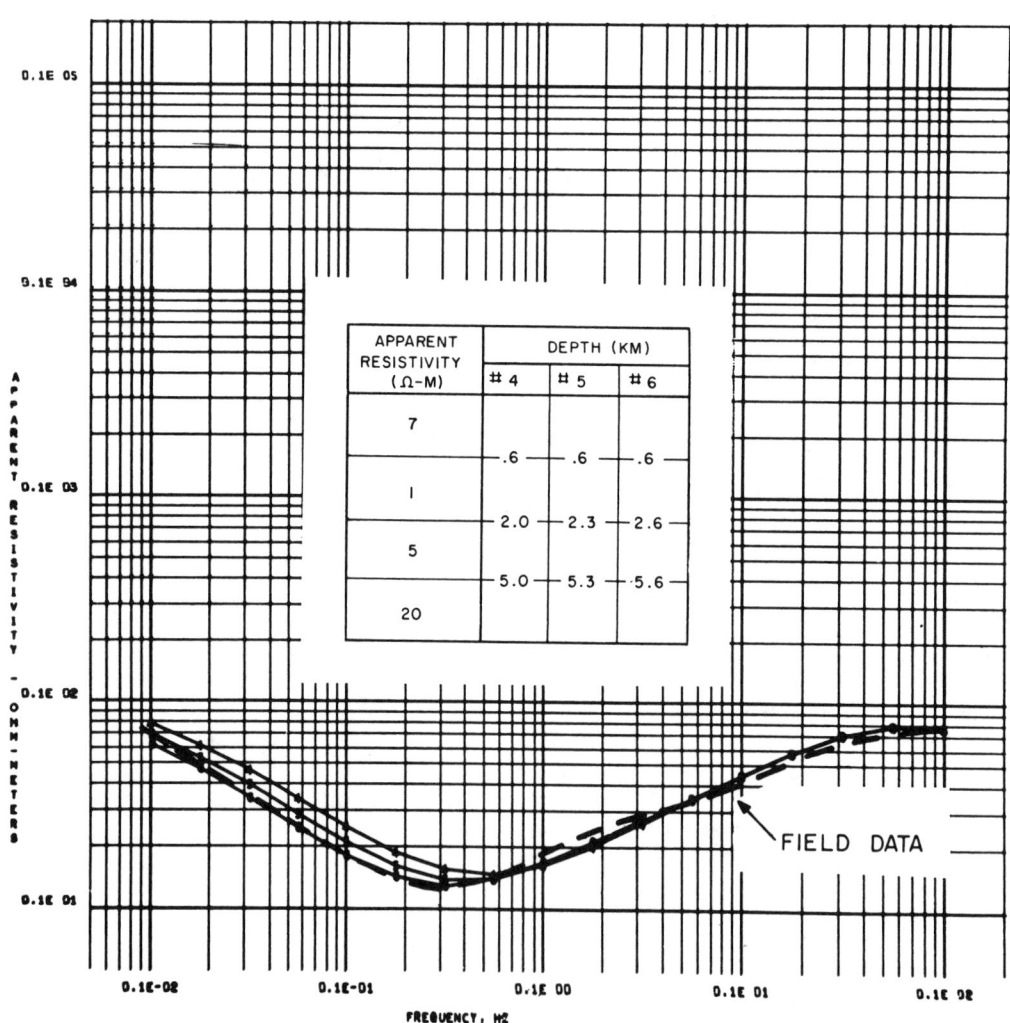

FIG. 19. Cut-and-try layered model fit, South Texas traverse. Model no. 6 fits best the dashed curve drawn through the field results.

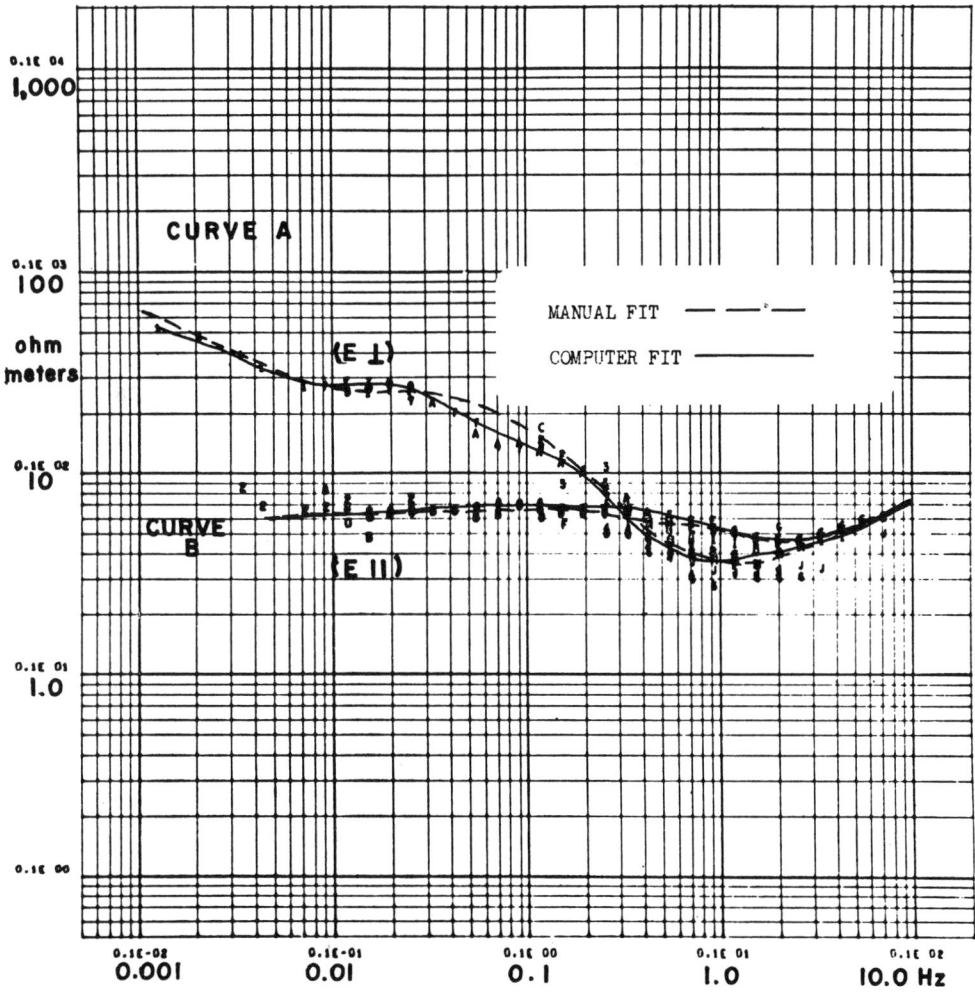

FIG. 20. Interpretation of apparent resistivity data from Oklahoma, showing manually fitted and computer-fitted model curves. Corresponding models are shown in Figure 21.

computing three-dimensional models, but no examples have thus far appeared.

CURVE MATCHING

Most layered model curve matching to date has been done manually, by cut-and-try methods. Numerous sets of computed model curves are available (Yungul, 1961; and Srivastava, 1967) but efficient computer programs for arbitrary horizontal layering are common. Time-sharing systems have been found very useful and inexpensive in this application. An example of a cut-and-try fit is shown in Figure 19.

In addition to the cut-and-try technique, direct computer curve fitting techniques are available.

The more common of these use least squares fitting, where the computer attempts to find a model which fits every data point, so as to produce the smallest sum of squared errors (Wu, 1968; Patrick and Bostick, 1969). In the first reference, only apparent resistivity curves are matched, whereas in the second, both apparent resistivity and phase can be fitted.

In least squares methods the number of parameters found (layer thicknesses plus layer resistivities) cannot exceed the number of data points, and all data points used as computer input are usually weighted equally; that is, there is no way to distinguish good data from poor data. More significant, there is no simple, objective way to tell how good the data are. Indirect indicators are smoothness, predictability, and consistency with time; but these do not distinguish between good one-dimensional data and good three-dimensional data, where the dimensionality refers to the model used to fit the data. The generalized inverse technique, suggested first (in this application) by T. R. Madden and developed by Harter and Madden (in preparation), has the virtue of being able to ignore bad data points, of indicating when the data are otherwise inade-

quate, and of showing when the model itself is inadequate.

In both manual and direct inversion, the choice of starting parameters can be critical, so that any prior knowledge must be used as early as possible.

To compare the results of curve matching by manual cut-and-try methods with direct (generalized inverse) computer fitting, data were analyzed from a site in the Anadarko Basin. The computer output apparent resistivities are shown in Figure 20. Two curves, the manually fitted

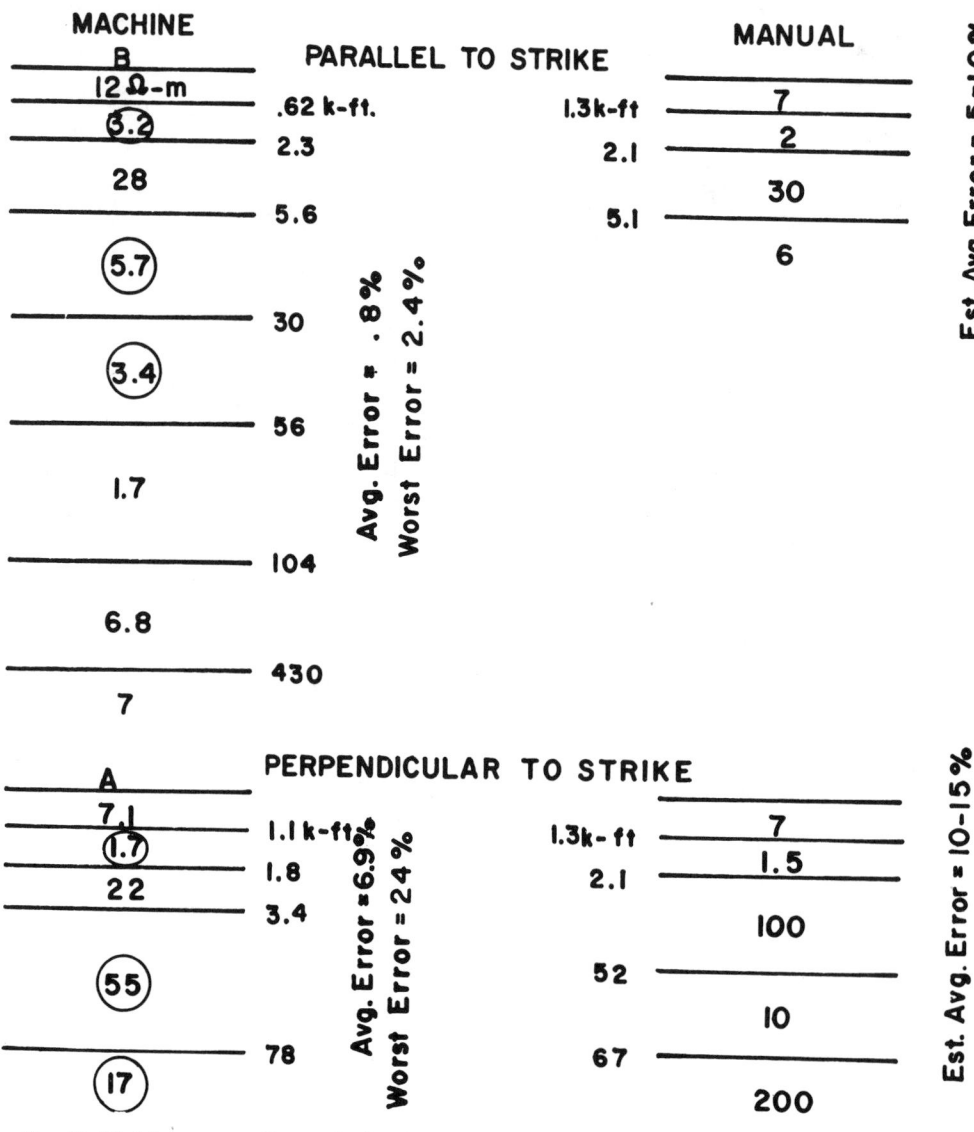

FIG. 21. Models corresponding to the fitted curves of the previous figure. Errors for the manually fitted curves are estimated.

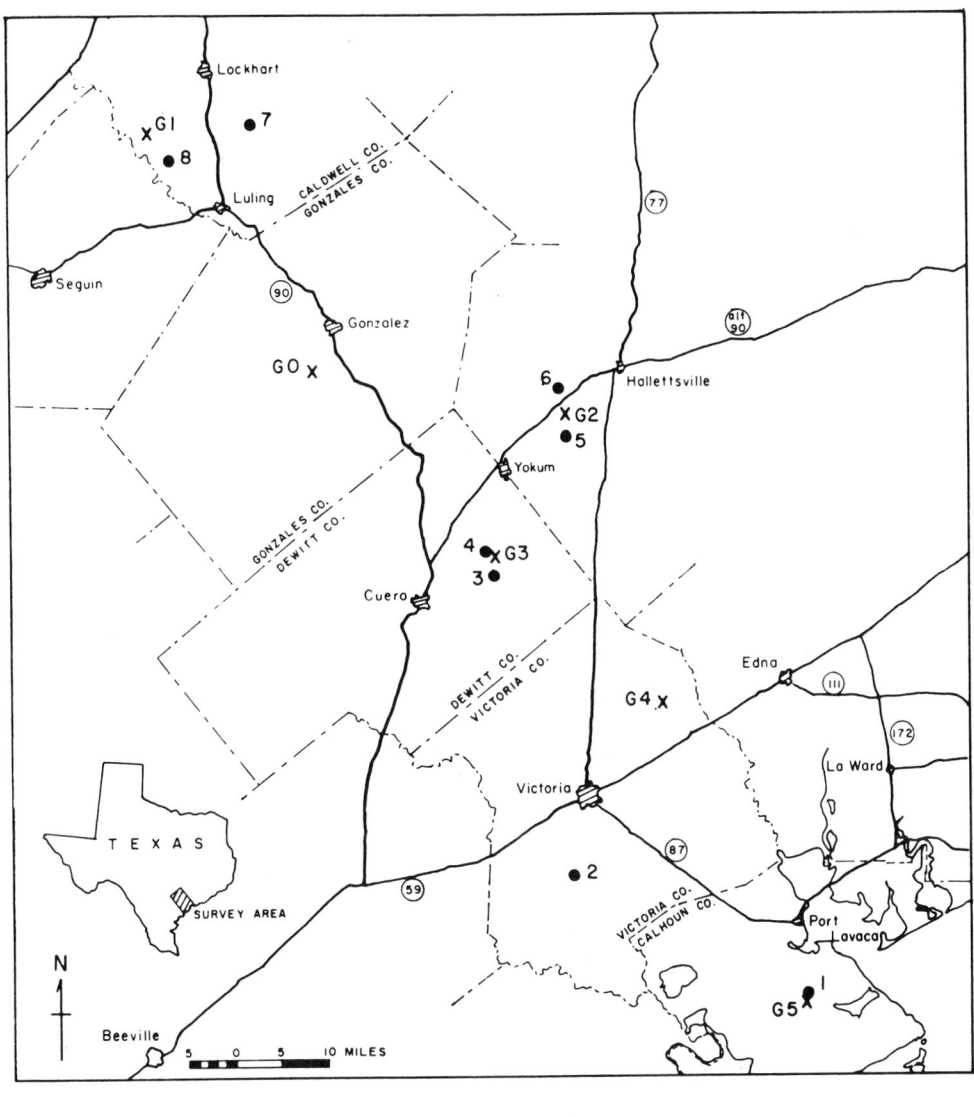

X = MT Site
● = Well Log (deep)

Fig. 22. Location map, South Texas traverse, showing sites and deep wells used for control.

and the direct fitted, are superposed on each set of points; and the resulting models are compared in Figure 21. No error is calculated for the manually fitted curves; in this case, the average error appears to be 10–15 percent for the $E_\perp$ data and 5–10 percent for the $E_\parallel$. The computer fit is obviously closer in each case, but the models are not very different in the upper few thousand feet. Considering the apparent anisotropy, the deeper portions of the models are remarkably consistent with the results to the north, presented in a later section. They indicate the bottom of horizontal layering to be at about 3400 ft subsurface.

The circled values on the machine-fitted models, Figure 21, indicate layers whose parameters are individually best resolved and most stable. These are unlikely to differ much if other starting models are used, for example.

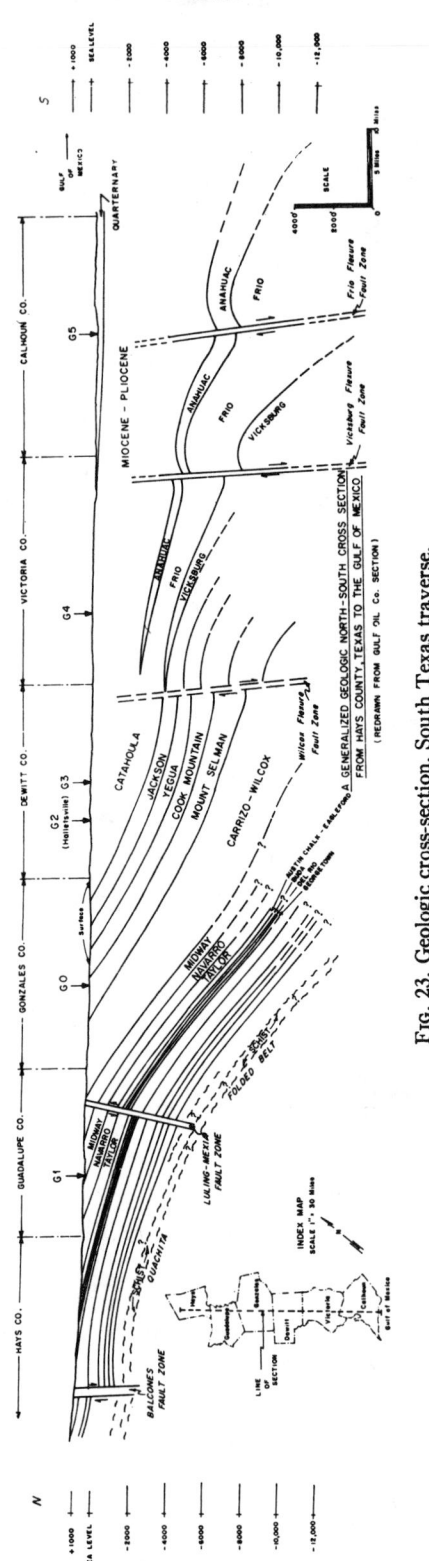

FIG. 23. Geologic cross-section, South Texas traverse.

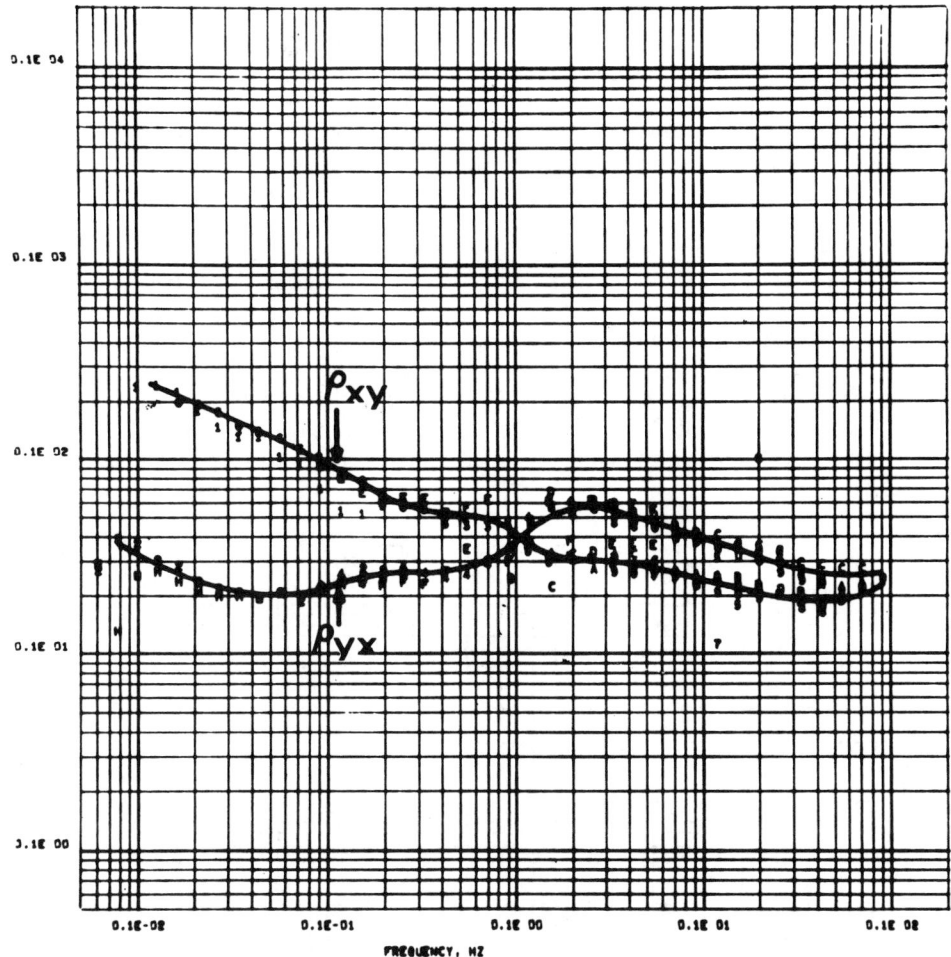

Fig. 24. Rotated apparent resistivities, site G-1, South Texas.

Two examples of the results of field surveys are presented here. One is a wide-spacing reconnaissance traverse in South Texas. The other is an intermediate-spacing traverse in the Anadarko Basin. Both would be considered basin evaluations if they had been carried out in unknown areas. As it is, new data were developed which gave information about the deeper portions of both basins.

South Texas

The first of these surveys (Figure 22) trends northwestward for 115 miles, from Port Lavaca, Texas, on the Gulf Coast to a point 9 miles southwest of Lockhart, Texas. Five magnetotelluric

sites were set up along the traverse. One additional site, G-2, was set up 16 miles to the northeast of site G-3 to examine the continuity of data off the traverse. Word et at (1969) have continued MT measurements in the up-dip direction, but their traverse is offset about 40 miles to the northeast.

This area, which is relatively well known geologically, is an important one to the oil industry of the United States. With the good geologic control, it was possible to set up sites such that each succeeding site to the southeast showed the effect of later deposition. Therefore, the traverse enabled us to see, on a gross scale, a magnetotelluric cross-section of the Gulf Coast. We also attempted to detect the presence of high pressure shales. This area was selected partly because it is open-ended to the Gulf, so that conditions and results ought to resemble those offshore.

The traverse line extended from the flank of the Ouachita folded belt on the northwest, southeastward across the tremendous deposits of Gulf Coast sediments (Figure 23). The line was along the San Marcos arch between the Houston and Rio Grande embayments. The stratigraphic units strike northeast and dip to the southeast. Thus, the traverse extended in a down-dip direction from G-1 to G-5. In general, from northwest to southeast the sites encountered younger sediments and a thickening of the older sediment. Numerous facies changes within the units exist.

The traverse crossed a series of the "down-to-the-coast" fault/flexure zones.

Results are discussed first for each site and then for the entire traverse. All apparent resistivities shown have been rotated by equation (26). These data were analyzed before the tipper and direct curve-fitting computations had been started. The vertical magnetic component was analyzed in terms of coherency ratio and phase relative to each of the horizontal component. From these a rough tipper was estimated mentally.

Site G-1

The rotated apparent resistivity curves of Figure 24 indicated that 2–10 ohm-m values predominate beneath this site. Increases in apparent

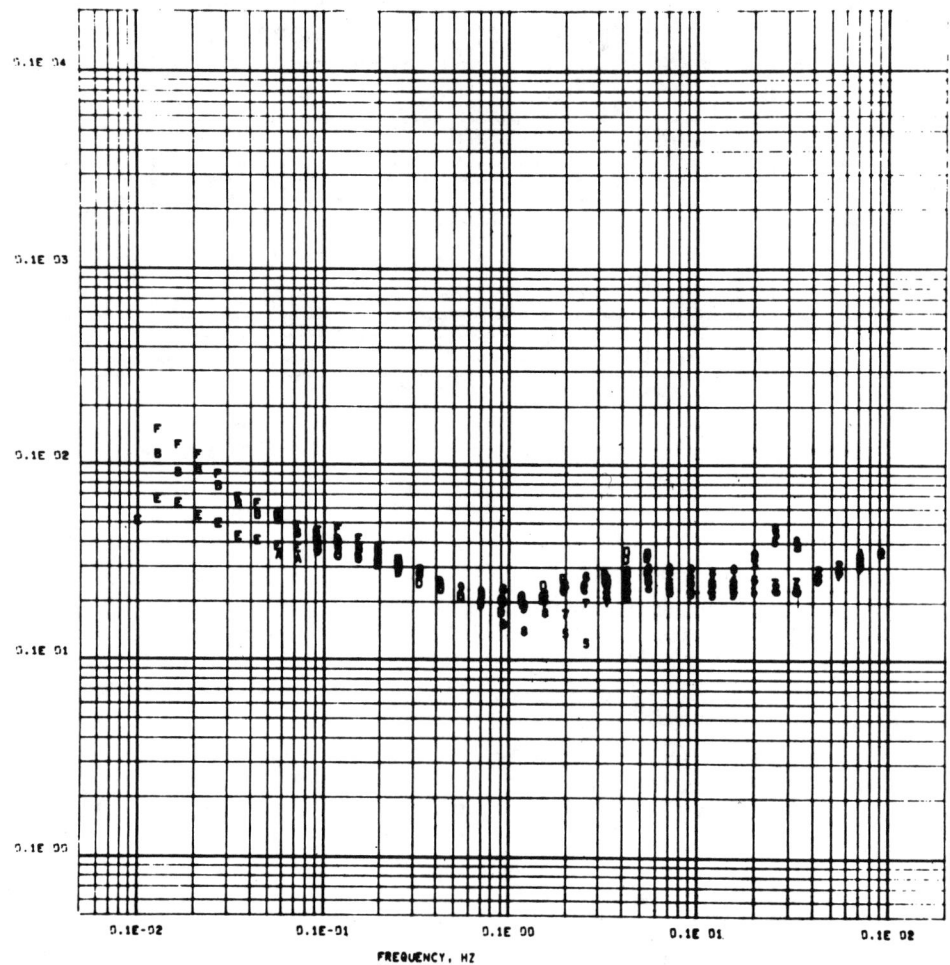

Fig. 25. Rotated apparent resistivities, site G-0, South Texas.

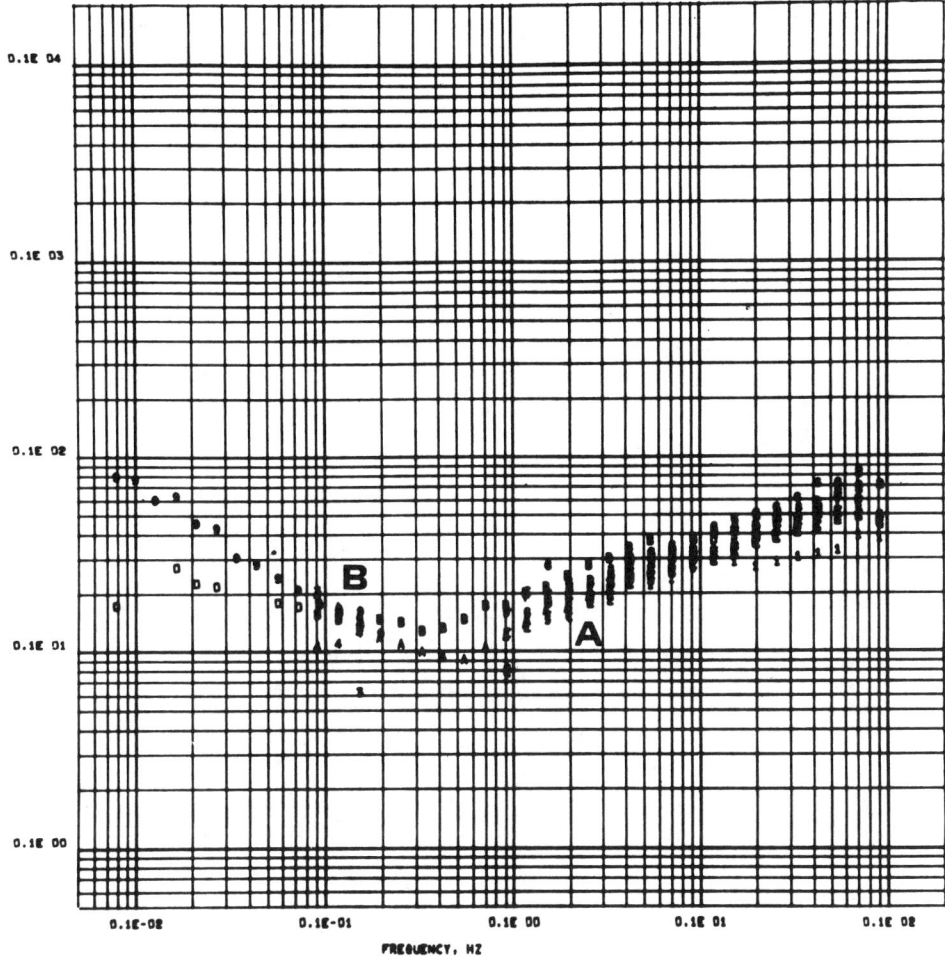

FIG. 26. Rotated apparent resistivities, site G-2, South Texas. E∥ is curve B.

anisotropy with decreasing frequency could be the result of increasing dips, of increasing resistivity contrasts between layers, or both. Pronounced lateral conductivity changes at depths or horizontal distances of 3–5 miles would give similar effects.

Strike at shallow depth is either ENE or NNW, with a 90 degree ambiguity which requires another site for resolution.[1] ENE is believed to be more likely of the two directions. At greater depths, the strike direction is very nearly northward.

Near-surface dips are minor. The data indicate

[1] Choose one of the directions as strike and set another site several miles away along that direction. If the choice is correct the two sets of results will be very similar. If not, a difference will be observed.

that resistivity may increase slightly toward the east at shallow depth, but below 3000–4000 ft, it increases strongly toward the west. Plot criteria are skew less than 0.2 and predictability greater than 0.95.

Site G-0

Although the rotated apparent resistivities at G-0 (Figure 25) were similar to those at G-1, apparent anisotropy (i.e., the gap between sets of points) is much smaller at G-0. Another major difference between the two sites is the greater thickness of moderately conductive material at G-0. These two observations are reflected in the smaller vertical magnetic field component H_z and the smaller skew at G-0.

Apparent electrical strike direction is definite

from 0.05 hz downward, varying from N30W at 0.05 hz, to N50W at 0.005 hz and back to N30W at 0.001 hz, with the 90 degree ambiguity throughout. Regional geology suggests that true strike is at 90 degrees to these directions. Resistivity in the uppermost several thousand feet appears to increase eastward.

Sites G-2 and G-3

As shown on the location map, G-2 is 16 miles off the traverse line to the NE from G-3, along what was believed to be regional strike. Results at the two sites differed in some important features, with those from G-2 being the more complex (Figure 26). Apparent anisotropy at G-2 is better developed, rotation angles are more definite, and skew is generally larger except at the lowest fre-

quencies. Below 4000–5000 ft at G-3 (Figure 27) there is a pronounced resistivity increase to the NW. This may continue northeastward past G-3, but is not as clearly indicated there. According to the phase information (not shown), the low frequency apparent resistivities at G-3 rise to larger values than those at G-2 before starting to decrease again. Hence a thicker, more resistive section is expected at depths of 25,000–30,000 ft beneath G-3.

Apparent electrical strike is N70W to N85W at G-2, and probably indicates a real strike direction of N5E to N20E. The apparent electrical strike at G-3 alternates randomly between these same two values. For a shallow strike direction of N5–10E, G-2 should be projected farther to the SE on the interpreted resistivity cross-section, although its

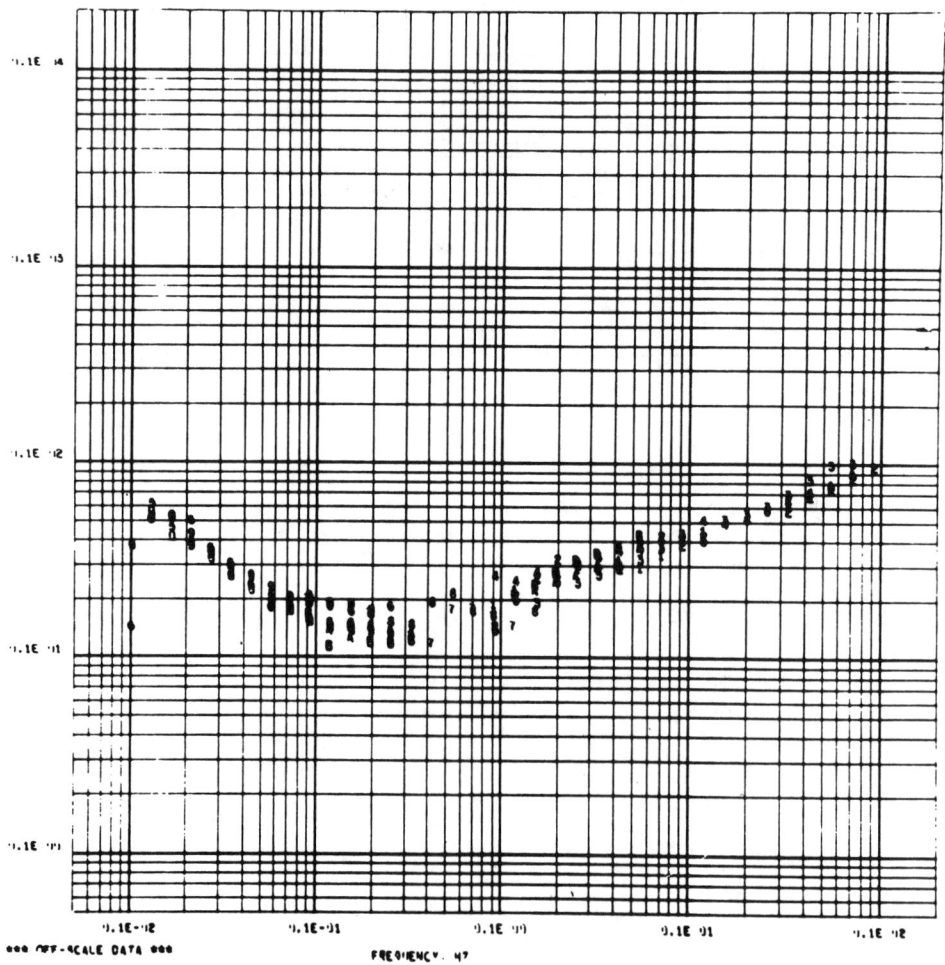

FIG. 27. Rotated apparent resistivities, site G-3, South Texas.

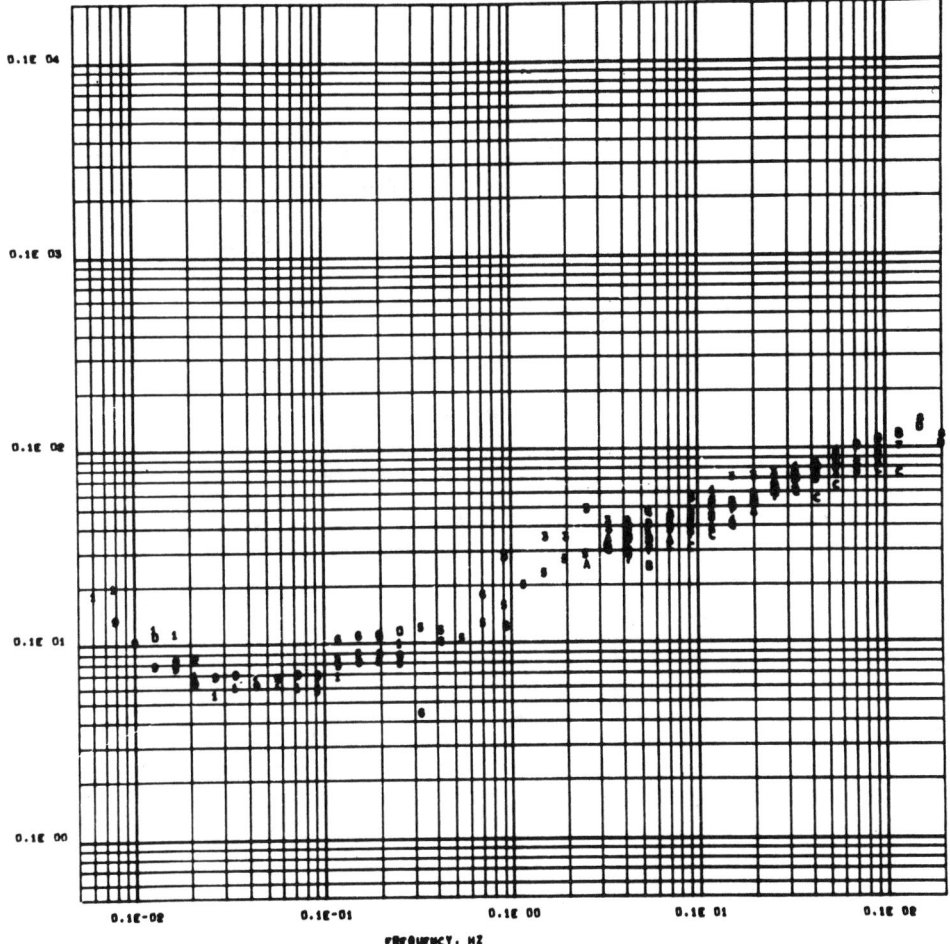

Fig. 28. Rotated apparent resistivities, site G-4, South Texas.

projection is correct for the deeper strikes. Plot criteria are skew less than 0.2 and predictibility greater than 0.85 (G-2) and 0.95 (G-3).

Site G-4

Apparent resistivities (Figure 28) decrease very gradually before rising sharply at 0.002 hz. Apparent anisotropy is minor; apparent electrical strike direction is poorly defined but changes smoothly with frequency from N20W (low frequency) to N10E (high frequency). Noise is large and few low frequency points of ρ_{21} are passable. Plot criteria are skew less than 0.5 and predictability greater than 0.85.

Site G-5

Anisotropy is about the same order as the scatter in the data, about ± 5 percent over most of the frequency range (Figure 29). Rotation angles are poorly defined except at the highest frequencies.

The most unusual aspect (not shown here) of the G-5 data are large values of H_z at the higher frequencies, decreasing linearly with decreasing frequency and closely correlated with H_y. The H vector in the y-z plane is tilted downwards to the east, indicating a north–south conductor lying near the site and to the west of it. This may be caused by an unseen pipeline. An alternative explanation involves near-surface salt-water invasion to the west, possibly as a result of locally enhanced porosity. The University of Texas data at Port Aransas to the south and west of G-5 have high frequency apparent resistivity less than 1 ohm-m (Smith, 1968). Plot criteria are skew less than 0.5 and predictability greater than 0.95.

Traverse

Layered models were matched to each of the $E_\parallel$ apparent resistivity curves by trial and error; and a correlation between stations was carried out, with the result shown in Figure 30. These correlations are obviously open to question and adjustment in view of the large spacings between sites.

The interpreted resistivity section shows three ranges of values: 0.5–3 ohm-m, isotropic; 4–10 ohm-m, isotropic; and >2 ohm-m, anisotropic. The last category is predominant at depth, particularly in the north-west near the thrust. The anisotropy was interpreted as due to dipping or

monoclinally folded beds of alternating resistive and conductive materials. Bands of graphitic or black shaly materials, which could provide the lower resistivity value, have been observed in and beneath the thrust zone, both in East Texas and in the exposed Precambrian near Llano. Going from NW to SE, these materials become buried beneath an increasing thickness of the conductive isotropic rocks and, in particular, by a zone of nearly isotropic 0.6–2 ohm-m resistivity. Minor apparent anisotropy, indicating minor dips with small true anisotropy (probably interbedded sands and shales), persists through the entire traverse. The computed resistivity rotation angles and skews suggest that these dips are due to minor local structure, and that it might be possible by careful model work to interpret strike

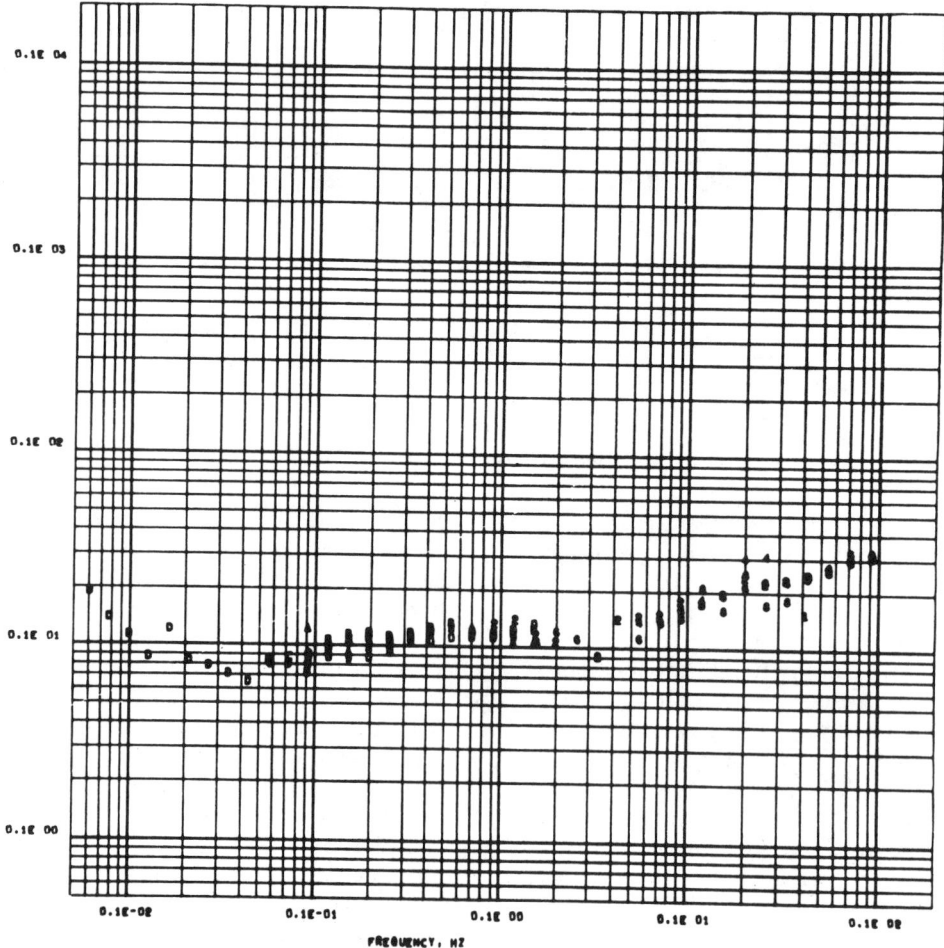

FIG. 29. Rotated apparent resistivities, site G-5, South Texas.

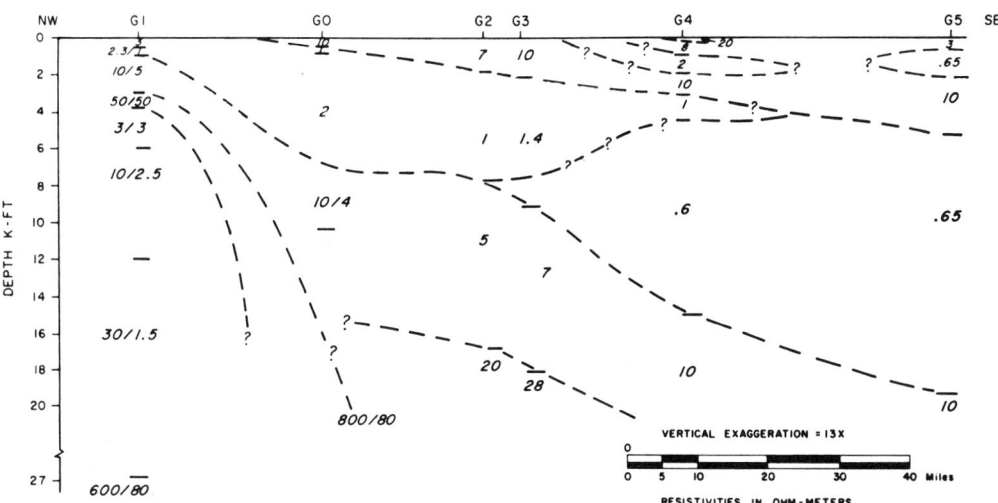

FIG. 30. Interpreted resistivity cross-section, South Texas traverse.

and dip variations with depth beneath the site. The resemblance between this figure and the generalized geologic section (Figure 23) is unmistakable.

There are some interesting differences which appear to be significant. For example, the tendency of the resistivity boundaries to be more nearly horizontal than the geological boundaries appears to be real, and may be a result of water and clay distribution and of compaction. Resistivity in the Eocene decreases systematically seaward, possibly because of a systematic increase in clay content. It has been suggested that the 10 ohm-m material near the surface indicates fresh water invasion.

One of the targets chosen was high-pressure shale, i.e., a shale having a resistivity substantially less than 1 ohm-m, which was known to be present in large amounts beneath G-4 and virtually absent beneath G-2. The apparent resistivities at G-4 and G-5 could not be matched without the thick 0.6 ohm-m zones. No amount of reinterpretation can modify the basic result. The method is unable to resolve the fine structure of the interbedded sands within the overpressured zone, but is able to map features beneath it.

The limited contrasts and simple structure permit us to evaluate resistivity with what is still a surprising amount of resolution. In such situations MT takes on the aspect of a regional stratigraphic tool.

Upon completion of the interpretation, logs

from various drill holes along the traverse were obtained and compared with the interpreted section. The location of the drill holes relative to the sites is shown in Figure 22. The comparison of electric log information and interpreted structure is shown in Figure 31.[2] For this example, the electric logs were subdivided by hand into segments, each segment being given the average conductivity for its depth range. A segment then represents a layer. Overall, the agreement is quite good. Disagreement at D.H. 6 (bottom) is within the limits of accuracy; D.H. 7 is further from G-1 than is D.H. 8 and probably reflects lateral variations in near-surface structure. The discrepancy between D.H. 3 and the interpreted section at G-3 may be related to the age of the log (1949).

A more detailed interpretation was carried out at one location, while testing the direct model fitting program. The program uses the generalized matrix inversion program developed by R. Harter and T. R. Madden (in preparation). The results, comparing G-2 and D.H. 6, are summarized in the four curves of Figure 32. The heavy solid curve is the result of smoothing the induction log for D.H. 6 (No. 1 J. Orsak, Mobil Oil Co., 1960–61, mud resistivity=0.4 at 108 degrees) over steps of 10 percent of depth. The dashed curve is a portion of the four-layer model which was pre-

[2] Induction logs commonly use millimhos/meter (mmho/m) as units of measure. To convert, 1 mho/m =10³ mmho/m.

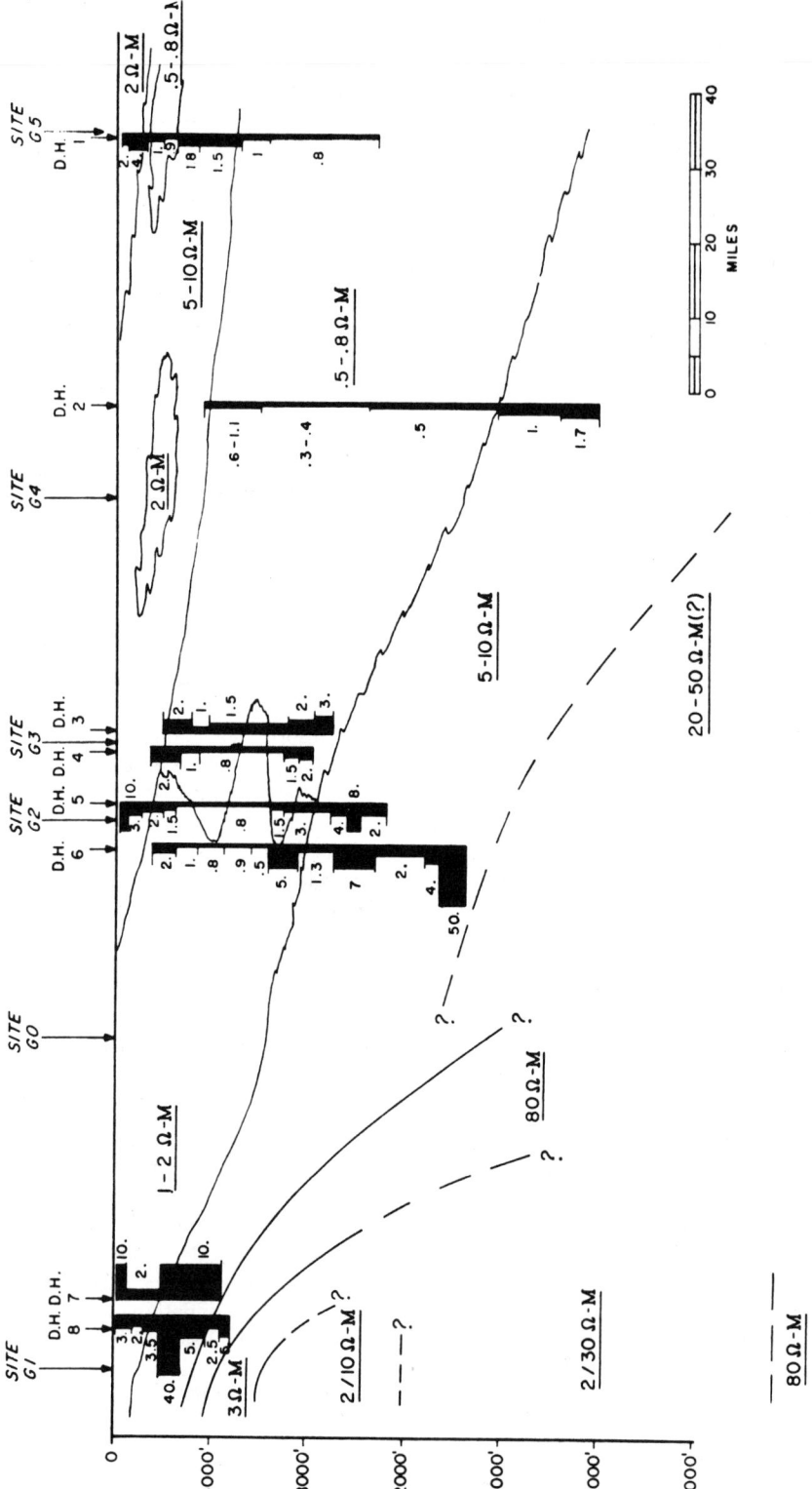

FIG. 31. Comparisons of digitized well logs with models fitted manually. Well locations are shown in Figure 22.

228

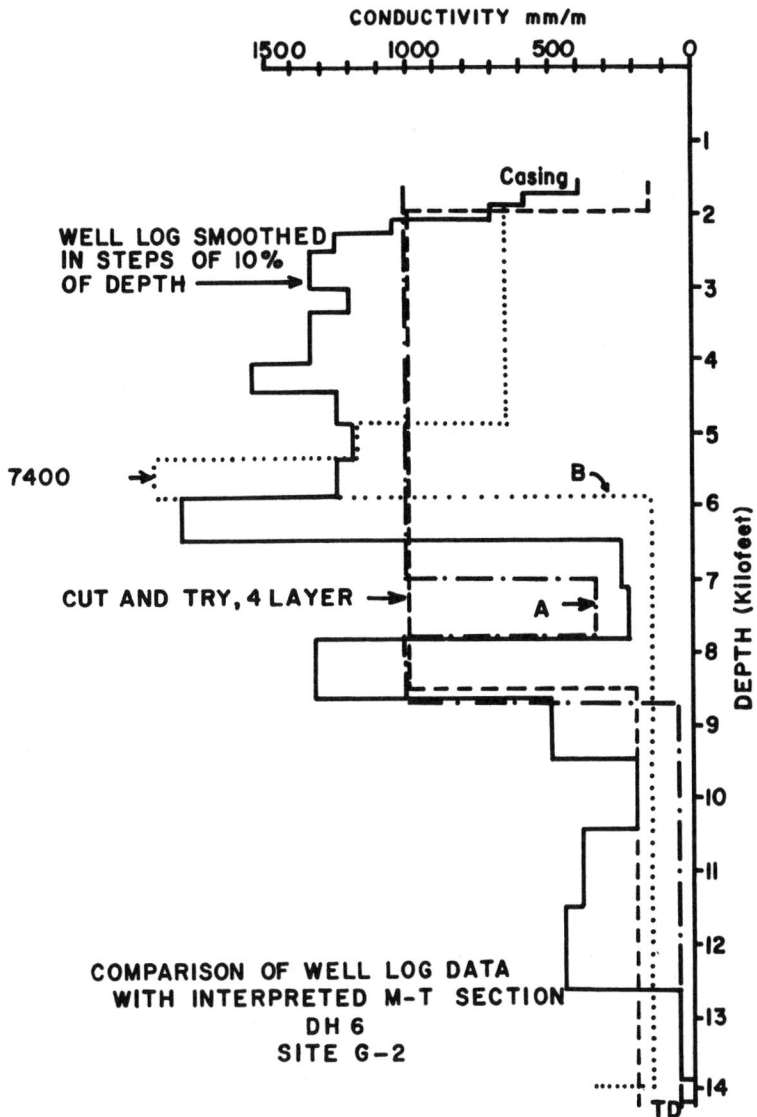

CONDUCTIVITY mm/m

DEPTH (Kilofeet)

WELL LOG SMOOTHED
IN STEPS OF 10%
OF DEPTH →

Casing

7400 →

B

CUT AND TRY, 4 LAYER →

A →

COMPARISON OF WELL LOG DATA
WITH INTERPRETED M-T SECTION
DH 6
SITE G-2

TD

FIG. 32. Detailed comparison of digitized well logs with direct computer interpretations at site G-2. Conductivity is in millimhos/meter.

viously fitted by cut-and-try for $E_\perp$ (curve A, Figure 26). It can be seen to fit the major features of the smoothed log rather well, especially the breaks near 2000 ft and 8500 ft. The significant low conductivity zone from 6500–8000 ft was missed, and conductivity was underestimated in the thick conductive zone from 2000–6500 ft. The two direct-fit models labeled A and B are for the two apparent resistivity curves at the site, B being a fit to $\rho_\parallel$. Six-layer models were used in

which both layer thicknesses and their resistivities were permitted to vary. These early inversion results were considered encouraging.

Anadarko Basin

The second field example is a traverse across a portion of the Anadarko Basin, Oklahoma. The basin has an area of approximately 35,000 square miles in western Oklahoma, the northern part of the Texas Panhandle, and southwestern Kansas.

In terms of its depth and volume, it is one of the major crustal features of North America (Basement Map of North America, 1967). Despite its long history of oil production, the Anadarko Basin is currently the site of intensive exploration and some of the deepest drilling in the world. A location map showing our traverse appears in Figure 33. A geological cross-section showing the general basin configuration is shown in Figure 34. This section was constructed from data which are not as recent or as closely controlled as might be desired, but which seem to be the best available.

At the southern end of the traverse are the Wichita Mountains, consisting of Cambrian rocks of the Wichita granite group. Immediately to the north where the basement is still near-surface,

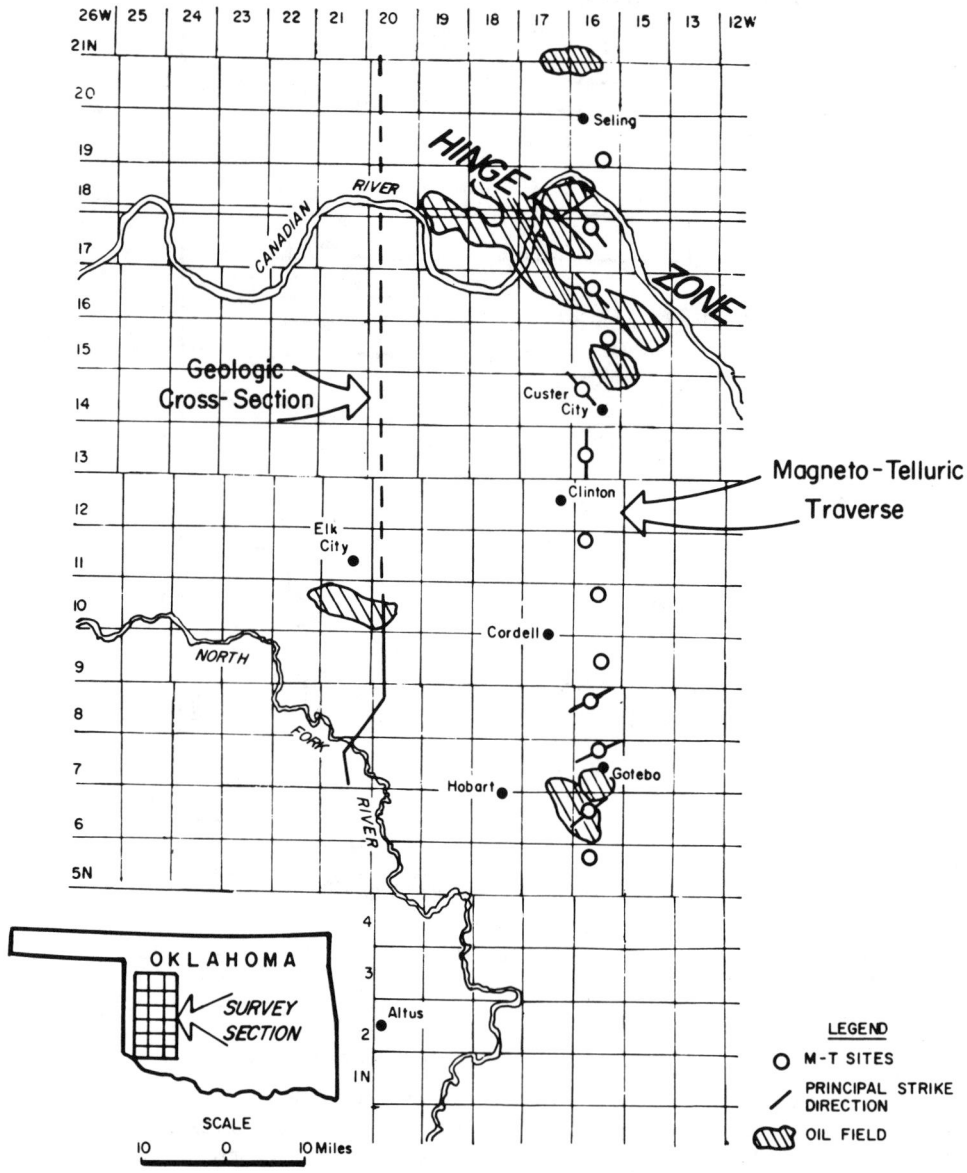

FIG. 33. Location map, Anadarko Basin traverse, Oklahoma, Ranges 20W–21W (after Lang, 1955 and Tulsa Geologic Society, 1951).

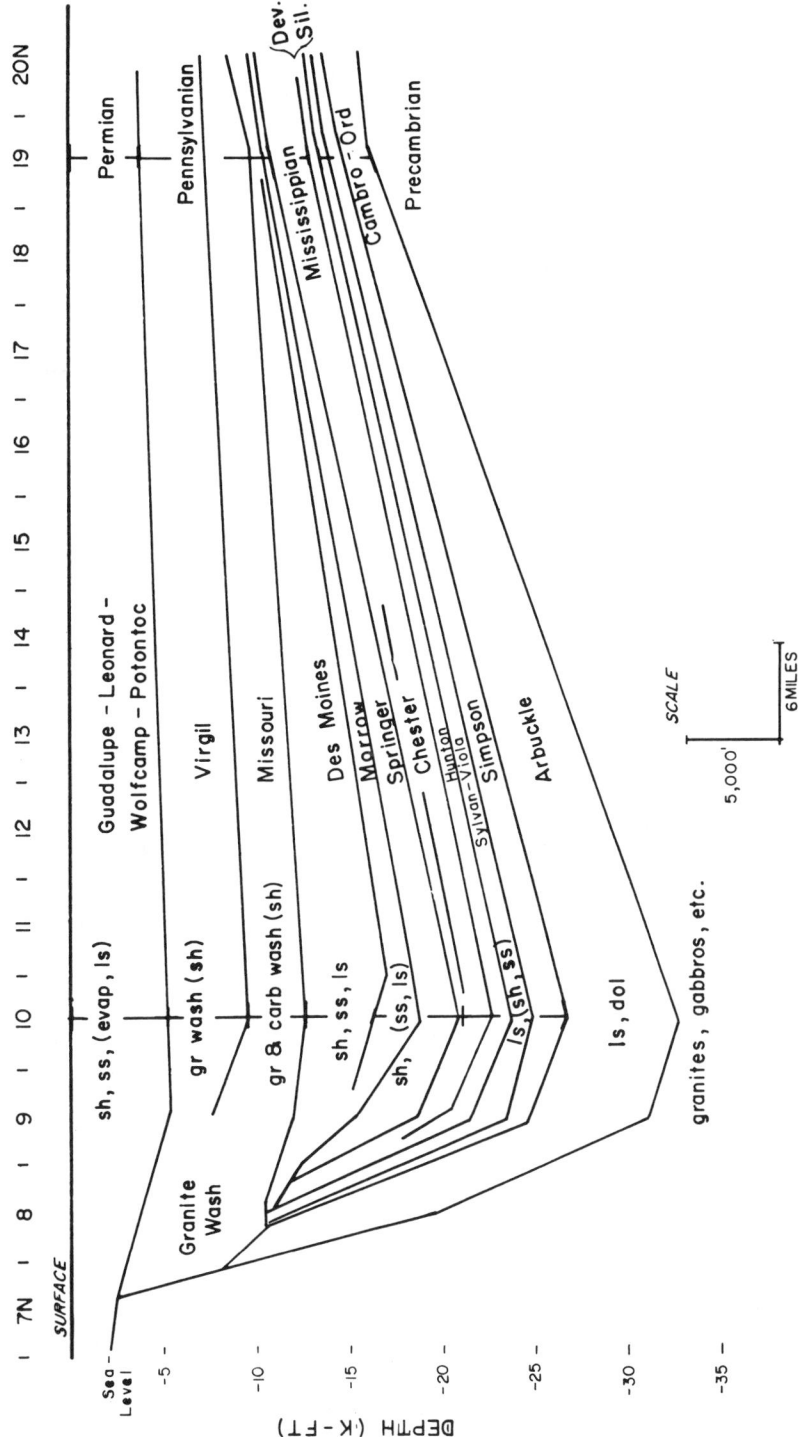

Fig. 34. Geologic cross-section 25 miles west of the Anadarko Basin traverse.

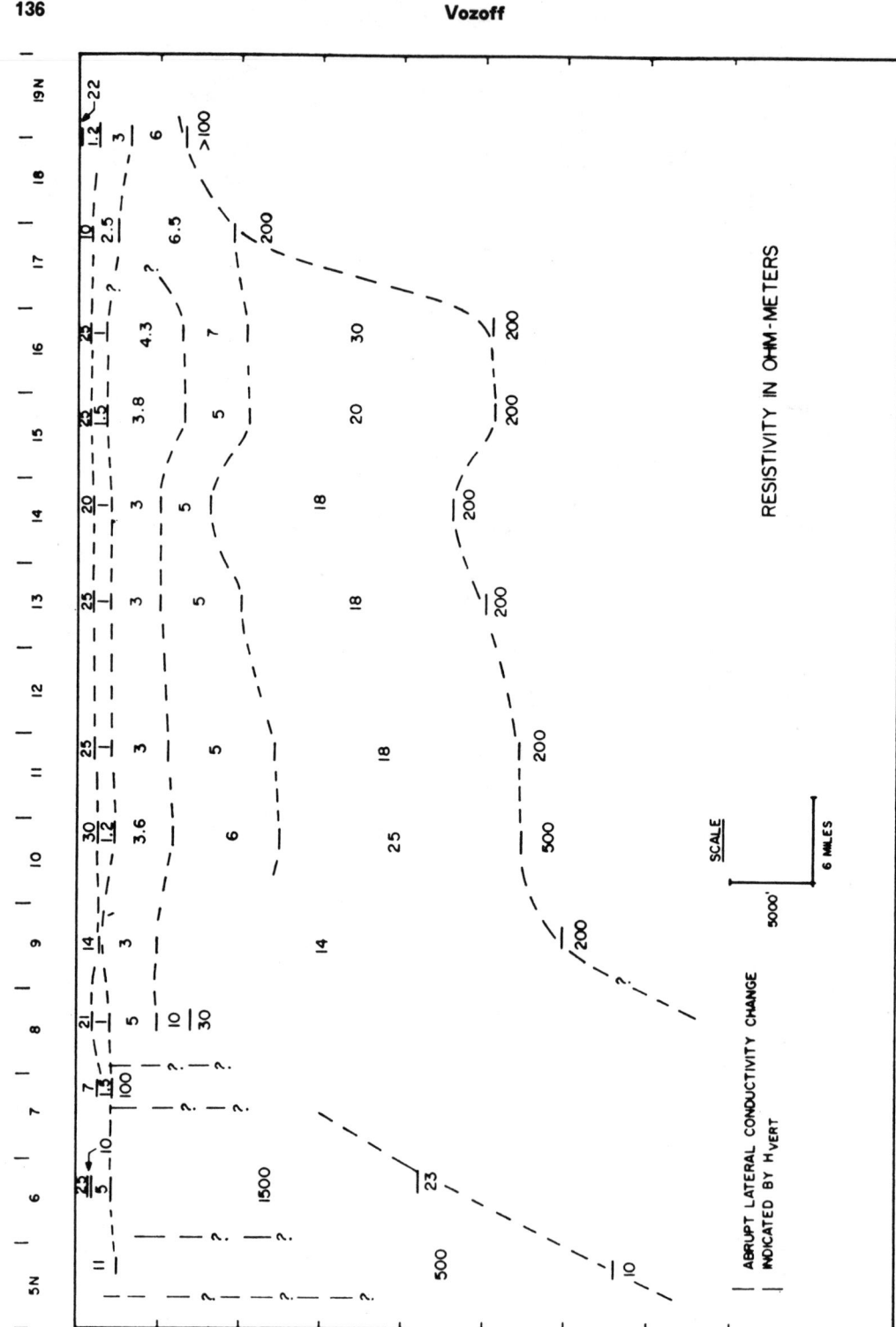

FIG. 35. Interpreted resistivity cross-section, Anadarko basin traverse.

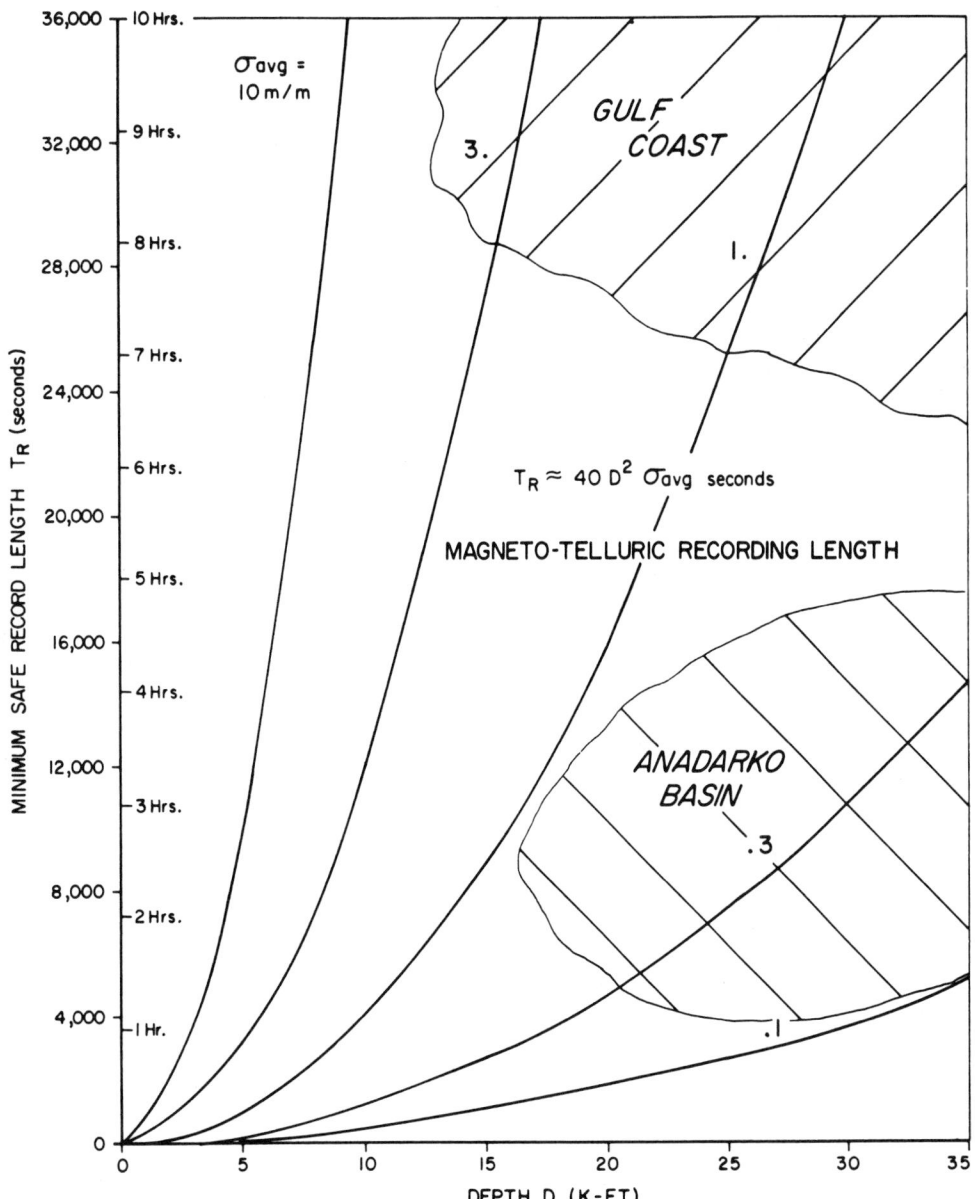

FIG. 36. Recording times under various conditions of section conductivity encountered in typical sedimentary basins.

basement rocks to a depth of approximately 8000 ft are layered Cambrian gabbros of the Raggedy Mountain group, a thick, layered intrusion made up of gabbros, anorthosite, and diorite. Where present, the Wichita granite group overlies the Raggedy Mountain gabbro as a thin veneer.

Moving from south to north, the first major structural feature encountered is the Meers Fault, where the basement drops abruptly from near-surface to a depth of approximately 10,000 ft. North of the Meers Fault is an area of complex faulting culminating in the Mountain View Fault, beyond which point the basement is at a depth of more than 35,000 ft. (The faults are not shown on this section.) Sediment thickness then remains fairly constant for a distance of approximately 20 km when basement begins a shallow upward slope toward the Central Kansas uplift. North of the Meers Fault, the basement consists of layered rhyolite flows of the Carlton rhyolite group of Cambrian age. This has been inferred from drill holes penetrating through the Arbuckle group. The Carlton rhyolite group is thought to be underlain by a fairly thin granite zone and then bedded basaltic flows.

The sedimentary sequence is summarized on the cross-section. At the top are Permian red shales and sandstones with subsidiary evaporites and limestones. The major part of the section is Pennsylvanian, which may exceed 15,000 ft at its thickest. It includes substantial granite and carbonate washes, particularly at the south, underlain by shales with decreasing amounts of sands and limes. The underlying Mississippian consists primarily of limestones with subsidiary shale and sandstone beds, while the Cambro-Ordovician Sylvan, Viola, Simpson, and Arbuckle are primarily carbonates. The basal Reagan sandstone, not shown in the section, is very thin and intermittent, thickening to an estimated 100 ft in basement topographic lows.

Cutting diagonally across the traverse from NW to SE is a hinge zone, which crosses the traverse in the region of Twp 17 N. Facies changes in the Pennsylvanian are substantial and well-documented here, as well as farther south where the washes gradually terminate away from the mountains. The thick Arbuckle is also reported to contain major facies changes from limestone to dolomites, and from impermeable to very porous zones. From discussions with several geological specialists in this region, it appears to be generally believed that the faulting at the south end of the basin is of an overthrust nature, with the crystalline rocks overthrust to the north.

An intensive effort was made to estimate the resistivities of the rocks in the basin from subsurface logs, but this effort was unsuccessful because of the extremely conductive drilling mud used in the area. The values which were estimated apply only south from Twp 9 N. These are:

Virgil and Missouri Washes	10–25	ohm-m
Mississippian and lower		
Pennsylvanian	2–5	ohm-m
Arbuckle	50–250	ohm-m
Basement	100–5000	ohm-m
	(locally 5 ohm-m).	

The field program was, to the best of our knowledge, the first undertaken commercially in North America, fieldwork starting in early 1967. For purposes of objectivity, three geologists were involved independently in fitting layered models to the apparent resistivities. Each interpreted several nonadjacent sites, and the results were assembled and correlated by a fourth person, with the result shown in Figure 35.

North of Twp 7 N there was insufficient anisotropy to justify two-dimensional models. Such models were fitted to the more complex southern zone and resulted in some significant reinterpretation of the subsurface geological information there. For example, the trace of the large Mountain View Fault was shifted southward from its mapped location by over a mile.

The resistivity values correlate directly with rock type. Resistivities in the 100–200 ohm-m range correlate with low porosity carbonates and with some of the (probably factured) granites. Values greater than 1000 ohm-m seem to indicate more competent crystalline rocks. The 18–30 ohm-m resistivities at depth show the moderately porous limestone and sandstones, while the compactive effect of burial may drive the resistivities of limey and sandy shales into this range as well (McCrossan, 1961).

Considerable resemblance is evident between the electrical and geological cross-sections. Where they differ, the electrical section is in closer agreement with present geologic concepts of the basin. The Permian consists of three major subdivisions: a thousand feet or so of 10–30 ohm-m material at the surface, a similar thickness of 1 ohm-m material, and then 2000–3000 ft of 3–4 ohm-m rock. The Virgil and Missouri are made up of 6–30 ohm-m material in the granite wash facies and of a consistent 5–7 ohm-m material farther north. The remainder of the Pennsylvanian, the Mississippian, and portions of the Arbuckle comprise a massive thickness, with resistivities averaging about 20 ohm-m, within which little detail can be resolved. Beneath this mass, resistivity rises again by an order of magnitude. The low-resistivity values at depth at the south end strongly suggest that the resistive rocks forming the Wichita Mountains have been thrust over a large thickness of more conductive rocks. The resistivities of the latter are typical of the Lower Pennsylvanian and Mississippian rocks in the basin proper. The major lateral breaks of the electrical cross-section bear a one-to-one correlation with the faults of Ham et al (1964). The abrupt resistivity change at depth in the hinge zone (Townships 16N–17N) may represent the effects of the extensive facies changes known to occur here (e.g., see Cambridge, 1970).

The smaller features are of questionable significance. For example, the apparent structure beneath Twp 14 at 25,000 ft can certainly be smoothed out with negligible effect on curve fit by minor changes in the shallower parameters. The structure at 10,000 ft could likely be reduced in size by similar reinterpretation, although it would require more effort to do so.

In this basin, the bulk resistivity is seen to increase with depth, from 1 ohm-m near the surface, to a few hundred ohm-meters in the basement. The apparent resistivity curves typically rise gradually with decreasing frequency over a very wide frequency range. Matching that portion of the curves requires only that resistivity increase at a well-defined average rate with depth. Neither the boundaries between layers nor the average resistivity within each layer can be very accurately determined in that depth range. A model made up of many thin layers, each more resistive than the one above, can be constructed that leads to practically the same result as the model consisting of a few thick layers.

The limits of interpretability of any set of data are shown clearly by two of the computer curve matching techniques now in use, the Monte Carlo searches (Greenfield, personal communication) and the generalized inversion technique (Harter and Madden, in preparation). Obviously, supplementary information regarding some parameters can be extremely useful in narrowing these limits for the remaining unknowns.

ECONOMICS

An MT operation involves field support, interpretive data processing, and logistical support. In a region of road access, a field crew would consist of 6–8 men, two of whom operate the equipment while the rest install and retrieve induction coils, porous pots, and wire. If surveying, line-cutting, and station clearing are necessary, more men may be needed if no recording time is to be lost in waiting for site preparation. Data processing requires part or full time of one man, depending on the number of sites per day. Minimum recommended recording times per station for

various regions are shown in Figure 36. Another major cost factor is the amount of geophysicist-geologist time spent on planning and interpretation. This can be limited in a routine investigation once it has been established that no unusual problems exist, but may require nearly full time for junior and senior interpreter in the interpretation stages of a nonroutine problem. Total costs in areas of road access have averaged about $2000 per site on a one site per day basis. Thus, the cost of the entire South Texas traverse was about $12,000, while that of the Anadarko Basin traverse was about $25,000.

Where logistics are more difficult, their costs can completely overwhelm those of the basic survey. Here the relatively small MT crew and the near-portability of the equipment lead to logistics costs which are much smaller than those of, say, a reflection seismic crew in the same area.

CONCLUSIONS

The results presented here clearly show the vast amount of significant geological information which the MT method can yield through careful application. Equipment and computer programs now available are functional, practicable, and generally reliable; but they could benefit from additional development. A higher speed (albeit cruder) reconnaissance capability, as practiced in the USSR (Yungul, 1971), leads to more favorable consideration of the method when rapid coverage at low cost is more important than great accuracy. Better display techniques would be of great help in comprehending and explaining the results.

Improvements continue to be made. For example, the accuracy and convenience of two-dimensional modeling can now be greatly enhanced by using the computer to space the networks which it then solves. Nevertheless, other approaches to modeling, such as those of Parry and Ward (1971) and Hohmann (1971), should be pursued, as should the development of analog models and analytical solutions for their obvious advantages. Research planned or in progress may in the next few years permit MT measurements to be made offshore and on unstable surfaces where they are not now possible.

With the range of questions that arise in the application of MT and the variety of tools which are available to the interpreter, each survey demands that decisions be made based on subjective, preliminary evaluation. The correct decision is often critical to the technical and economic success of the survey. However, both our total understanding of the MT technique and our total knowledge of the gross electrical properties which it measures are still very limited. Hence, in order to obtain the best possible results from the method, planning and interpretation of surveys will probably have to continue to rely on specialists for some time to come. The alternative could be another 25 years of ignominy in petroleum geophysics for the electrical methods.

The magnetotelluric method has advanced tremendously in the past five years as a result of improved instrumentation, computer analysis, and (especially) interpretation. By and large, the method is well suited for mapping the broad features of porosity distribution and poorly suited for mapping fine detail. Its major potential contribution to oil and gas exploration appears to be in early stages of basin evaluation. An important secondary application is in exploration of areas which are unusually difficult to explore by conventional seismic means, such as areas of near-surface volcanic or metamorphic rocks or of very thick sands or gravels.

ACKNOWLEDGMENTS

The lessons and results reported here derive from the efforts of a highly professional team. Dr. G. H. Hopkins was responsible for the equipment; Dr. B. J. Woznick, for the data analysis programs; and Dr. J. N. Galbriath, Jr., for the data editing, data bookkeeping, and model-plotting routines. Messrs. A. Orange and H. S. Lahman ably supervised the field operations and data processing, and Mr. D. Halpin brought the essential outlook of the structural geologist to the interpretation group. Mr. R. Harter contributed significantly to development of the generalized inversion technique and also did much of its coding.

The novel analytical and interpretive tools which we used in this work were developed in the incisive researches of Prof. T. R. Madden and a group of outstanding students, notably, C. M. Swift, Jr., R. J. Greenfield, P. Nelson, and T. Cantwell, at MIT.

Drs. R. L. Caldwell and G. L. Hoehn suggested many improvements of presentation in early versions of this material. I am also indebted to the Rev. James Skehan, S.J., for his assistance, and to Dr. S. H. Yungul for a detailed reading and comments.

Funding for about half of this project was provided by a group of two major oil companies, while the remainder was sponsored by Geoscience, Inc., of Cambridge, Massachusetts.

REFERENCES

Cagniard, L., 1953, Basic theory of the magneto-telluric method of geophysical prospecting: Geophysics, v. 18, p. 605.

Cambridge, T. R., 1970, Potential of Marmaton oil production: Oil and Gas J., v. 68, p. 179.

Flawn, Peter, T., Chairman, 1967, Basement Map of North America: Basement Rock Project Committee, AAPG and USGS, Washington, D. C., Scale—1: 5,000,000.

Ham, W. E., Denison, R. E., and Merritt, C. A., 1964, Basement rocks and structural evolution, southern Oklahoma: Oklahoma Geol. Surv., Bull. 95.

Harthill, N., 1968, The CSM test area for electrical surveying methods: Geophysics, v. 33, p. 675.

Heirtzler, J. R., and Davidson, M. J., 1967, Synoptic measurements of geomagnetic field data: Tech. Report No. 1, CU-1-67 Nonr 4259(05), Lamont Geological Observatory (AD 665037).

Hohmann, G. W., 1971, Electromagnetic scattering by conductors in the earth near a line source of current: Geophysics, v. 36, p. 101.

Hopkins, G., 1965, Instrumentation for geofield measurements: EERL, Univ. Texas, Rep. No. 138.

Keller, G. V., 1968, Electrical prospecting for oil: Quart. Colo. School of Mines, v. 63, p. 38.

Kunetz, G., 1969, Traitement et interpretation des sondages magneto-telluriques: Rev. de l'IFP, v. 25, p. 685.

Lang, R. C., III, 1955, A geologic cross section from the Wichita Mountains to the Elk City pool: Shale Shaker Digest, p. 5.

Madden, T., and Nelson, P., 1964, A defense of Cagniard's magneto-telluric method: Geophysics Lab. MIT Project NR-371-401 rep.

Madden, T., and Swift, C. M. Jr., 1969, Magnetotelluric studies of the electrical conductivity structure of the crust and upper mantle: AGU Monograph 13, The Earth's Crust and Upper Mantle, p. 469.

McCrossan, R. G., 1961, Resistivity mapping and petrophysical study of Upper Devonian inter-reef calcareous shales of Central Alberta, Canada: Bull. AAPG, v. 45, p. 441.

Neves, A. S., 1957, The magneto-telluric method in two-dimensional structures: Dept. of Geology and Geophysics, MIT, Ph.D. thesis.

Nisenoff, M., 1969, Superconducting magnetometers with sensitivities approaching 10^{-10} gauss: Paper presented at C.N.E.S. Conference on Low Magnetic Fields, Paris.

Parry, J. R., and Ward, S. H., 1971, Electromagnetic scattering from cylinders of arbitrary cross-section in a conductive half-space: Geophysics, v. 36, p. 67.

Patrick, F. W., and Bostick, F. X., 1969, Magnetotelluric modeling techniques: EERL, Univ. Texas, Tech. Rep. No. 59.

Rikitake, T., 1966, Electromagnetism and the earth's interior: Amsterdam, Elsevier Publ. Co.

Sims, W. E., and Bostick, F. X., Jr., 1969, Methods of magnetotelluric analysis: EGRL Tech. Rep. No. 58, Univ. of Texas at Austin.

Smith, H. W., 1968, The magneto-telluric method—

field examples, Texas: Engineering Ext. Dept., Univ. Calif., Berkeley, Course notes.

Srivastava, S. P., 1967, Magnetotelluric two- and three-layer master curves: Dom. Obs. Publ., v. 35, no. 7, Canada Dept. of Energy, Ottawa.

Swanson, Donald C., 1967, Major factors controlling Anadarko Basin production: World Oil, p. 81.

Swift, C. M., Jr., 1967, A magnetotelluric investigation of an electrical conductivity anomaly in the southwestern United States: Ph.D. thesis, MIT.

Tikhonov, A. V., 1950, Determination of the electrical characteristics of the deep strata of the earth's crust: Dokl. Akad. Nauk, v. 73, p. 295.

Tulsa Geological Society (Hendricks, et al), 1951, Possible future petroleum provinces of North America, Mid-Continent region: AAPG, Tulsa.

Vozoff, K., and Ellis, R. M., 1966, Magnetotelluric measurements in southern Alberta: Geophysics, v. 31, p. 1153.

Vozoff, K., and Swift, C. M., Jr., 1968, Magneto-telluric measurements in the North German Basin: Geophys. Prosp., v. 16, p. 454.

Waeselynck, M., 1967, Etude de structures anticlinales ou synclinales par la methode magneto-tellurique: Bull. Centre Rech. Pau—S.N.P.A., v. 1, p. 417.

Wait, J. R., 1962, Theory of magneto-telluric fields: J. Res. N.B.S., v. 66D, p. 509.

Word, D. R., Smith, H. W., and Bostick, F. X., Jr., 1969, An investigation of the magnetotelluric tensor impedance method: EGRL Tech. Rep. No. 82, U. of Texas at Austin.

Wu, Francis T., 1968, The inverse problem of magneto-telluric sounding: Geophysics, v. 33, p. 972.

Yungul, S. H., 1961, Magnetotelluric sounding three-layer interpretation curves: Geophysics, v. 26, p. 465.

———— 1971, Personal communication.

Reprinted by permission of the Society of Exploration
Geophysicists from *Geophysics,* v. 49, no. 9 (1984), p.
1517-1533.

Magnetotelluric responses of three-dimensional bodies in layered earths

Philip E. Wannamaker*, Gerald W. Hohmann‡, and Stanley H. Ward*

ABSTRACT

The electromagnetic fields scattered by a three-dimensional (3-D) inhomogeneity in the earth are affected strongly by boundary charges. Boundary charges cause normalized electric field magnitudes, and thus tensor magnetotelluric (MT) apparent resistivities, to remain anomalous as frequency approaches zero. However, these *E*-field distortions below certain frequencies are essentially in-phase with the incident electric field. Moreover, normalized secondary magnetic field amplitudes over a body ultimately decline in proportion to the plane-wave impedance of the layered host. It follows that tipper element magnitudes and all MT function phases become minimally affected at low frequencies by an inhomogeneity.

Resistivity structure in nature is a collection of inhomogeneities of various scales, and the small structures in this collection can have MT responses as strong locally as those of the large structures. Hence, any telluric distortion in overlying small-scale extraneous structure can be superimposed to arbitrarily low frequencies upon the apparent resistivities of buried targets. On the other hand, the MT responses of small and large bodies have frequency dependencies that are separated approximately as the square of the geometric scale factor distinguishing the different bodies. Therefore, tipper ele-

ment magnitudes as well as the phases of all MT functions due to small-scale extraneous structure will be limited to high frequencies, so that one may "see through" such structure with these functions to target responses occurring at lower frequencies.

About a 3-D conductive body near the surface, interpretation using 1-D or 2-D TE modeling routines of the apparent resistivity and impedance phase identified as transverse electric (TE) can imply false low resistivities at depth. This is because these routines do not account for the effects of boundary charges. Furthermore, 3-D bodies in typical layered hosts, with layer resistivities that increase with depth in the upper several kilometers, are even less amenable to 2-D TE interpretation than are similar 3-D bodies in uniform half-spaces. However, centrally located profiles across geometrically regular, elongate 3-D prisms may be modeled accurately with a 2-D transverse magnetic (TM) algorithm, which implicitly includes boundary charges in its formulation. In defining apparent resistivity and impedance phase for TM modeling of such bodies, we recommend a fixed coordinate system derived using tipper-strike, calculated at the frequency for which tipper magnitude due to the inhomogeneity of interest is large relative to that due to any nearby extraneous structure.

INTRODUCTION

Magnetotelluric (MT) measurements are sensitive to the resistivity structure of the earth, potentially to depths exceeding 100 km (Cagniard, 1953; Swift, 1967; Word et al., 1971; Vozoff, 1972; Larsen, 1981; Wannamaker, 1983). Recent advances in instrumentation and data processing (e.g., Gamble et al., 1979; Weinstock and Overton, 1981; Stodt, 1983) have enabled procurement of precise tensor MT data. However, the skills necessary to translate these measurements into trustworthy models of subsurface resistivity have been slow in developing.

MT data are, strictly speaking, responses from three-dimensional (3-D) resistivity structure in the earth, but traditionally they have been interpreted using 1-D and sometimes 2-D model structures (for example, Petrick et al., 1977; Stanley et al., 1977; Rooney and Hutton, 1977). This tradition has arisen because 3-D modeling routines require considerable computing resources to handle complex earth structure, re-

Manuscript received by the Editor July 12, 1983; revised manuscript received March 15, 1984.
*Earth Science Laboratory, University of Utah Research Institute, 391 C Chipeta Way, Salt Lake City, UT 84108.
‡Dept. of Geology and Geophysics, University of Utah, Salt Lake City, UT 84108.

1517

sources not readily available. This shortcoming produces a lack of concensus on the interpretive errors which occur when 1-D and 2-D computational aids are used in 3-D areas.

We favor keeping the interpretation of observations as simple as possible. This philosophy underlies the major purposes of this paper, which are as follows: first, develop magnetotelluric theory for 3-D bodies buried in otherwise 1-D media to establish the fundamental controls on observed responses; and second, investigate the utility of 1-D and 2-D algorithms for interpreting 3-D geology. The latter goal is achievable only through rigorous, three-dimensional model studies, which we perform using the computer program of Wannamaker et al. (1984).

MAGNETOTELLURIC THEORY FOR THREE-DIMENSIONAL BODIES IN LAYERED EARTHS

Here we pursue the essentials that determine the magnetotelluric signatures of three-dimensional bodies. Special attention is paid to understanding the low-frequency limits of MT responses. By incorporating this understanding with standard EM scaling concepts, in our model study we can propose means of discriminating the response of a large resistivity anomaly, which may represent an exploration target, from that of a small one, which may constitute mere extraneous, overlying structure.

Electromagnetic field relations

A 3-D body in the earth is a source of scattered electric and magnetic fields. Establishing relations between the incident plane-wave fields and the scattered and total fields, as well as exploring the behavior of these relations as frequency varies, is an important step toward resolving the roles of the inhomogeneity and the host layering in creating anomalous MT response functions.

Governing equations.—Source-free versions of Maxwell's equations (Harrington, 1961) describe the total, plane-wave induced electric and magnetic fields (**E**, **H**) as a function of position **r** about a 3-D body in a layered medium (Figure 1). In the manner of Wannamaker et al. (1984), (**E**, **H**) can be decomposed into an incident set ($\mathbf{E}_i$, $\mathbf{H}_i$) which are the plane-wave fields and a secondary set ($\mathbf{E}_s$, $\mathbf{H}_s$) contributed by the inhomogeneity. The secondary fields are specified in terms of a scattering current $\mathbf{J}_s = (\hat{y}_b - \hat{y}_j)\mathbf{E}_b$, where $\hat{y}_b$ and $\hat{y}_j$ are the admittivities of the body and of layer j containing the body, while $\mathbf{E}_b$ is the total electric field within the body. If the scattering current within the inhomogeneity is known, the secondary EM fields elsewhere can be given by integral equations (Wannamaker et al., 1984).

Equation (7) of Wannamaker et al. (1984) describes the source of the secondary electric field with a volume current component $\hat{z}\mathbf{J}_s$, where $\hat{z}$ is taken to be the impedivity of free space, and a free-charge component

$$\frac{-1}{\hat{y}_j} \nabla \cdot \mathbf{J}_s.$$

The free charge preserves continuity of normal total current density, but in doing so makes the normal total electric field discontinuous (Price, 1973). Furthermore, since $\mathbf{E}_i$ is continuous, $\nabla \cdot \mathbf{E}_s = \nabla \cdot \mathbf{E}$.

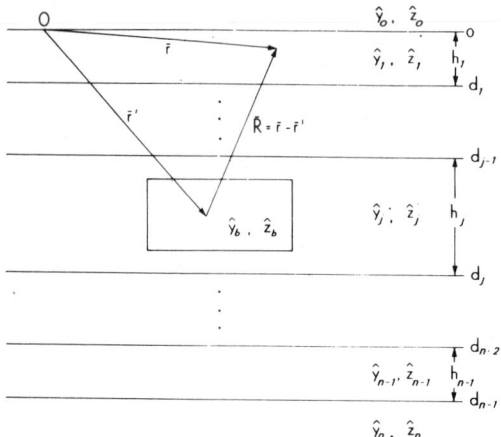

FIG. 1. Section view showing physical properties of a 3-D body in a layered earth.

Tensor field relations.—Considering the linearity of the governing equations of Wannamaker et al. (1984), we postulate

$$\mathbf{E}^0(\mathbf{r}) = \mathbf{E}_i^0 + [\tilde{\mathbf{P}}_s^0(\mathbf{r})] \cdot \mathbf{E}_i^0 \qquad (1)$$

and

$$\mathbf{H}^0(\mathbf{r}) = \mathbf{H}_i^0 + [\tilde{\mathbf{Q}}_s^0(\mathbf{r})] \cdot \mathbf{E}_i^0, \qquad (2)$$

where superscript zero indicates we are considering **r** at the surface of the earth over which $\mathbf{E}_i^0$ is constant. $[\tilde{\mathbf{P}}_s^0(\mathbf{r})]$ and $[\tilde{\mathbf{Q}}_s^0(\mathbf{r})]$ are 3×2 normalized tensors representing the scattered field unique for a specified 3-D body, layered host and frequency, of the form

$$[\tilde{\mathbf{P}}_s^0(\mathbf{r})] = \begin{bmatrix} [\tilde{\mathbf{P}}_{hs}^0(\mathbf{r})] \\ \text{-----} \\ [\tilde{\mathbf{P}}_{vs}^0(\mathbf{r})] \end{bmatrix} = \begin{bmatrix} P_{xx}^0 & P_{xy}^0 \\ P_{yx}^0 & P_{yy}^0 \\ \text{-----} \\ P_{zx}^0 & P_{zy}^0 \end{bmatrix}, \qquad (3)$$

and

$$[\tilde{\mathbf{Q}}_s^0(\mathbf{r})] = \begin{bmatrix} [\tilde{\mathbf{Q}}_{hs}^0(\mathbf{r})] \\ \text{-----} \\ [\tilde{\mathbf{Q}}_{vs}^0(\mathbf{r})] \end{bmatrix} = \begin{bmatrix} Q_{xx}^0 & Q_{xy}^0 \\ Q_{yx}^0 & Q_{yy}^0 \\ \text{-----} \\ Q_{zx}^0 & Q_{zy}^0 \end{bmatrix}. \qquad (4)$$

Note that we have subdivided the scattered field tensors into horizontal subtensors, $[\tilde{\mathbf{P}}_{hs}^0(\mathbf{r})]$ and $[\tilde{\mathbf{Q}}_{hs}^0(\mathbf{r})]$, and vertical subtensors, $[\tilde{\mathbf{P}}_{vs}^0(\mathbf{r})]$ and $[\tilde{\mathbf{Q}}_{vs}^0(\mathbf{r})]$, which pertain, respectively, to horizontal and vertical electric and magnetic field components induced by the incident electric vector $\mathbf{E}_i^0$. Discrete, approximate versions of $[\tilde{\mathbf{P}}_s^0(\mathbf{r})]$ and $[\tilde{\mathbf{Q}}_s^0(\mathbf{r})]$ are computed by Wannamaker et al. (1984). Hermance and Thayer (1975), Larsen (1975, 1977, 1981), Klein and Larsen (1978), Stodt et al. (1981), and Hermance (1982) also considered E- and H-field tensor approaches.

The incident fields at the surface are related through

$$E_i^0 = [\tilde{\mathbf{Z}}_\ell] \cdot H_i^0, \qquad (5)$$

(Cagniard, 1953; Ward, 1967, p. 117–124), with the layered earth impedance

$$[\tilde{\mathbf{Z}}_\ell] = \begin{bmatrix} 0 & Z_\ell \\ -Z_\ell & 0 \end{bmatrix}. \qquad (6)$$

Using equations (1) through (5), and with $[\tilde{\mathbf{I}}]$ the 2×2 identity tensor, the horizontal fields become

$$E_h^0(\mathbf{r}) = \{[\tilde{\mathbf{I}}] + [\tilde{\mathbf{P}}_{hs}^0(\mathbf{r})]\} \cdot E_i^0, \qquad (7)$$

and

$$H_h^0(\mathbf{r}) = \{[\tilde{\mathbf{I}}] + [\tilde{\mathbf{Q}}_{hs}^0(\mathbf{r})] \cdot [\tilde{\mathbf{Z}}_\ell]\} \cdot H_i^0. \qquad (8)$$

Low-frequency conditions.—At low frequencies, such that wavelengths in the host layers are long compared to the observation distance **R** from the body (Figure 1), the Helmholtz equations of Wannamaker et al. (1984) asymptote to Laplace's and Poisson's equations. In particular, note that the volume current source term $\tilde{z}\mathbf{J}_s$ in equation (7) of Wannamaker et al. (1984) vanishes. This term, to first order, is proportional to frequency so that $E_s^0(\mathbf{r})$ and thus $[\tilde{\mathbf{P}}_s^0(\mathbf{r})]$ at low frequencies are determined solely by the free charge. However, equation (7) of Wannamaker et al. (1984) shows that the charge density is intimately associated with the E-field inside the inhomogeneity. Also, the E-field interior to the body defines $\mathbf{J}_s$, which in turn provides the source for the secondary magnetic field (op. cit.). We thus require frequencies sufficiently low so that wavelengths within the inhomogeneity are long compared to the size of the inhomogeneity before it is strictly valid to treat the secondary EM fields as though induced by a zero-frequency incident electric field. When both exterior and interior long-wavelength criteria are satisfied, then $[\tilde{\mathbf{P}}_s^0(\mathbf{r})]$ and $[\tilde{\mathbf{Q}}_s^0(\mathbf{r})]$ will be essentially real and independent of frequency.

In conclusion, with free charges $E^0(\mathbf{r})$ near a 3-D body remains anomalous to arbitrarily low frequencies. Such anomalous behavior due to this charge, when it occurs about conductive bodies, is referred to as current gathering by various investigators (e.g., Berdichevskiy and Dmitriev, 1976). Boundary charges, however, do not enter into $H^0(\mathbf{r})$ as frequency falls. Over an arbitrarily layered earth, it is not difficult to show that $|Z_\ell|$ decreases monotonically with decreasing frequency (see Cagniard, 1953). In particular, for a uniform half-space, we have

$$|Z_\ell| = \left| \frac{\omega\mu_0}{k_1} \right| \simeq \left| \frac{\omega\mu_0}{\sigma_1} \right|^{1/2}, \qquad (9)$$

indicating that $[\tilde{\mathbf{Z}}_\ell]$ in this case varies as $\omega^{1/2}$ (ω is angular frequency). Hence, even though $[\tilde{\mathbf{Q}}_s^0(\mathbf{r})]$ possesses a nonzero, low-frequency limit, $H_s^0(\mathbf{r}) = [\tilde{\mathbf{Q}}_s^0(\mathbf{r})] \cdot [\tilde{\mathbf{Z}}_\ell] \cdot H_i^0$ will vanish as frequency approaches zero.

Tensor magnetotelluric quantities

The tensor field relations we have specified may be used to construct MT functions. In doing so, the relative contributions of the body and the layered host to anomalous MT functions become evident. Studies of MT functions over single bodies, and in particular the low-frequency asymptotes of such quantities, are required before considering multiple bodies.

Impedance tensor.—The impedance tensor $[\tilde{\mathbf{Z}}(\mathbf{r})]$, defined by

$$E_h^0(\mathbf{r}) = [\tilde{\mathbf{Z}}(\mathbf{r})] \cdot H_h^0(\mathbf{r}), \qquad (10)$$

where $[\tilde{\mathbf{Z}}(\mathbf{r})]$ is of the form

$$[\tilde{\mathbf{Z}}(\mathbf{r})] = \begin{bmatrix} Z_{xx} & Z_{xy} \\ Z_{yx} & Z_{yy} \end{bmatrix}, \qquad (11)$$

can be formulated by substituting equations (7) and (8) into equations (5). One obtains

$$[\tilde{\mathbf{Z}}(\mathbf{r})] = \{[\tilde{\mathbf{I}}] + [\tilde{\mathbf{P}}_{hs}^0(\mathbf{r})]\} \cdot [\tilde{\mathbf{Z}}_\ell] \cdot \{[\tilde{\mathbf{I}}] + [\tilde{\mathbf{Q}}_{hs}(\mathbf{r})] \cdot [\tilde{\mathbf{Z}}_\ell]\}^{-1}. \qquad (12)$$

As frequency approaches zero, equation (12) reduces to

$$[\tilde{\mathbf{Z}}(\mathbf{r})] \simeq \{[\tilde{\mathbf{I}}] + [\tilde{\mathbf{P}}_{hs}^0(\mathbf{r})]\} \cdot [\tilde{\mathbf{Z}}_\ell]. \qquad (13)$$

All four elements Z_{ij} of $[\tilde{\mathbf{Z}}(\mathbf{r})]$ are related to Z_ℓ by real constants, so that in this low-frequency limit a Hilbert transform relates magnitude and phase of Z_{ij} (Kunetz, 1972; Boehl et al., 1977). Also, it is easy to show from equation (13) that the impedance ellipticity (Word et al., 1971) approaches zero as frequency becomes low (and see Ting and Hohmann, 1981).

The apparent resistivities at low frequencies are, from equation (13),

$$\rho_{xx} \simeq \frac{1}{\omega\mu_0} |Z_\ell|^2 \cdot |P_{xy}^0|^2, \qquad (14a)$$

$$\rho_{xy} \simeq \frac{1}{\omega\mu_0} |Z_\ell|^2 \cdot |1 + P_{xx}^0|^2, \qquad (14b)$$

$$\rho_{yx} \simeq \frac{1}{\omega\mu_0} |Z_\ell|^2 \cdot |1 + P_{yy}^0|^2, \qquad (14c)$$

and

$$\rho_{yy} \simeq \frac{1}{\omega\mu_0} |Z_\ell|^2 \cdot |P_{yx}^0|^2 \qquad (14d)$$

for a 3-D body. Like the Z_{ij}, all ρ_{ij} are distorted to arbitrarily low frequencies by boundary charge effects and are related to $\rho_\ell = 1/\omega\mu_0 |Z_\ell|^2$ by positive constants as given in equation (14). If interpreted assuming a 1-D model structure, apparent resistivity soundings distorted in this manner by a nearby 3-D body will yield model resistivities in error by a factor ρ_{ij}/ρ_ℓ and model layer thicknesses in error by $\sqrt{\rho_{ij}/\rho_\ell}$ (Larsen, 1977, 1981).

However, since the P_{ij}^0 become real as frequency approaches zero, the phases of all Z_{ij} (i.e., ϕ_{ij}) asymptote to the phase of the layered host impedance ϕ_ℓ and are no longer affected by the inhomogeneity. Nevertheless, this does not mean that the parameters of the layered host can be recovered through 1-D inversion of the impedance phase sounding alone. It is well-known in the literature (for example, Cagniard, 1953), and can be inferred from equations (13) and (14), that a specific impedance phase sounding can correspond to an infinite number of apparent resistivity soundings, and thus an infinite number of layered resistivity structures.

Let us now briefly consider a 2-D inhomogeneity, whose strike direction corresponds to the x coordinate axis. An x-oriented incident electric field induces only x-oriented secondary E-fields about such a structure, so that the total electric field parallels all resistivity contacts and no boundary charges exist. This is the transverse electric (TE) mode of wave polarization (Swift, 1967). At low frequencies for the TE mode, neither volume currents nor boundary charges pertain as sources for a secondary electric field; therefore, P_{xx}^0 becomes zero and ρ_{xy} in equation (14b) approaches ρ_ℓ.

Thus, if all near-surface extraneous structure were 2-D, then 1-D inversion of apparent resistivity and impedance phase soundings identified as TE would yield models of any deep resistivity layering below that are relatively free from distortion due to such structure (Word et al., 1971; Vozoff, 1972; Berdichevskiy and Dmitriev, 1976). On the other hand, if resistivity structures possess finite strike lengths as do the models we will compute, then 1-D inversions in such instances potentially may suffer serious error.

Similarly, a y-oriented incident electric field causes only y-oriented secondary E-fields over a 2-D body. Such a field polarizaton defines the transverse magnetic (TM) mode. However, since the incident electric field for this mode is normal to resistivity contacts in the earth, boundary charges will be induced as sources for secondary E-fields, P_{yy}^0 will have a nonzero value to arbitrarily low frequencies and ρ_{yx} remains defined by equation (14c). Because of boundary charges, the TM mode in the case of conductive bodies exhibits vertical current-gathering (Park et al., 1983).

Tipper transfer tensor.—The tipper transfer function is defined by

$$[\mathbf{H}_z^0(\mathbf{r})] = [\tilde{\mathbf{K}}_z(\mathbf{r})] \cdot \mathbf{H}_h^0(\mathbf{r}), \qquad (15)$$

(Word et al., 1970), in which

$$\tilde{\mathbf{K}}_z(\mathbf{r}) = [K_{zx} \quad K_{zy}]. \qquad (16)$$

In terms of scattered field tensors, we have

$$[\tilde{\mathbf{K}}_z(\mathbf{r})] = [\tilde{\mathbf{Q}}_{vs}^0(\mathbf{r})] \cdot [\tilde{\mathbf{Z}}_\ell] \cdot \{[\tilde{\mathbf{I}}] + [\tilde{\mathbf{Q}}_{hs}^0(\mathbf{r})] \cdot [\tilde{\mathbf{Z}}_\ell]\}^{-1}. \qquad (17)$$

The low-frequency asymptotes of the tipper elements are

$$K_{zx} \simeq -Z_\ell Q_{zy}^0 \qquad (18a)$$

and

$$K_{zy} \simeq Z_\ell Q_{zx}^0. \qquad (18b)$$

Note that K_{zx} and K_{zy} are related to Z_ℓ by real constants, so their magnitudes approach zero as frequency approaches zero (Word et al., 1970). In addition, phases of K_{zx} and K_{zy} approach ϕ_ℓ as frequency falls, since $[\tilde{\mathbf{Q}}_{vs}^0(\mathbf{r})]$ becomes real.

The tipper (Vozoff, 1972) has a magnitude given by

$$|T| = [|K_{zx}|^2 + |K_{zy}|^2]^{1/2}. \qquad (19)$$

From equation (18), at low frequencies, the tipper likewise becomes zero as $\omega \to 0$. Because of Z_ℓ, T and $[\tilde{\mathbf{K}}_z(\mathbf{r})]$ contain a good deal of information about the layered host.

THREE-DIMENSIONAL MAGNETOTELLURIC MODEL STUDY

The foregoing theoretical concepts are now exemplified using two specific resistivity inhomogeneities. The first is a small-scale, shallow, conductive body. This might represent a single zone of the hydrothermal alteration which is common at geothermal resource areas of the western United States (e.g., Sandberg and Hohmann, 1982) or some other near-surface heterogeneity (SEG Workshop on Magnetotellurics, Las Vegas, 1983). Such heterogeneity can reside directly over exploration

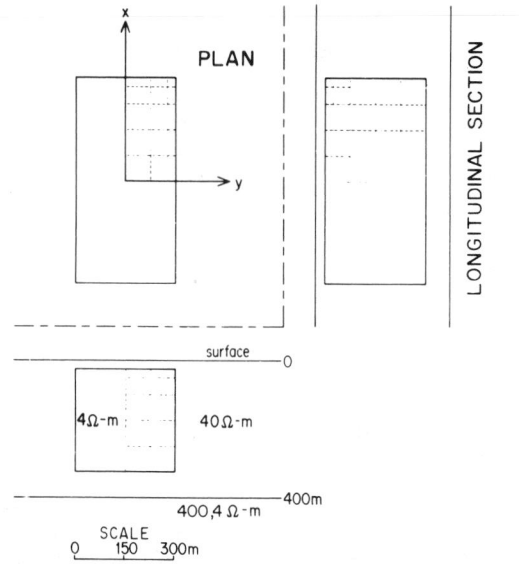

FIG. 2. Prismatic 3-D body in a two-layered earth used to represent small-scale, shallow structure. Dashes outline the discretization of the conductor into rectangular cells, shown only for the right half of the body in section and the upper right-hand quadrant in plan. Basal half-space resistivities of 400 and $4\ \Omega \cdot m$ are considered.

targets and can be extremely variable over distances of only a few hundred meters. This presents a grave sampling problem for MT measurements. Even if present 3-D modeling algorithms could accommodate this complex heterogeneity, which they cannot, it would be prohibitively expensive to record sufficient MT data to delineate its response. Thus, it becomes important to make increased use of MT response functions that are relatively insensitive to this sort of extraneous structure.

The second 3-D body we examine is intended to portray a sedimentary basin such as is frequently encountered in the Basin and Range province (Stewart, 1980; Eaton, 1982). These basins sometimes can be targets of investigations while at other times are of only secondary interest. Our particular model derives from study of the Milford Valley in southwestern Utah (Ward et al., 1978; Cook et al., 1981). However, gravity surveys suggest that the podiform geometry and overall dimensions of our model apply to the sedimentary fill of most grabens of the northern Basin and Range (Erwin and Berg, 1977; Erwin and Bittleston, 1977; Cook et al., 1981; Healy et al., 1981). Apart from relatively coarse-grained alluvium deposited from erosion of surrounding mountain ranges, many grabens contain large amounts of especially conductive, Pleistocene lacustrine clays (Stewart, 1980; Hintze, 1980). Also, we have studied the effects on MT response functions of a regional resistivity host to our basin model, a host which reflects tectonic setting and which may be close to one-dimensional to depths of tens of kilometers (Brace, 1971; Wannamaker, 1983).

We have disregarded inhomogeneities which are resistive compared to their layered hosts. Alteration by hydrothermal

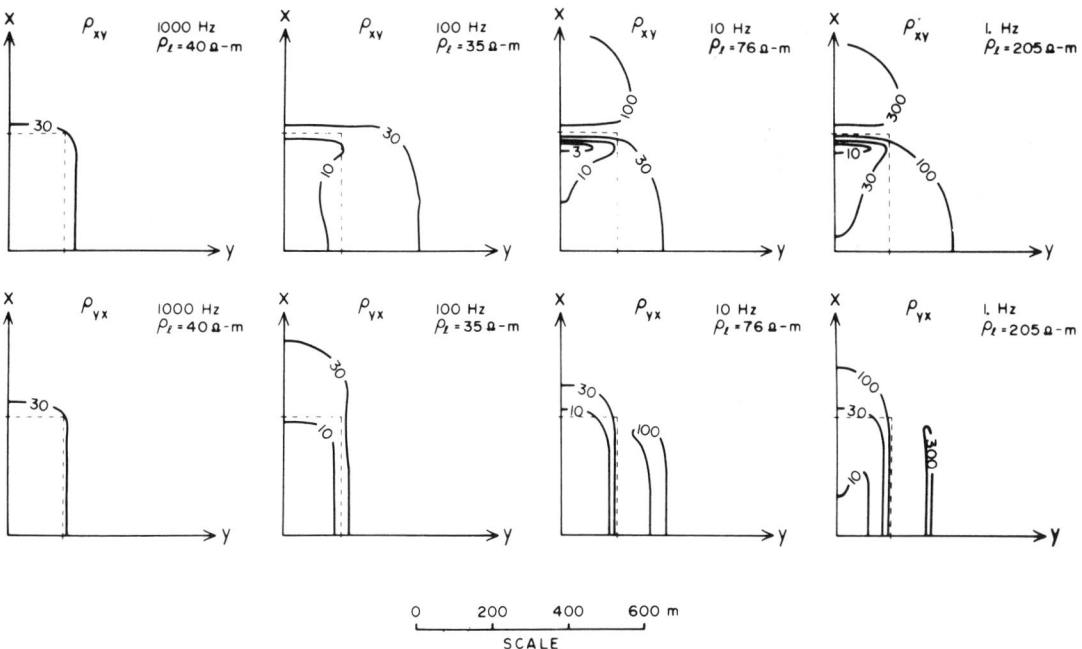

FIG. 3. Multifrequency plan maps of tensor apparent resistivities ρ_{xy} and ρ_{yx} over upper right-hand quadrant of the inhomogeneity of Figure 2. The body outline in plan is shown with dashes, the basal half-space resistivity is 400 $\Omega \cdot$ m and contour values are in $\Omega \cdot$ m. Also, the frequency and the value of the layered earth apparent resistivity ρ_ℓ are given in the upper right-hand corner of each plot.

brines discharging through rock is relatively conductive, although alteration by brines discharging through alluvium may be relatively resistive (e.g., Gamble et al., 1981). However, we consider distributions of sediments, due to their comparatively high porosity and clay contents, in general to be conductive structures in relatively resistive hosts (also Porath, 1971a; Hermance, 1982). Sedimentary basins form through erosion with collection of resulting detritus in topographic lows. In the northern Basin and Range, this concept is supported by the extensive gravity surveys cited previously. For areas where the existence of resistive bodies can be demonstrated, future model simulations certainly would be worthwhile.

The following MT responses were computed using the algorithm of Wannamaker et al. (1984), designed to model 3-D bodies in arbitrarily layered hosts with plane-wave incident fields. This algorithm is an extension of that developed by Ting and Hohmann (1981), which simulates the MT responses of 3-D structures in uniform half-spaces. Per Jones and Vozoff (1978), MT quantities are derived from total fields computed using two independent polarizations of $\mathbf{E}_t^0$. All calculations in the study were performed on the Prime 400 Series minicomputer of the Earth Science Laboratory of the University of Utah Research Institute.

The response of a small-scale structure

In Figure 2, a small, shallow conductive prism appears in a layer of overburden upon a resistive basement. The scattering current within the body was approximated by 48 rectangularly

prismatic cells to a quadrant. Contoured MT response functions shown next, with coordinate axes paralleling the axes of symmetry of the body, were derived from 92 variously spaced receiver points per quadrant and required about 4 hours CPU time for each frequency on the Prime 400. One hour CPU time on the Prime 400 compares to less than 20 s on a CRAY-1 without optimal vectorization.

Apparent resistivities and impedance phases.—The apparent resistivity signatures produced by our small-scale structure are displayed in Figure 3. Especially at the lower frequencies of 1. and 10 Hz where equation (14) becomes accurate and current-gathering is of particular importance, the anomalies are roughly electric dipolar in nature (Stratton, 1941, p. 431–434, p. 563–573), with undershoots and overshoots with respect to ρ_ℓ occurring over the ends of the body for ρ_{xy} and over the sides for ρ_{yx}. Note also at the lower frequencies that the anomalies are greater than those at 100 and 1 000 Hz. Boundary charges cause apparent resistivities to vary spatially by a factor of nearly 100, which is much higher than the body-host layer contrast, although such extremes are due partly to the abrupt nature of the resistivity contacts of the model and may be subdued for diffuse boundaries. It is most important, however, to realize that current gathering into conductive structure similar to our model will produce strong apparent resistivity anomalies that actually increase to a low-frequency asymptote as frequency falls. The results complement the study of Berdichevskiy and Dmitriev (1976), who considered a great variety of elliptically shaped, near-surface inhomogeneities but who con-

243

Wannamaker et al.

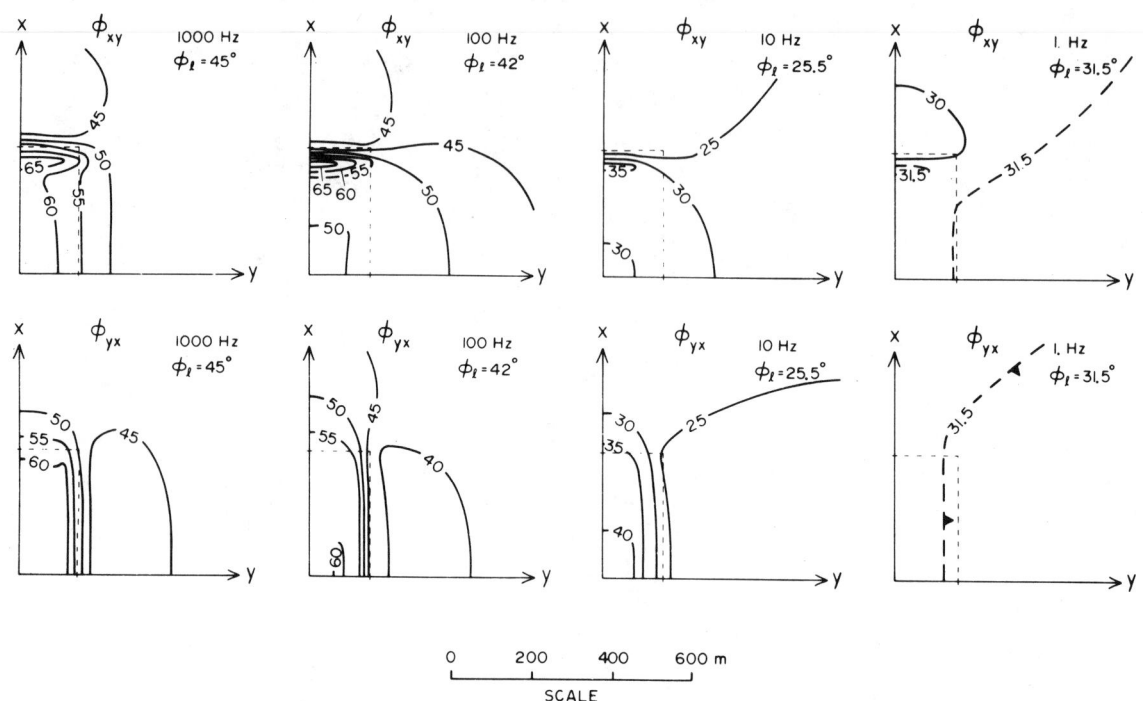

FIG. 4. Multifrequency plan maps of tensor impedance phases ϕ_{xy} and ϕ_{yx} over upper right-hand quadrant of the inhomogeneity of Figure 2. The body outline in plan is shown with dashes, the basal half-space resistivity is 400 $\Omega \cdot$ m and contour values are in degrees. Also, the frequency and the value of the layered earth impedance phase ϕ_ℓ are given in the upper right-hand corner of each plot.

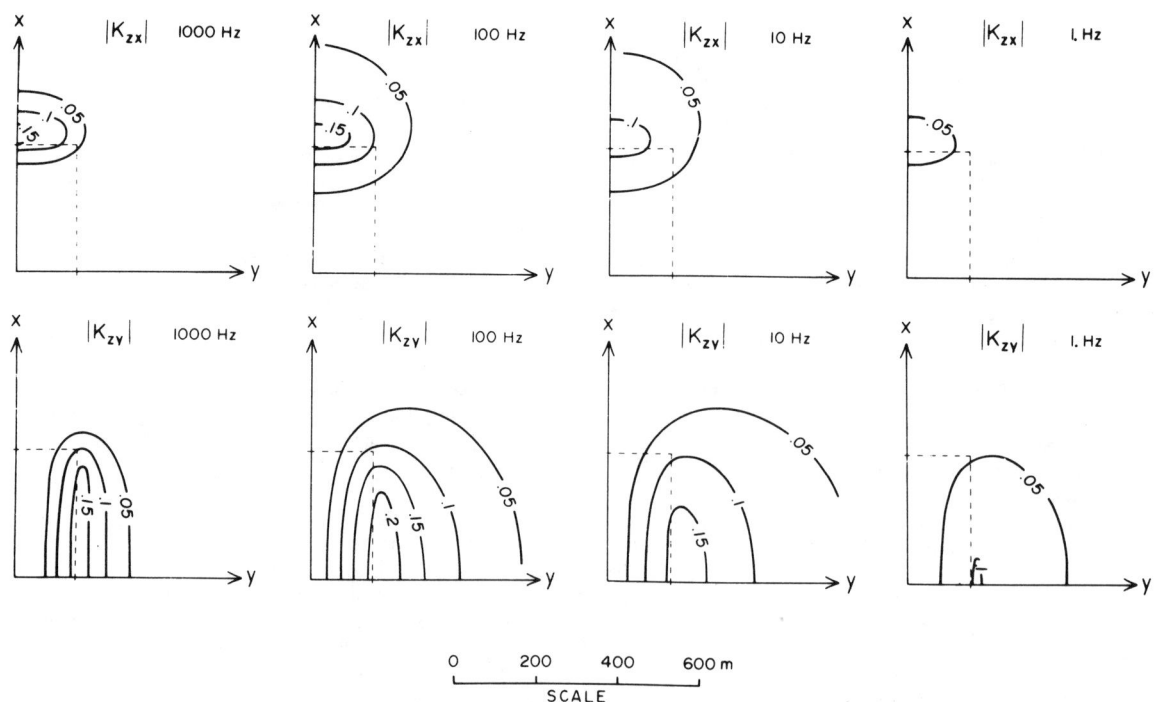

FIG. 5. Multifrequency plan maps of tipper element magnitudes $|K_{zx}|$ and $|K_{zy}|$ over upper right-hand quadrant of the inhomogeneity of Figure 2. The body outline in plan is shown with dashes, the basal half-space resistivity is 400 $\Omega \cdot$ m, and contour values are dimensionless.

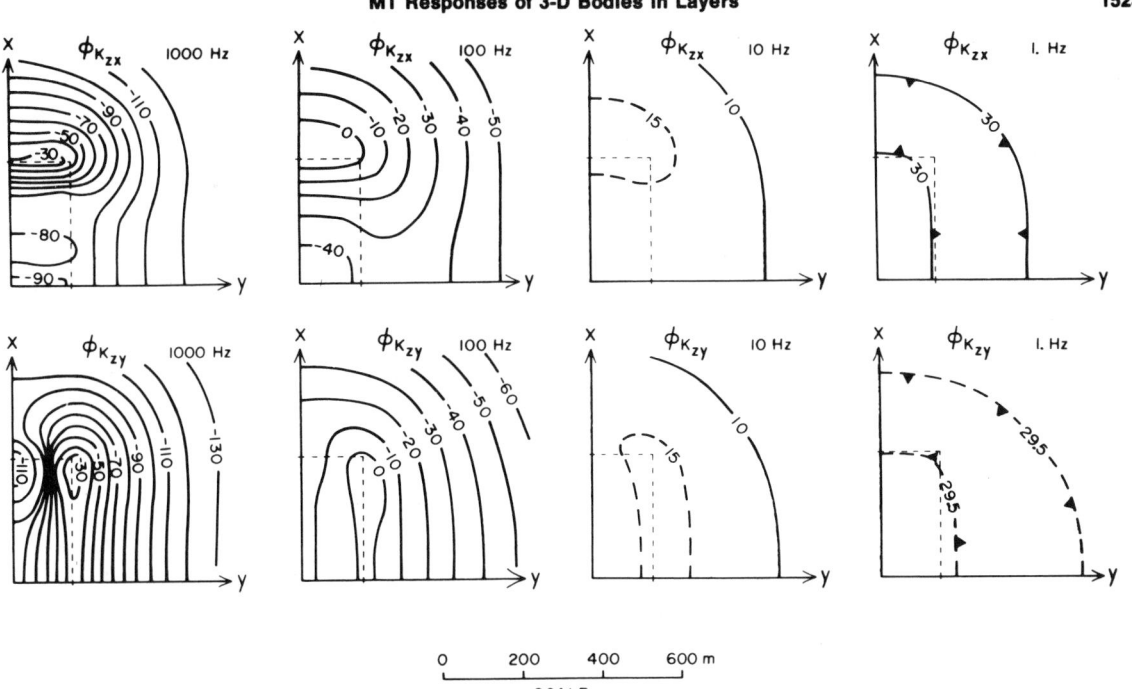

FIG. 6. Multifrequency plan maps of tipper element phases over upper right-hand quadrant of the inhomogeneity of Figure 2. The body outline in plan is shown with dashes, the basal half-space resistivity is 400 Ω · m, and contour values are in degrees.

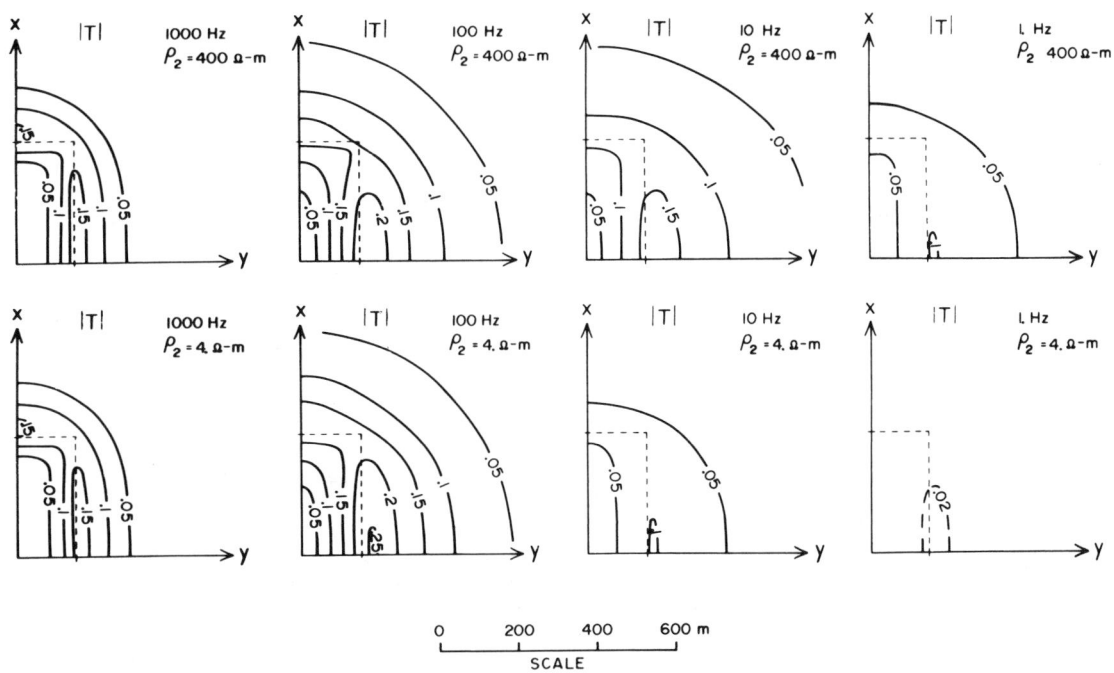

FIG. 7. Multifrequency plan maps of tipper magnitude | T | over upper right-hand quadrant of the inhomogeneity of Figure 2. The body outline is shown in plan with dashes and the contours are dimensionless. Basal half-space resistivities of 400 and 4 Ω · m, identified in the upper right-hand corner of each plot, have been considered in the upper and lower rows of diagrams.

fined their attention to only the low-frequency limits of the distortions due to such inhomogeneities.

Behavior of the impedance phase is entirely different from that of the apparent resistivities, as seen in Figure 4. At 100 and 1 000 Hz, departures may appear in excess of 20 degrees from the layered host phase ϕ_ℓ, which is labeled in the upper right corner of each panel of the figure. At 1. Hz, on the other hand, the secondary electric field is essentially in phase with the incident E-field and the total and incident H-fields are very nearly equal, so that impedance phase values deviate less than 3 degrees from ϕ_ℓ. Impedance phase anomalies due to our model of small-scale, extraneous structure peak at high frequencies, certainly in excess of 100 Hz, and contribute negligibly to observed phase responses below 1. Hz.

Tipper transfer function.—Peak amplitudes of around 0.15 for $|K_{zx}|$ and 0.20 for $|K_{zy}|$ at 100 and 1 000 Hz over this conductor are illustrated in Figure 5. At 1. Hz, because of the appearance of Z_ℓ in equation (18), values of $|K_{zy}|$ have decreased markedly and barely reach 0.10. Complicated anomalies in the phase of elements of $[\tilde{\mathbf{K}}_z(\mathbf{r})]$ are seen in Figure 6 at 100 and 1 000 Hz. At distance from the body at these higher frequencies, the rather uniform spacing of the phase contours represents an outwardly propagating secondary wavefront from this essentially electric dipole scatterer. Beyond several hundred meters, however, the contour spacing broadens, indicating we are approaching the far-field where secondary wavefronts become transverse electromagnetic (TEM) to z. Phases at 1. Hz, as foretold by equation (18), have approached the layered earth impedance phase $\phi_\ell = 31.5$ degrees. In conclusion, responses in both magnitude and phase of elements of $[\tilde{\mathbf{K}}_z(\mathbf{r})]$ due to this sort of small-scale structure are most important at frequencies above 100 Hz.

Figure 7 illustrates the effect of the layered host on $[\tilde{\mathbf{K}}_z(\mathbf{r})]$. At 100 and 1 000 Hz, anomalies in tipper for a conductive basement of 4 $\Omega \cdot$m are essentially identical to those for a resistive basement of 400 $\Omega \cdot$m, since at these high frequencies, $[\tilde{\mathbf{Q}}_s^0(\mathbf{r})]$ and $[\tilde{\mathbf{Z}}_\ell]$ are insensitive to ρ_2. Significant differences are apparent at 1. Hz, however, with peak anomalies for $\rho_2 = 400$ $\Omega \cdot$m being greater by a factor of about $4\frac{1}{2}$ than those for $\rho_2 = 4$ $\Omega \cdot$m. This factor is close to the ratio of the magnitudes of the layered earth impedance at 1. Hz for the two layered hosts, as explained by equation (18). One-dimensional hosts with layer resistivities that increase with depth, through their effect on $[\tilde{\mathbf{Z}}_\ell]$ as discussed with equation (9), tend to prolong the anomalies in $[\tilde{\mathbf{K}}_z(\mathbf{r})]$ to lower frequencies than do layered hosts that become more conductive with depth.

The response of sedimentary basins

An accumulation of Basin and Range graben alluvial fill is depicted in Figure 8 as a large, plate-like inhomogeneity. Although outcropping in nature, this model valley is buried 500 m to obtain accurate results with the integral equations algorithm (Ting and Hohmann, 1981). A model one-dimensional host for this basin, similar to those of Brace (1971) or Wannamaker (1983), also is illustrated in Figure 8. The scattering current in this model was approximated by 110 rectangularly prismatic cells per quadrant and as many as 195 receiver points per quadrant were used to construct upcoming plots. Results required about 20 hours CPU time for each frequency on the Prime 400.

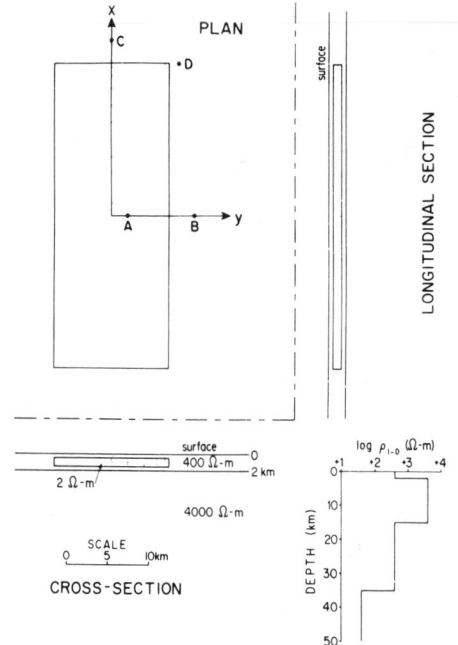

FIG. 8. Plate-like 3-D body in a four-layered earth representing a sedimentary basin in the Basin and Range tectonic province of the western United States. Dashes outline the discretization of the conductor into rectangular cells, shown only for the right half of the body in section and the upper right-hand quadrant in plan. The 1-D resistivity host for this basin model is shown toward the lower right and four points at which sounding curves were computed are given in the plan view.

Regional current gathering.—To begin, we study the widespread distortion of electric fields in the vicinity of our model of a sedimentary basin. Figures 9 and 10 contain plan views of total electric field polarization ellipses at 0.032 Hz over one quadrant of the basin for x- and y-directed polarizations of $\mathbf{E}_i^0$. This is a fairly low frequency for this scale of structure and the response is largely near-field or galvanic, so the ellipses are almost linear.

A clear display of regional current gathering appears in Figure 9, showing the undershoot-to-overshoot, electric dipolar behavior of the ellipses expected over the end of a 3-D body for this orientation of $\mathbf{E}_i^0$. With the incident field directed along the x-azis, a regional depression of $\mathbf{E}_h^0(\mathbf{r})$ occurs to the side of the basin in the y-direction. Outside the corner of the prism, very large values of $\mathbf{E}_h^0(\mathbf{r})$ occur, locally exceeding twice the incident E-field and indicating current convergence from a large volume of the host toward the smaller end of the basin.

The electric field ellipses in Figure 10 for a y-oriented incident E-field show behavior complementary to those in Figure 9. The electric dipolar character of the ellipses is evident over the side of the body for this polarization of $\mathbf{E}_i^0$, in particular causing a regional amplification of $\mathbf{E}_h^0(\bar{\mathbf{r}})$ to the side of the basin in the y-direction. Very small E-fields appear directly over the model for both polarizations of $\mathbf{E}_i^0$, with the y-directed incident field giving a somewhat more extreme anomaly.

Specific effects of the layering upon regional current gathering can be demonstrated in longitudinal section views. In

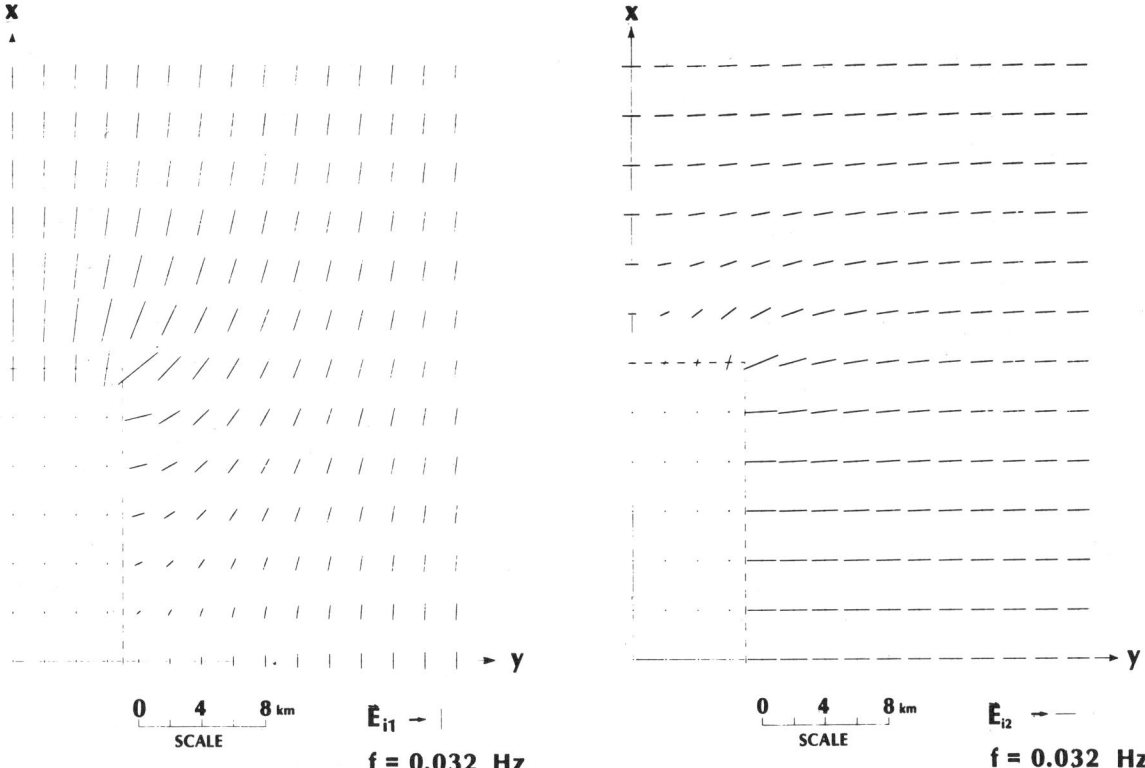

FIG. 9. Plan view of total *E*-field polarization ellipses over upper right-hand quadrant of basin model at 0.032 Hz for an *x*-directed incident field. The magnitude of the linearly polarized $\mathbf{E}_i^0$ is shown in the lower right-hand corner of the diagram.

FIG. 10. Plan view of total *E*-field polarization ellipses over upper right-hand quadrant of basin model at 0.032 Hz for a *y*-directed incident field. The magnitude of the linearly polarized $\mathbf{E}_i^0$ is shown in the lower right-hand corner of the diagram.

Figure 11, the basin model resides in a uniform half-space of 400 Ω · m and total electric field polarization ellipses have been computed, again at 0.032 Hz, to a depth in excess of 20 km. The approximately electric dipolar character of $\mathbf{E}^0(\mathbf{r})$ in longitudinal section is apparent, where boundary charges result in current from great depth gathering up to the valley. As in the plan views involving the layered host, very large values of total *E*-field are seen near the ends of the plate. The undershoot-to-overshoot behavior of $E_{xs}^0(\mathbf{r})$ drawn at the top of the figure is another view of this phenomenon. At lateral distances near 40 km from the prism's center, $|E_{xs}^0(\mathbf{r})|$ has decayed to about 5 percent of E_{xi}^0.

In Figure 12, the model is returned to its four-layered host and *E*-field polarization ellipses are again calculated. In the uppermost 400 Ω · m layer containing the basin model the ellipses are nearly horizontal and attenuate with distance much more slowly than in the case of the half-space host. The plot of $E_{xs}^0(\mathbf{r})$ in the upper part of the figure supports this, with $|E_{xs}^0(\mathbf{r})|$ at lateral distances near 40 km from the plate's center exceeding 20 percent of $|E_{xi}^0|$. To comprehend this relatively slow attenuation, one must realize that secondary currents induced in the 400 Ω · m layer about the basin have difficulty penetrating the more resistive 4 000 Ω · m medium, and hence can decay geo-

metrically most easily in just the *x-y* plane (Wannamaker et al., 1984). By contrast, secondary currents about the basin in the uniform 400 Ω · m half-space can die away readily in the *z*-direction as well. Hence, the surface anomaly in Figure 12 is observed to be important to greater distances from the body than is that in Figure 11.

However, despite the fact that secondary current flow about the basin in the four-layered earth has difficulty penetrating the 4 000 Ω · m medium, electric field ellipses in this most resistive layer in Figure 12 exhibit much stronger vertical components than those at comparable depths in the 400 Ω · m half-space in Figure 11. This phenomenon is due to the preservation of the normal component of current density across layer interfaces, which in turn means that the vertical component of secondary electric field experiences a step jump by a factor of ten going from the 400 Ω · m layer down to the 4 000 Ω · m layer (note there is no vertical component of the primary field). Nevertheless, we emphasize that values of secondary current density in the 4 000 Ω · m material are actually much smaller than values at corresponding positions in the less resistive, 400 Ω · m half-space host. Confirming this is the very flat nature of the ellipses in Figure 12 in the deep 400 Ω · m layer extending from 15 to 35 km, showing that deep regional current flow is essentially insu-

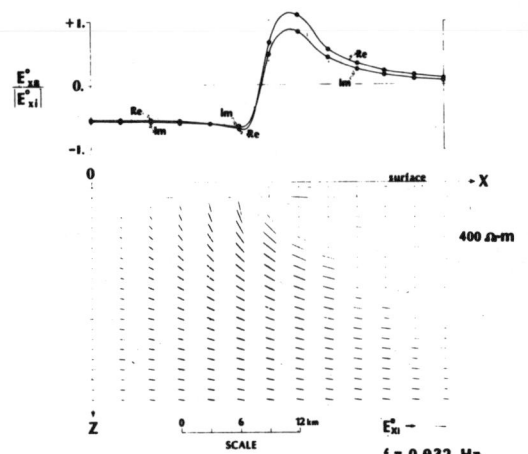

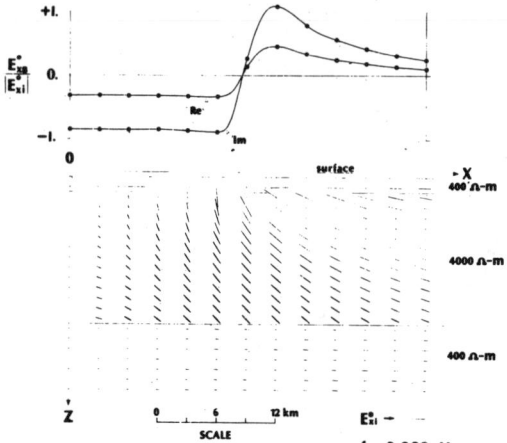

FIG. 11. The basin model is enclosed in a 400 Ω·m half-space for this section view of total E-field polarization ellipses through one half of the basin at 0.032 Hz. The incident field is linearly polarized in the x-direction and its magnitude at the surface is shown in the lower right-hand corner of the diagram. At the top of the figure are profiled real and imaginary components of $\mathbf{E}_{xs}^0(\mathbf{r})$ normalized by $|\mathbf{E}_{xi}^0|$.

FIG. 12. The basin model is returned to the four-layered host for this section view of total E-field polarization ellipses at 0.032 Hz. The incident field is again linearly polarized in the x-direction and its magnitude at the surface is shown in the lower right-hand corner of the diagram. At the top of the figure are profiled real and imaginary components of $\mathbf{E}_{xs}^0(\mathbf{r})$ normalized by $|\mathbf{E}_{xi}^0|$.

lated by the 4 000 Ω·m layer from the effects of the conductive basin.

Effects of the layered host upon electric field anomalies carry over to the apparent resistivity signatures (Ranganayaki and Madden, 1980). In Figure 13, ρ_{xy} and ρ_{yx} have been plotted for a frequency of 0.032 Hz along the y-axis of the 3-D basin model, with both half-space and layered hosts being considered. To the side of the basin within either host, an apparent resistivity anisotropy in excess of an order of magnitude is observed; it diminishes with distance from the body. With the half-space host, neither ρ_{xy} nor ρ_{yx} depart more than 10 percent from the 1-D sounding ρ_ℓ beyond distances of about 25 km from the center of the basin. However, the apparent resistivity anomalies about the basin in the layered host exist to considerably greater distances than this. With the layered sequence, ρ_{xy} and ρ_{yx} do not lie within 10 percent of the ρ_ℓ at this frequency until about 60 km from the center of the basin, which is well off our diagram. We conclude that in the interpretation of MT soundings in the Basin and Range province, not only must one be aware of the sedimentary basin immediately adjacent to the soundings, but perhaps also of basins at greater distances.

Magnetotelluric strike estimations.—Principal coordinate directions of tensor MT response functions provide measures of preferred geoelectric trends. In Figure 14 are plotted off-diagonal impedance polar diagrams as well as tipper-strike directions for a variety of receivers over the basin model at 0.032 Hz. Our definition of tipper-strike is the x-axis for which $|K_{zy}|$ is maximized. The length of the tipper-strike bars is proportional to tipper magnitude. Since $|T|$ is zero directly over the model, no tipper-strike can be defined here.

From the impedance polar diagrams, one sees that principal axes of $[\mathbf{Z}(r)]$ occur approximately every 90 degrees, so that

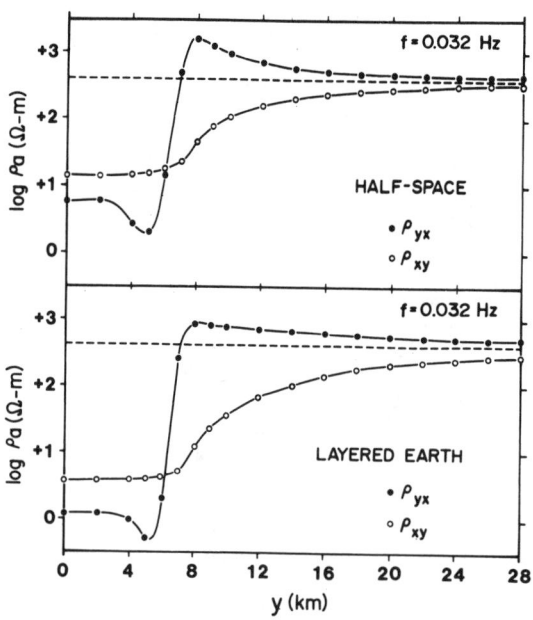

FIG. 13. Profiles of ρ_{xy} and ρ_{yx} along the y-axis over the 3-D basin model computed for a frequency of 0.032 Hz. The response over the basin in a uniform 400 Ω·m half-space appears in the top half of the diagram for comparison to the response over the basin in the layered host of Figure 8. The 1-D host apparent resistivity ρ_ℓ also has been plotted with dashes for reference for both the half-space and the layered earth.

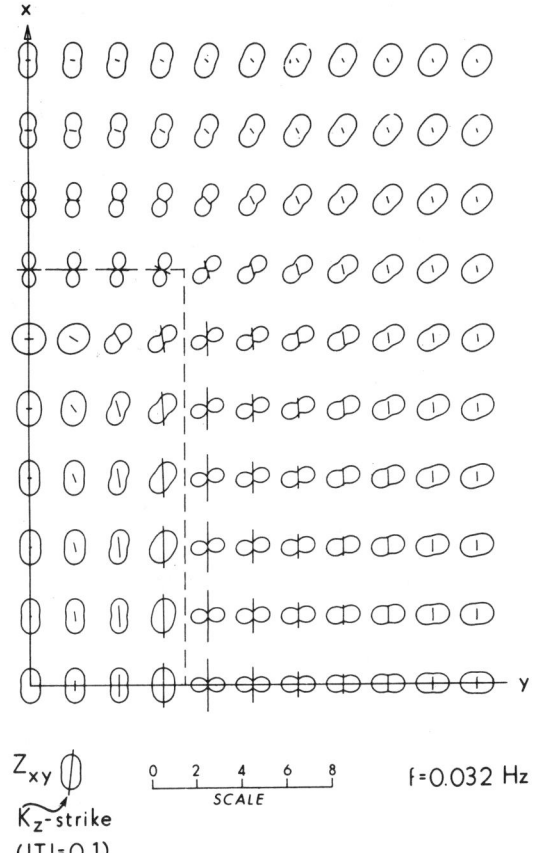

FIG. 14. Plan view of tipper-strike and off-diagonal impedance polar diagrams over the upper right-hand quadrant of the basin model for a variety of receivers at 0.032 Hz. Length of the tipper-strike bars is proportional to magnitude of tipper, exemplified by a value of $|T| = 0.1$ in the key.

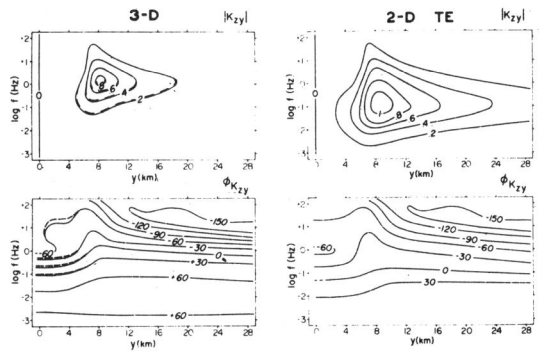

FIG. 15. Pseudosections of magnitude and phase of K_{zy} for profiles at $x = 0$ (solid contours) and $x = 9$ (dashed contours) over the 3-D basin model compared to 2-D TE pseudosections for corresponding model of infinite strike length. Pseudosections commence at $y = 0$ over the center of the basin and extend in the y-direction. Contours of magnitude and phase are, respectively, dimensionless and in degrees.

they alone cannot give unique strike directions (Word et al., 1970; Vozoff, 1972). Tipper-strike parallels more closely the true limits, both sides and ends, of the valley than does the principal axis of $[\tilde{Z}(\mathbf{r})]$ closest to tipper-strike (cf., Jones and Vozoff, 1978). Principal axes of the impedance also seem to be a less stable indicator of geoelectric trends just inside the basin. Over single inhomogeneities such as that of Figure 8, we conclude that tipper-strike is somewhat superior to the principal axes of the impedance due to its closer conformity to true geoelectric trends, its greater stability and its lack of a 90 degree ambiguity. Nevertheless, both types of trend estimator show reasonably close agreement overall, especially to the side of the valley, and thus are considered largely equivalent for single simple bodies.

Tipper pseudosections.—Multifrequency calculations of tipper element K_{zy} were performed along two traverses over the 3-D basin and are compared in Figure 15 to calculations over a 2-D model of identical cross-section (Rijo, 1977; Stodt, 1978). For display in pseudosection form, log frequency and the y-axis serve as ordinate and abscissa for contour plots of amplitude and phase (Vozoff, 1972). The 3-D traverses reside along the line $x = 0$, over the prism's center, and along $x = 9$ km, halfway to the prism's end. For the pseudosection displays, the x-y coordinates of the MT response functions parallel the plate symmetry axes. In this manner, K_{zy} over the 3-D model can be compared to the 2-D transverse electric responses.

Pseudosection contours were constructed from computations at every half decade in frequency, from 0.001 to 100 Hz. However, we distrust the numerical accuracy of the 3-D calculations above 10 Hz for this large body, although the discretization we have chosen is as fine as is practical with our computer. Since 3-D and 2-D values of all MT functions presented have converged by the point that frequency has risen to 10 Hz, 3-D contours at higher frequencies were derived by extrapolation using 2-D results.

In Figure 15, agreement in $|K_{zy}|$ between the profiles at $x = 0$ and at $x = 9$ km over the 3-D body is close at all frequencies. However, a significant departure from the 2-D transverse electric responses occurs below 3 Hz, with the 3-D values at lower frequencies being greatly subdued by comparison. To understand this 3-D anomaly, consider the regional depression of $E_i^0(\mathbf{r})$ occurring over and beside the valley when E_i^0 is x-oriented (Figure 9). The depression results overwhelmingly from boundary charges on the ends of the 3-D prism and reaches a maximum at low frequencies. These charges effect a reduction of J_s within, and hence $[\tilde{Q}_{vs}^0(\mathbf{r})]$ over, the 3-D basin. In contrast, the lack of charges on the 2-D body allows a strong 2-D TE response in $|K_{zy}|$ over a broader frequency range.

The magnitude of K_{zy} over the 3-D basin model peaks at a frequency near 1. Hz. Both 2-D and 3-D amplitudes of $|K_{zy}|$ decay rapidly at frequencies below 0.03 Hz in Figure 14, a result of the resistivity of the layered host decreasing at depths beyond 15 km, so that $|Z_\ell|$ in equation (18) attenuates quickly at these lower frequencies. Due to this phenomenon, and in light of the strength of the 2-D response, we conclude that large amplitudes of tipper will occur over long, high contrast bodies in hosts that have layer resistivities increasing with depth.

Agreement between 2-D and 3-D phases of K_{zy} in Figure 15 is limited to quite high frequencies, above 10 Hz for this model. The inclination of phase contours away from the valley above about 1. Hz is another view of outwardly traveling secondary waves, and corresponds to the rather uniform spacing of contours on single frequency plan maps such as Figure 6. At low frequencies for both 2-D and 3-D valley models, phases of K_{zy} approach the phase of the layered host impedance ϕ_ℓ, which is near 60 degrees at 0.003 Hz.

The separation in frequency of the responses of $[\tilde{\mathbf{K}}_z(\mathbf{r})]$ for the basin model and for the small-scale structure, apparent by comparison of Figures 5, 6, and 15, is fundamentally a demonstration of EM scaling in MT responses (Stratton, 1941, p. 488–490; Grant and West, 1965, p. 478–482). In the concept of EM scaling, small inhomogeneities can have MT responses as strong as those of large ones. However, provided corresponding resistivities and aspect ratios are similar, the small and large body responses have frequency dependencies that are separated as the square of the geometric scale factor distinguishing the different bodies. Specifically, then, examination of observed tipper may be a quick and easy means of identifying the presence of any large, buried resistivity anomaly, constituting perhaps an exploration target, beneath small-scale extraneous structure. For estimating geoelectric trends of basin sediments like the model we have considered, tipper-strike should be computed in the frequency range 0.1 to 1. Hz to ensure that the basin response in $[\tilde{\mathbf{K}}_z(\mathbf{r})]$ is strong while that of any small-scale, extraneous structure is attenuated.

Nevertheless, it is possible that resistivity structure of a scale significantly greater than that of typical Basin and Range graben sediments, for instance resulting from regional thermotectonic perturbations common to this province (Eaton, 1982), can dominate tipper amplitudes at frequencies of 0.01 Hz or below (consider Porath, 1971b). This underscores the need for care in defining the term "regional structure" and thus to impose a selective weighting with respect to frequency of one's tipper-strike estimates (compare Gamble et al., 1982). Given the dependence of tipper responses on both the properties of the body and of the layered host, use of an algorithm like that of Wannamaker et al. (1984) is recommended for assessment of this weighting.

Apparent resistivity and impedance phase pseudosections.—As was done for the tipper, apparent resistivities ρ_{xy} and ρ_{yx} as well as impedance phases ϕ_{xy} and ϕ_{yx} have been displayed in pseudosection form in Figures 16 and 17. A strong discrepancy between 2-D TE and corresponding 3-D responses, designated ρ_{xy} (3-D) and ϕ_{xy} (3-D), for our basin model is illustrated again in Figure 16, this time for frequencies below about 1. Hz. This apparent resistivity over and adjacent to the 3-D plate decreases as frequency falls relative to the apparent resistivity of the layered host ρ_ℓ, which the 3-D response approaches at large distances from the valley. Correspondingly, anomalously high values of ϕ_{xy} (3-D) appear in the vicinity of the body, although the discrepancy between this quantity and the host impedance phase ϕ_ℓ is less than 5 degrees outside the valley below 0.03 Hz and less than 5 degrees anywhere below 0.003 Hz. In contrast, ρ_{xy} (2-D) falls with respect to ρ_ℓ, while ϕ_{xy} (2-D) surpasses ϕ_ℓ only above 0.3 Hz with just the opposite behavior at lower frequencies.

This 3-D anomaly may be explained again in terms of boundary charges, acting on ρ_{xy} and ϕ_{xy} through equations (12) and (14b). Such charges do not occur in the 2-D body for this transverse electric mode, so that the wave equation for $\mathbf{E}_s(\mathbf{r})$ approaches the homogeneous Laplace's equation at lower frequencies and there is a diminishing contribution by the secondary E-field to the anomalous ρ_{xy} and ϕ_{xy}. In fact, the 2-D TE responses below about 0.1 Hz result predominantly from a strong secondary H-field, which remains important as governed by equations (8) and (12) until frequencies under 0.001 Hz for the 2-D basin model. On the other hand, over the 3-D model at frequencies less than 0.1 Hz, $\mathbf{H}_{hs}^0(\mathbf{r})$ is much smaller than that over the 2-D counterpart. Once more, boundary charges severely depress the scattering current within, and thus $[\tilde{\mathbf{Q}}_{hs}^0(\mathbf{r})]$ over, the 3-D body relative to the 2-D structure. Hence, ρ_{xy} and ϕ_{xy} for the 3-D basin model arrive at low-frequency asymptotes around 0.003 Hz.

If our model is representative of graben sedimentary fill in the Basin and Range, then 2-D transverse electric modeling

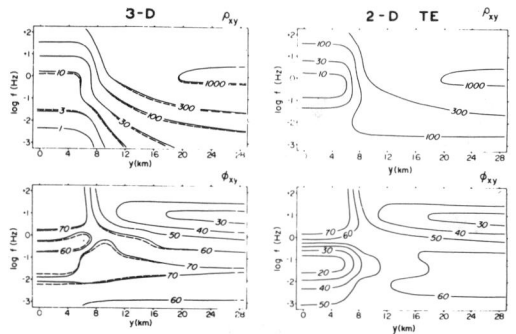

FIG. 16. Pseudosections of ρ_{xy} and ϕ_{xy} for profiles at $x = 0$ (solid contours) and $x = 9$ km (dashed contours) over the 3-D basin model compared to 2-D TE pseudosections for corresponding model of infinite strike length. Pseudosections commence at $y = 0$ over the center of the basin and extend in the y-direction. Contours of ρ_{xy} and ϕ_{xy} are in $\Omega \cdot$ m and degrees.

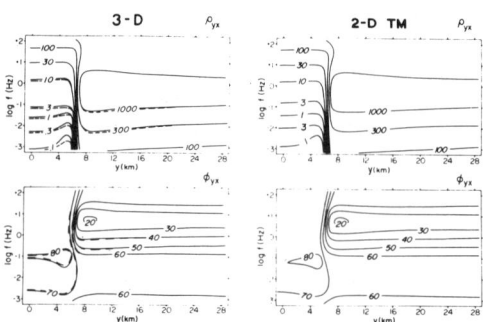

FIG. 17. Pseudosections of ρ_{yx} and ϕ_{yx} for profiles at $x = 0$ (solid contours) and $x = 9$ km (dashed contours) over the 3-D basin model compared to 2-D TM pseudosections for corresponding model of infinite strike length. Pseudosections commence at $y = 0$ over the center of the basin and extend in the y-direction. Contours of ρ_{yx} and ϕ_{yx} are in $\Omega \cdot$ m and degrees.

algorithms may result in erroneous resistivity cross-sections when applied to apparent resistivities and impedance phases identified as TE in this province. In attempting to replicate the 3-D response in Figure 16 with a 2-D TE routine, one would need to place false low resistivities at some depth below the true basin. Further errors through use of 2-D TE programs will likely occur if small-scale, near-surface structure such as we have simulated in Figure 2 is widespread (for example, see Wannamaker et al., 1978).

Given the size of the basin model, the frequencies at which 2-D TE and corresponding 3-D MT responses agree closely may seem surprisingly high. Wannamaker et al. (1984), however, explained that the length a 3-D body must have to achieve agreement with 2-D TE components depends strongly on the 1-D host, with bodies in layers overlying resistive basements needing to be longer than bodies overlying conductive basements. Layered host resistivities in nature increase with depth in the upper several kilometers (Brace, 1971; Wannamaker, 1983), exacerbating the difficulties with 2-D TE interpretations.

In contrast to the responses depicted in Figure 16, anomalies in ρ_{yx} and ϕ_{yx} in Figure 17 are essentially the same for the two 3-D traverses and the 2-D profiling at all frequencies. Some minor differences right over the valley at lower frequencies will be explained with the presentation of sounding curves. The reasons for this agreement are twofold. First, no secondary *H*-field exists for the 2-D TM mode (Swift, 1967), and there is only an insignificant contribution by $H_s^0(\mathbf{r})$ to the corresponding 3-D response. Second, boundary charges on the sides of the basin are included in both 2-D TM and 3-D formulations. These boundary charges in turn lead to current gathering into the sides of the 2-D and 3-D basin models. As was demonstrated for the 3-D model in Figure 10, such current gathering is manifested in the development of a crudely dipolar variation in the electric field over the basin toward lower frequencies. Low-frequency asymptotes to ρ_{yx} and ϕ_{yx} are reached by about 0.01 Hz outside the basin model. In contrast, these asymptotes directly over the model do not occur until about 0.001 Hz due to the tendency of the secondary and incident electric fields to cancel here.

As was the case for the tipper, the frequency dependence of both ρ_{xy} and ρ_{yx} as well as anomalous values of both ϕ_{xy} and ϕ_{yx} for the basin model occur over a much lower frequency range than for the small-scale, near-surface model of Figure 2. This result is a further manifestation of EM scaling in MT responses. Thus, examination of observed impedance phases may be another straightforward means of identifying the presence of any large, buried resistivity anomaly, constituting perhaps an exploration target, beneath small-scale extraneous structure.

We infer from Figure 17 that accurate cross-sections of earth resistivity may be interpreted from profiles of MT measurements across elongate, geometrically regular 3-D bodies using a 2-D transverse magnetic algorithm. For 2-D TM modeling of field data, we recommend tipper-strike for specifying a fixed coordinate system for the MT response functions. Tipper-strike is a stable, unambiguous strike estimator that conforms closely to true geoelectric trends. Use of $[\tilde{\mathbf{K}}_z(\mathbf{r})]$ to define the coordinate axes will result in pseudosections that are essentially identical to those in Figures 16 and 17, with the possible exception of field points close to the center of the model. These latter axes may be constrained to be consistent with the majority of the

strike estimates. For single simple bodies such as our valley model, however, principal axes of $[\tilde{\mathbf{Z}}(\mathbf{r})]$ probably are satisfactory also for this purpose.

On the other hand, telluric distortions in any near-surface extraneous structure such as modeled in Figure 2 may cause scatter from sounding to sounding in the principal axes of $[\tilde{\mathbf{Z}}(\mathbf{r})]$ from their orientations over the basin alone. The resulting nonuniform coordinate system leads to variations specifically in the impedance phase ϕ_{yx}, and in the frequency dependence of the apparent resistivity ρ_{yx}, of high spatial frequency occurring at middle to low angular frequencies. These variations in turn are due to the sedimentary basin, which has induced a broad anisotropy in both ϕ_{xy} and ϕ_{yx}, and in the frequency dependence of both ρ_{xy} and ρ_{yx}, at middle to low frequencies. A nonuniform coordinate system mixes secondary field components of differing phases in differing proportions from sounding to sounding, thus leading to inconsistent complex quantities. These rapid spatial variations in ϕ_{yx} and in the frequency dependence of ρ_{yx}, since they occur at middle and low frequencies, would be difficult if not impossible to fit with a 2-D TM algorithm. As discussed previously, tipper-strike remains relatively uniform in the face of small-scale extraneous structure and therefore is superior to the principal axes of the impedance for coordinate system definition.

It should be emphasized that use of a 2-D transverse magnetic algorithm as we propose it is no interpretive panacea. For example, consider a buried 3-D dike-like conductive inhomogeneity elongated in the *x*-direction. If the dike is thin compared to its depth, the measured response in ρ_{yx} and ϕ_{yx} may be very weak (Vozoff, 1972), even though it can be modeled with a 2-D TM routine. On the other hand, if the strike extent of the dike is large compared to its depth, a strong response in ρ_{xy} and ϕ_{xy} may exist. However, this response requires a full 3-D interpretation including the effects of current gathering. Also, apart from employing MT strike estimations in transverse magnetic mode identification, this 2-D TM approach can make no use of the tipper.

Furthermore, if a 3-D structure is highly irregular in geometry, then application of a 2-D TM algorithm can be hazardous (Hermance, 1982). In such instances, however, we doubt the validity of 2-D TE approaches as well. To us, the gravity surveying of the northern Basin and Range cited previously suggests that most graben sedimentary fill is podiform, with limited electrical connection between basins. MT interpretation certainly should be approached on a case-by-case basis to avoid errors due to unwarranted generalizations. Nevertheless, 2-D TM modeling can be exploited to the fullest given careful MT survey design at the outset utilizing independent constraints such as gravity and geology.

Apparent resistivity and impedance phase soundings.—To explore the applicability of 1-D interpretation schemes for 3-D data, in Figures 18 through 21 we present principal apparent resistivity and impedance phase soundings for positions A, B, C, and D of Figure 8. In a manner similar to that for the pseudosections, the transverse electric mode has been associated with ρ_{xy} and ϕ_{xy} by selecting the principal direction of the impedance which is closest to tipper-strike as the *x*-axis of the principal coordinate system (Word et al., 1970; Vozoff, 1972). Definition of modes in this manner is common practice although strictly speaking, the EM fields scattered by 3-D bodies do not decouple into TE and TM modes (Swift, 1967). Note in Figure 14 that this principal axis swings, from a direction paralleling the x-axis of symmetry of the basin model for points A and B, counterclockwise about 40 degrees for site D until finally it parallels the y-axis of symmetry of the basin for site C.

At point A over the basin model, principal apparent resistivities and impedance phases for the 3-D body are compared in Figure 18 to a 1-D response which would occur if the 2 $\Omega \cdot$ m basin sediments had no lateral bounds, i.e., formed a continuous layer. Both ρ_{xy} and ρ_{yx} for the 3-D body fall increasingly below the 1-D sounding as frequency falls, while both ϕ_{xy} and ϕ_{yx} generally exceed the 1-D response at lower frequencies. For point A, ρ_{xy} and ϕ_{xy} are less anomalous than ρ_{yx} and ϕ_{yx}, because the resistivity boundaries normal to the incident electric field in the former case are more distant relative to their dimensions. Still, all the 3-D signatures depart substantially from the 1-D response. As explained with the pseudosections, free charges on the boundaries of the 3-D model cause a progressive depression of the electric fields for both modes over the basin as frequency decreases.

Also in Figure 18, the 3-D responses are compared to results over the 2-D model of identical cross-section. The 2-D TE response is quite similar to the 1-D sounding over most of the

frequency range, but it shows some departure toward lower frequencies. Due to the boundary charges, the difference between the 2-D TE and corresponding 3-D responses is much more striking. On the other hand, the 2-D TM results compare well with ρ_{yx} (3-D) and ϕ_{yx} (3-D) over the whole frequency range, apart from a minor discrepancy mainly in amplitude below about 0.1 Hz. We stress that is difficult to achieve accuracy right over a shallow, high-contrast model since the secondary E-field for which we solve nearly cancels the incident field here. We are especially suspicious of the irregular 3-D behavior around 0.1 Hz. Alternately, a small amount of current gathering in the horizontal plane from the layered host may explain at least partially the difference in these 2-D TM and 3-D results. Nevertheless, very small changes in basin resistivity would result in substantial changes in ρ_{yx} and ϕ_{yx}, either for the 3-D or 2-D bodies, so that in reality these discrepancies are of little consequence.

At point B beside the basin, principal apparent resistivities and impedance phases for the 3-D model are compared in Figure 19 to the response of the 1-D regional resistivity profile in the absence of the basin. The 3-D response in ρ_{xy} falls increasingly below the response of the regional host ρ_ℓ for frequencies below about 10 Hz while that in ϕ_{xy} exceeds ϕ_ℓ,

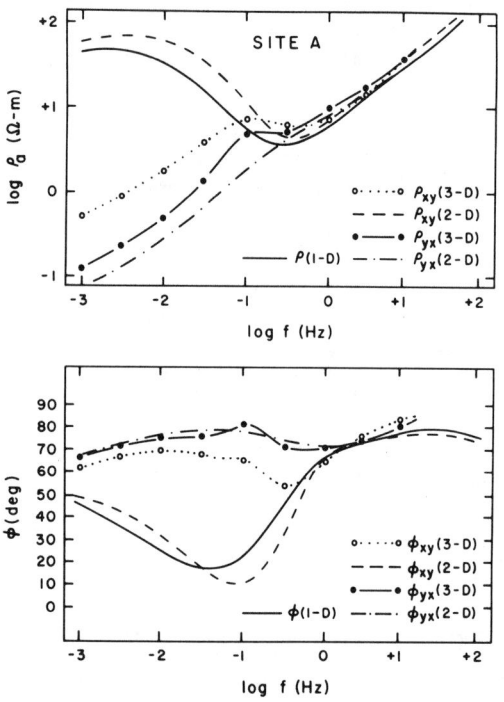

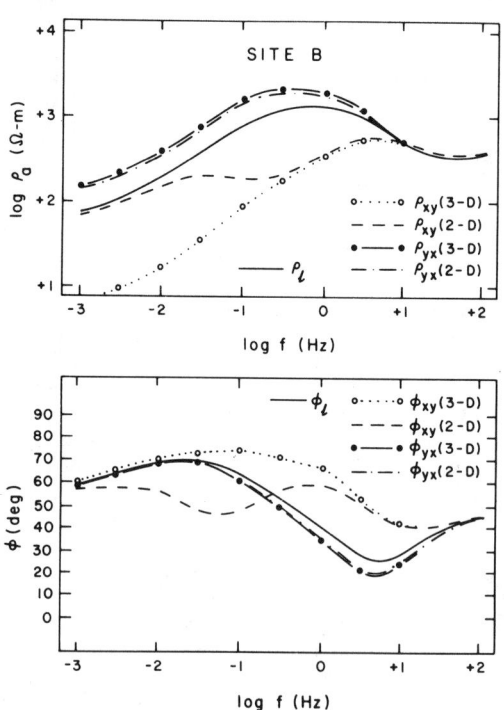

FIG. 18. Principal apparent resistivities and impedance phases ρ_{xy} and ϕ_{xy} (dotted curves, identified as TE) and ρ_{yx} and ϕ_{yx} (dashed curves, identified as TM), for point A over the basin model of Figure 8. The 3-D responses were calculated every half decade in frequency, as shown by the open and solid circles. For comparison, the 1-D response which would occur if the basin had no lateral bounds is shown with solid curves. Also presented are the results for a 2-D basin model of identical cross-section.

FIG. 19. Principal apparent resistivities and impedance phases for point B over the basin model of Figure 8. Plotting convention is the same as that in Figure 16. For comparison, the 1-D response of the regional host to the basin model shown in Figure 8 is plotted with solid curves. Also presented are the results for a 2-D basin model of identical cross-section.

especially over the middle frequency range. On the other hand, the 3-D response ρ_{yx} increases beyond ρ_ℓ while ϕ_{yx} is less than ϕ_ℓ, again especially at mid-frequencies. For point B, ρ_{yx} and ϕ_{yx} are less anomalous than ρ_{xy} and ϕ_{xy} for the 3-D body. As explained also with the pseudosections, free charges on the ends of the 3-D basin model depress electric fields both over and adjacent to the basin model for the TE mode. However, charges on the sides of the model result in an undershoot-to-overshoot behavior in the electric field causing depression of ρ_{yx} over the basin but amplification of ρ_{yx} to the side with respect to ρ_ℓ. These 3-D results are similar to those presented by Park et al. (1983).

The 3-D results at site B have been compared as well to those beside the 2-D model. The 2-D TE response here differs a great deal from ρ_ℓ and ϕ_ℓ, showing an apparent resistivity minimum near 0.2 Hz along with a maximum in ϕ_{xy} (2-D) at somewhat higher frequencies. Unlike the 3-D values which decrease monotonically, the 2-D TE apparent resistivities approach the layered host sounding at low frequencies. Again, however, the 2-D TM signatures very much resemble the corresponding 3-D calculations over the entire frequency range. We are not certain how much of the apparent difference is due to numerical inaccuracy and how much is 3-D current gathering, but the difference is very small anyway.

Principal apparent resistivities and impedance phases shown in Figures 20 and 21 for sites C and D outside the 3-D basin model qualitatively are similar to those at site B. However, in contrast to responses at B, ρ_{yx} and ϕ_{yx} at C and D appear essentially as anomalous as ρ_{xy} and ϕ_{xy} when compared to the regional profile sounding. For site D in particular, current gathers from large volumes of the host toward the corner of the basin model resulting in large electric fields and large values of ρ_{yx}. Furthermore, an apparent resistivity anisotropy exceeding two orders of magnitude has occurred here.

Note especially that the transverse electric apparent resistivity defined in the conventional manner is depressed relative to the response of the 1-D host alone everywhere over and exterior to the conductive 3-D body to arbitrarily low frequencies. In light of the concept of EM scaling, such depression can be as severe locally about small conductive bodies as about large ones. Given this depression, we are skeptical of many published models of layered resistivity structure in the Basin and Range and other extensional regimes. We believe many such models have erroneously shallow depths to interfaces and erroneously low values of layer resistivities (see also Porath, 1971a; Wannamaker, 1983).

To be sure, our skepticism is based upon the assumption that upper crustal lateral inhomogeneities, for the greater part, are conductive compared to their environs. It has been argued previously that this is the case for sedimentary distributions.

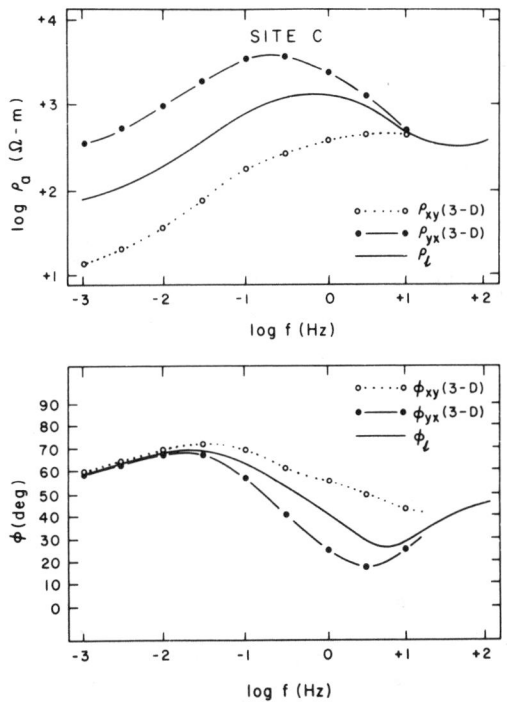

FIG. 20. Principal apparent resistivities and impedance phases for point C of the end of the basin model of Figure 8 compared to the 1-D response of the regional host. Plotting convention is the same as that in Figure 18.

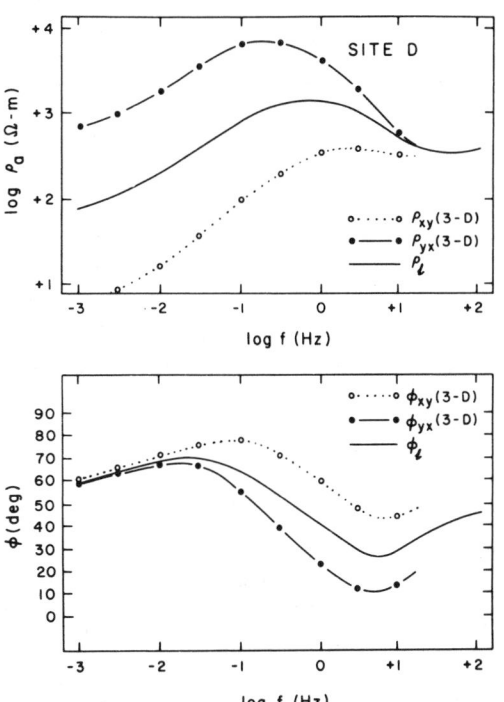

FIG. 21. Principal apparent resistivities and impedance phases for point D outside the corner of the basin model of Figure 8 compared to the 1-D response of the regional host. Plotting convention is the same as that in Figure 18.

However, even if resistive inhomogeneities were as common as conductive ones, our perception of a bias toward low resistivity at depth probably still is valid. This bias occurs because, for a given body-host layer resistivity contrast, the upward shift in the TE apparent resistivity about a resistive body is less than the downward shift in this function about a conductive body (Berdichevskiy and Dmitriev, 1976).

CONCLUSIONS

Resistivity structure in nature is an ensemble of inhomogeneities of different scales, and the small structures in this collection may have MT responses as strong as those of the large ones. Any telluric distortion in overlying, small-scale extraneous structure will be superimposed upon the apparent resistivities measured over buried 3-D targets to arbitrarily low frequencies. However, the responses of the small and large bodies have frequency dependencies that are separated approximately as the square of the geometric scale factor distinguishing the different structures. Also, the magnitudes of the tipper elements as well as the phases of all MT functions due to a particular body are significant only over a finite frequency range, i.e., they are band-limited. Thus, these quantities may allow one to "see through" small-scale extraneous structure to observe the signature of a buried target if the scales of the two types of inhomogeneity differ sufficiently.

Relative to the TE response of a 2-D body of identical cross-section, the apparent resistivity identified as TE by conventional means over and around a confined 3-D conductive body suffers a widespread depression that is increasingly pronounced toward lower frequencies. This depression of the 3-D response results from current gathering. Interpretation of such a 3-D response using 1-D or 2-D TE modeling routines would infer erroneously low resistivities at depth below the true inhomogeneity. From this we conclude that many published models of deep resistivity derived from MT have experienced a bias toward shallow, low resistivities.

Fortunately, our model studies have shown that centrally located profiles of apparent resistivity ρ_{yx} and impedance phase ϕ_{yx} across elongate, geometrically regular 3-D prisms can be modeled accurately with a 2-D TM algorithm. Boundary charges are included in both 3-D and 2-D TM formulations. To define pseudosections for transverse magnetic modeling, we recommend employment of a fixed coordinate system based on tipper-strike. Since $[\tilde{\mathbf{K}}_z(\mathbf{r})]$ is band-limited, one may choose an optimal frequency for defining a tipper-strike to minimize the contributions of secondary structures much smaller or, for that matter, much larger than the target.

ACKNOWLEDGMENTS

We are grateful for the constructive criticisms of John F. Hermance, William A. SanFilipo, William R. Sill, and C. M. Swift, Jr. Further technical assistance came from Sandra Bromley, Doris Cullen, Terry Killpack, Carleen Nutter, and Joan Pingree.

This research was supported by DOE/DGE contract DE-AC07-80ID12079 and NSF contract EAR 8116602.

REFERENCES

Berdichevskiy, M. N., and Dmitriev, V. I., 1976, Basic principles of interpretation of magnetotelluric curves, *in* Geoelectric and geothermal studies: A. Adam, Ed., Akademini Kiado, 165–221.

Boehl, J. E., Bostick, F. X., Jr., and Smith, H. W., 1977, An application of the Hilbert transform to the magnetotelluric method: Elec. Geophys. Res. Lab. Rep., Univ. of Texas at Austin, 98 p.

Brace, W. F., 1971, Resistivity of saturated crustal rocks to 40 km based on laboratory measurements, *in* The Structure and Physical Properties of the Earth's crust: J. G. Heacock, Ed., AGU Mono. 14, 243–256.

Cagniard, L., 1953, Basic theory of the magnetotelluric method of geophysical prospecting: Geophysics, **18**, 605–635.

Cook, K. L., Adhidjaja, J. I., and Gabbert, S. C., 1981, Complete Bouguer gravity anomaly and generalized geology map of Richfield $1° \times 2°$ quadrangle, Utah: Map 59, Utah Geological and Mineral Survey, Salt Lake City.

Eaton, G. P., 1982, The Basin and Range province: origin and tectonic significance: Ann. Rev. Earth Plan. Sci., **10**, 409–440.

Erwin, J. W., and Berg, J. C., 1977, Bouguer gravity map of Nevada-Reno sheet: Map 58, Nevada Bureau of Mines and Geology.

Erwin, J. W., and Bittleston, J. C., 1977, Bouguer gravity map of Nevada-Millett sheet: Map 53, Nevada Bureau of Mines and Geology, Reno.

Gamble, T. D., Goubau, W. M., and Clarke, J., 1979, Magnetotellurics with a remote reference: Geophysics, **44**, 53–68.

Gamble, T. D., Goubau, W. M., Goldstein, N. E., Miracky, R., Stark, M., and Clarke, J., 1981, Magnetotelluric studies at Cerro Prieto: Geothermics, **10**, 169–182.

Gamble, T. D., Goubau, W. M., Miracky, R., and Clarke, J., 1982, Magnetotelluric regional strike: Geophysics, **47**, 932–937.

Grant, F. S., and West, G. F., 1965, Interpretation theory in applied geophysics: McGraw-Hill Book Co., Inc., 584 p.

Harrington, R. F., 1961, Time-harmonic electromagnetic fields: McGraw-Hill Electrical and Electronic Engineering Series, McGraw-Hill Book Co., Inc., 480 p.

Healey, D. L., Snyder, D. B., and Wahl, R. R., 1981, Bouguer gravity map of Nevada-Tonopah Sheet: Map 73, Nevada Bureau of Mines and Geology.

Hermance, J. F., 1982, The asymptotic response of three-dimensional basin offsets to magnetotelluric fields at long periods: The effects of current-channeling: Geophysics, **47**, 1562–1573.

Hermance, J. F., and Thayer, R. E., 1975, The telluric-magnetotelluric method: Geophysics, **40**, 664–668.

Hintze, L. F., 1980, Geologic map of Utah: Utah Geological and Mineral Survey, Salt Lake City, Utah.

Jones, F. W., and Vozoff, K., 1978, The calculation of magnetotelluric quantities for three-dimensional conductivity inhomogeneities: Geophysics, **43**, 1167–1175.

Klein, D. P., and Larsen, J. C., 1978, Magnetic induction fields (2-30 cpd) on Hawaii Island and their implications regarding electrical conductivity in the oceanic mantle: Geophys. J. Roy. Astr. Soc., **53**, 61–77.

Kunetz, G., 1972, Processing and interpretation of magnetotelluric soundings: Geophysics, **37**, 1005–1021.

Larsen, J. C., 1975, Low frequency (0.1-6.0 cpd) electromagnetic study of the deep mantle electrical conductivity beneath the Hawaiian Islands: Geophys. J. Roy. Astr. Soc., **43**, 17–46.

——— 1977, Removal of local surface conductivity effects from low frequency mantle response curves: Acta Geod. Geop. et Mont. Acad. Sci., **12**, 183–186.

——— 1981, A new technique for layered earth magnetotelluric inversion: Geophysics, **46**, 1247–1257.

Park, S. K., Orange, A. S., and Madden, T. R., 1983, Effects of three-dimensional structure on magnetotelluric sounding curves: Geophysics, **48**, 1402–1405.

Petrick, W. R., Pelton, W. H., and Ward, S. H., 1977, Ridge regression inversion applied to crustal resistivity sounding data from South Africa: Geophysics, **42**, 995–1005.

Porath, H., 1971a, A review of the evidence on low-resistivity layers in the earth's crust, *in* The structure and physical properties of the Earth's crust: J. G. Heacock, Ed., AGU Mono. **14**, 127–144.

——— 1971b, Magnetic variation anomalies and seismic low-velocity zone in the western United States: J. Geophys. Res., **76**, 2643–2648.

Price, A. T., 1973, The theory of geomagnetic induction: Phys. of the Earth and Plan. Int., **7**, 227–233.

Ranganayaki, R. P., and Madden, T. R., 1980, Generalized thin sheet analysis in magnetotellurics: an extension of Price's analysis: Geophys. J. Roy. Astr. Soc., **60**, 445–457.

Rijo, L., 1977, Modeling of electric and electromagnetic data: Ph.D. thesis, Univ. of Utah.

Rooney, D., and Hutton, V. R. S., 1977, A magnetotelluric and magneto-variational study of the Gregory Rift Valley, Kenya: Geophys. J. Roy. Astr. Soc., **51**, 91–119.

Sandberg, S. K., and Hohmann, G. W., 1982, Controlled-source audio-magnetotellurics in geothermal exploration: Geophysics, **47**, 100–116.

Stanley, W. D., Boehl, J. E., Bostick, F. X., Jr., and Smith, H. W., 1977, Geothermal significance of magnetotelluric soundings in the Snake River Plain—Yellowstone region: J. Geophys. Res., **82**, 2501–2514.

Stewart, J. H., 1980, Geology of Nevada: Nevada Bureau of Mines and Geology Special Pub. 4, 136 p.

Stodt, J. A., 1978, Documentation of a finite element program for solution of geophysical problems governed by the inhomogeneous 2D scalar Helmholtz equation: NSF Program Listing and Documentation, Univ. of Utah, 66 p.

———— 1983, Noise analysis for conventional and remote reference magnetotellurics: Ph.D. thesis, Univ. of Utah, 220 p.

Stodt, J. A., Hohmann, G. W., and Ting, S. C., 1981, The telluric-magnetotelluric method in two- and three-dimensional environments: Geophysics, **46**, 1137–1147.

Stratton, J. A., 1941, Electromagnetic theory: McGraw-Hill International Series in Pure and Applied Physics, McGraw-Hill Book Co., Inc., 615 p.

Swift, C. M., 1967, A magnetotelluric investigation of an electrical conductivity anomaly in the southwestern United States: Ph.D. thesis, Massachusetts Institute of Technology, 211 p.

Ting, S. C., and Hohmann, G. W., 1981, Integral equation modeling of three-dimensional magnetotelluric response: Geophysics, **46**, 182–197.

Vozoff, K., 1972, The magnetotelluric method in the exploration of sedimentary basins: Geophysics, **37**, 98–141.

Wannamaker, P. E., 1983, Resistivity structure of the Northern Basin and Range, in The role of heat in the development of energy and mineral resources in the Northern Basin and Range Province: Geothermal Resources Council, Special Rep., 13, 345–362.

Wannamaker, P. E., Sill, W. R., and Ward, S. H., 1978, Magnetotelluric observations at the Roosevelt Hot Springs KGRA and Mineral Mts., Utah: Geothermal Resources Council Trans., **2**, 697–700.

Wannamaker, P. E., Hohmann, G. W., and San Filipo, W. A., 1984, Electromagnetic modeling of three-dimensional bodies in layered earths using integral equations: Geophysics, **49**, 60–74.

Ward, S. H., 1967, Electromagnetic theory for geophysical application, in Mining geophysics, v. II: Soc. Expl. Geophys.

Ward, S. H., Parry, W. T., Nash, W. P., Sill, W. R., Cook, K. L., Smith, R. B., Chapman, D. S., Brown, F. H., Whelan, J. A., and Bowman, J. R., 1978, A summary of the geology, geochemistry and geophysics of the Roosevelt Hot Springs thermal area, Utah: Geophysics, **43**, 1515–1542.

Weinstock, H., and Overton, W. C., Jr., Eds., 1981, SQUID applications in geophysics: Soc. Expl. Geophys., 208 p.

Word, D. R., Smith, H. W., and Bostick, F. X., Jr., 1970, An investigation of the magnetotelluric tensor impedance method: Elec. Geophys. Res. Lab. Report 82, Elec. Res. center, Univ. of Texas at Austin, 264 p.

———— 1971, Crustal investigations by the magnetotelluric impedance method, in The structure and physical properties of the Earth's crust: J. G. Heacock, Ed., AGU Mono. 14, 145–167.